Optoelectronic Devices

oFor cona

Fo

Prentice Hall International Series in Optoelectronics

Consultant editors: John Midwinter, University College London, UK
Bernard Weiss, University of Surrey, UK

Optoelectronic Devices

S. Desmond Smith

Prentice Hall

London   New York   Toronto   Sydney   Tokyo   Singapore
Madrid   Mexico City   Munich

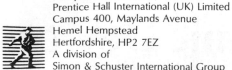

First published 1995 by
Prentice Hall International (UK) Limited
Campus 400, Maylands Avenue
Hemel Hempstead
Hertfordshire, HP2 7EZ
A division of
Simon & Schuster International Group

© Prentice Hall 1995

Typeset in 10/12 pt Times
by PPS Limited, London Road, Amesbury, Wilts.

Printed and bound in Great Britain by
Redwood Books, Trowbridge, Wiltshire

Library of Congress Cataloging-in-Publication Data

Available from the publisher

British Library Cataloguing in Publication Data

A catalogue record for this book is available from
the British Library

ISBN 0-13-143769-0

1 2 3 4 5   99 98 97 96 95

Contents

Preface

Optoelectronics is rapidly growing as a research interest as well as an industrial sector. The reason for such rapid growth and interest has stemmed from the potential advantages optoelectronics can offer in a very wide range of applications. In information technology the subject addresses the vital human–computer interface through the provision of displays. This subject, which of course extends to many areas of 2-D information presentation, will continue to develop in the future. Up to the present day it is still primarily achieved using the cathode ray tube display unit, but this has a number of disadvantages such as limited resolution and large weight, power and volume requirements. Progress is being made using several physical phenomena in order to provide displays with much higher resolution, smaller volumes and flicker-free pictures. Optical sensors allow the measurement of a wide range of physical variables and can often provide solutions where no other technology would be suitable.

Although not the main application discussed in this book, otpical fiber technology is another key area of development in which very large advances have already been made. The data communication rate of 'optical fiber' is substantially greater than that of conventional copper wire and it is feasible that in the future a variety of information (e.g. TV signals, videophones, interactive services, etc.) will be brought to the home using optical fibers. Optical detectors, essential components of any optoelectronic system, find applications in a myriad of technical problems such as sensors, night vision, telecommunications and video cameras. The fiber technology experience referred to above has already shown that light can be practically used for wide-band communication over long distances. There is however also the possibility of using optics in computing. Sequential binary digital processing suffers some disadvantages in requiring very high data rates. This in turn brings communication problems. Optics may alleviate these or even bring completely new perspectives to computational tasks such as real-time image recognition.

These are just a few of the key areas of optoelectronics, but it can clearly be seen why it is such an important and growing area. In fact, one Japanese estimate has suggested that by the year 2010, 20% of their GNP will come from optoelectronic devices and systems!

The level of presentation is suitable for graduate students in physics and engineering, as well as those working in research and in industry. The concept of this text began at the Optical Sciences Center, University of Arizona, Tucson, in 1980 when the author delivered an MSc course on the topic during a spell as Visiting Professor. It was quickly apparent that no existing book contained an adequate breadth or depth of basic physics and device mechanism explanation for the course.

In 1981 Heriot-Watt and St. Andrews Universities launched a joint MSc course on the same topic with an extensive scope. This course has enabled me to call upon a wide range of experience in the two universities and I have indeed adopted the basis of a number of these courses, under various chapter headings, constructed by the contributing staff members. The MSc course and this book aim to provide a clear practical understanding of the operation of a variety of optoelectronic devices, with adequate but not too detailed theoretical treatment (for deeper treatments the reader is referred to suitable references). I am indebted to Professor Brian Wherrett, who directs the Heriot-Watt course as well as to colleagues at St. Andrews.

Chapters 1 and 2 provide a basic introduction to the book and some background on the optical and electronic properties of materials (mainly crystalline solids and semiconductors), in preparation for the chapters that follow. Chapter 3 focuses on *pn* junction devices and devices that can be formed from variants of the *pn* junction. Chapter 4 looks at a variety of optically controlled devices that are either faster than or distinct from electronically controlled devices and mentions optical data storage. Chapter 5 concentrates on the science and technology of liquid crystal display devices, while Chapter 6 lays the foundations of crystal and nonlinear optics theory used later on. Chapter 7 considers a variety of fast optical modulators that allow control of the light beam. Chapter 8 discusses diode-pumped solid-state lasers which are growing in importance since they produce an all-solid-state laser with a wider range of applications than flashlamp-pumped solid-state lasers. Chapter 9 describes detectors, including array detectors, in both the visible and the infrared, and Chapter 10 deals with fiber optic waveguides and their applications, particularly for sensors.

With the wealth of source material available, as described above, my role has been both editor and author but I must, of course, take responsibility for the overall selection of material and for any errors.

Chapter 2, on the basic physics of solids (mostly semiconductors in this context), comes from Professor Carl Pidgeon's course at Heriot-Watt University and from *Infrared Physics*, by J.T. Houghton and myself. Chapter 3 is based on my own course from Tucson but has been considerably reconstructed, extended and updated with the help of Dr S.J. Adams. I would also like to thank Peter Selway and David Greene of BNR (Europe) for helpful criticisms and suggestions. On Chapter 5, I would acknowledge advice from Professor George Gray, FRS, of Merck Ltd. and Professor Ian Shanks, FRS, Chief Scientist at Thorn-EMI. Dr David Vass of Edinburgh University contributed a section on liquid-crystal-over-silicon spatial light modulators. Professor Andy Walker (Heriot-Watt University) essentially contributed Chapter 6 on crystal and nonlinear optics, whilst Chapter 8 on diode-pumped lasers was written by Professor Malcolm Dunn and Dr Bruce Sinclair from St. Andrews. Chapter 10 on optical fiber topics was a combination of efforts from Dr Jim S. Barton and Professor Julian D.C. Jones, both of Heriot-Watt.

I acknowledge the editorial assistance of the following: Kate Sugden, a student on the St. Andrews MSc course in 1991, who worked between January and September 1992 and began the activity; Steve Adams between September and December 1992, who continued the process; and Dharmesh Panchal (lately of Worcester College, Oxford) who filled in an enforced break in his MSc at Rochester from January to August 1993 and saw this activity of identification of appropriate material, editing and writing various sections through to delivery of a first draft to the publisher. Diagrams were prepared on computer by Stuart Fancey and Nigel Watson. Much encouragement and skill was provided by my secretary, Tillie Bell-Boulogne, who was ever present.

The incentive to produce this book was partly provided by the support of the SERC by way of providing studentships for the MSc course at the two Universities and also by the European Community Basic Research Action 'Workshop on Optical Information Technology', which stimulated much activity in this subject. I am grateful to Professor Peter Pusey and the Department of Physics at the University of Edinburgh for providing sabbatical facilities and thus aiding the completion of this book. I record my thanks to all the above.

S. Desmond Smith

1

Optoelectronic devices

1.1 Definition and ground plan

Optoelectronics is a topic of major significance in today's technical society. Optoelectronic devices are used in a variety of applications, such as light sources, modulators, display devices, communication systems and detectors. Display devices such as light-emitting diodes (LEDs) and liquid crystal displays (LCDs) are continually expanding their influence and improving in quality. Displays can be made with built-in logic circuits which interface with integrated circuits and computers. Optical fibers have become a practical way of transmitting information over long distances, with the semiconductor laser diode acting as the light source. Reading and writing optical audio and video memory disks also require diode lasers, as does laser printing. These applications continue to expand in volume and scope.

The conversion of optical into electrical signals and vice versa is fundamental to our use of optoelectronic technology. We receive much information optically and presently require electronics in the transmitting and processing functions, as in the video camera–television system, for example. Our discussion therefore includes devices capable of generating, modulating and detecting light, as well as elements for light transmission, optical memories and optical processing of data.

To restrict the scope of the book to fundamental principles of science and engineering, we concentrate more on components than on systems. We consider components used in display technology, optical communications, detectors (including those for arrays), devices required for advanced scientific measurements and for sensors for environmental or process control purposes (see Figure 1.1). Semiconductor devices are given prominence, due to their compatibility with integrated circuits. Liquid crystal technology is also discussed. A variety of further materials and mechanisms are mentioned where the above two technologies have not yielded adequate solutions to device requirements.

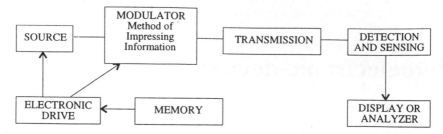

Figure 1.1 Schematic diagram of the different components of an optoelectronic system.

Most optoelectronic systems contain fundamental components as illustrated in Figure 1.1

A source of light is, of course, mandatory. There is a long technical history of this subject extending from the gas mantle, the tungsten lamp in vacuo, the fluorescent tube, to the various forms of present-day lasers. Efficiency in terms of light emission per unit of electrical power has always been important, as has size and cost. Industries have waxed and waned in spectacular manner according to technical advances. In this text we concentrate on semiconductor diode-based sources for the applications mentioned above.

Next, and a key theme of this text, is the means by which information is impressed onto the light beam. With our present basis of electronic information processing in computing, telecommunications and radiowave-based television, information is often transferred from 1-D (time sequential) to 2-D (optical). Thus all forms of modulators are fundamental and may themselves be 1-D (e.g. in telecommunications through fibers) or 2-D, such as the *spatial light modulator*, or SLM. Technologies for this purpose, which is of course the basis of displays, the essential man–machine interface, recur throughout the text.

Detection and sensing complete the set of functions of optoelectronic devices. Again, semiconductor technology is progressing rapidly, and shows considerable future possibility with some surprising new features arising from carrier exclusion and extraction impacting, particularly on detector performance.

The history of light sources provides an example of the impact of technology on industry. Lessons can be learned by looking backwards: a technology does not succeed until all its parts are in place; sometimes quite surprising (and simple) developments provide the necessary spur for a major leap forward. The impact of new and successful technical progress seems inevitable in a free market. We may expect such events to occur repeatedly with the present rate of progress in our subject. History shows, however, that predicting the time taken for a new invention to reach the production stage can be difficult (see Table 1.1).

The above discussion gives an idea of the scope of this text. An understanding of optoelectronic devices, whether from the point of view of a user of a complete system

Table 1.1 Typical induction times (i.e. time taken from discovery to production) of some optoelectronic products

Product	Induction time from laboratory basics to market (yrs)
Fluorescent tubes	70
Liquid crystal displays	90
Semiconductor lasers	35
Photodiode detectors	15
Optical fibers	10

or of a researcher or product developer engaged in advancing the technology, requires a combination of the necessary basic physics with a technical description of the function of the device. The basics begin, in Section 1.2, with a mathematical description of light waves that is applicable to all devices and effects throughout the book. This is followed by a treatment of guided waves that will be needed in Chapters 3 and 10. Section 1.3 introduces an important optical principle, spatial filtering, and some further principles of wide applicability concerned with modulators. To conclude this introductory chapter, Section 1.4 provides examples of some system issues, illustrated for the case of displays. The intention is to demonstrate that devices depend strongly on the capabilities of their supporting electronics, power supplies, etc., as well as their optical environments.

Chapter 2 treats the relevant physics of the largest class of materials from which devices are fabricated, i.e. crystalline solids, in particular semiconductors. The properties of electrons within these materials are covered under the heading 'band structure' and describe both the electronic and optical properties. The results in Chapter 2 are required particularly (and are quoted) in Chapter 3 (the pn junction), Chapter 4, and Chapter 9 (detectors). Readers who are fully familiar with band structure, the optical properties of semiconductors and electronic transport phenomena may omit this chapter. It serves, however, to simplify the operational descriptions in Chapters 3, 4 and 9, and appropriate sections are referenced when required.

Chapter 5 on liquid crystals and devices stands largely by itself and includes its own basic physics and chemistry. Chapters 6 and 7 revert to the separation of basic physics and device descriptions, whilst Chapter 8 brings together, in the modern topic of diode-pumped solid state lasers, material based on Chapters 3 and 6. Chapter 9 describes a selection of detectors of both practical and fundamental interest. It draws upon the physics and effects presented in Chapters 2 and 3 and extends to new principles of device design that have a bearing on the fundamentals of noise generation in semiconductors. Chapter 10 is necessarily selective, since specialized texts are available, particularly in the field of optical fiber telecommunications, but presents a review of the state of the art in optical fiber sensors and gives a summary of optical fiber communications technology.

1.2 Light beams

In order to discuss the properties of optoelectronic devices it is necessary to describe adequately the light beam itself in a mathematical form. This enables us to quantify all the changes in time and space that occur in the beam. In order to represent the light from a large low-powered optical system a *plane wave* is generally used.

1.2.1 The plane wave

A plane wave can be written in the form

$$\mathbf{E} = \mathbf{E}_0 \exp i(\omega t - \mathbf{k} \cdot \mathbf{r}) \tag{1.1}$$

which describes a wave with constant electric field amplitude and phase across the plane perpendicular to the propagation direction $\mathbf{k}$. This is *one* choice from the general wave solution derived from Maxwell's equations (2.12). The imaginary exponential time dependence can be chosen to be $+i\omega t$ or $-i\omega t$. There is a 'physics' sign convention opposite to an 'engineering' sign convention. This is arbitrary but the choice of signs of $\mathbf{k}$ and ω defines the *direction* of the travelling waves. We do not attempt to maintain uniformity, but conform to the common usage in the subtopic. However, this form has only limited practical application: to those cases distant from a point source or where the amplitude and phase do not vary across the dimensions of the device. In many cases, however, the light beam is either diverging or converging, especially where lasers are involved, and the plane wave equation approximation is not adequate.

1.2.2 The Gaussian beam

For many such cases a *Gaussian* description of the beam is often used especially with laser beams because of their modal nature. This is given in a mathematical form by

$$E(x, y, z, t) = E_0 \frac{w_0}{w(z)} \exp\left[-i(kz - \eta(z)) - r^2 \left(\frac{1}{w^2(z)} - \frac{ik}{2R(z)} - i\omega t \right) \right] \tag{1.2}$$

where
  w_0 is the minimum waist size (which occurs where the spot radius is a minimum)
  $w(z)$ is the waist size at any position z along the beam
  z_0 is the Rayleigh range (i.e. half the confocal parameter)

$$z_0 = \frac{\pi w_0^2 n}{\lambda}$$

$$r^2 = x^2 + y^2$$

$$\eta(z) = \tan^{-1} \frac{z}{z_0}$$

$$R = z\left(1 + \frac{z_0^2}{z}\right),$$ where R is the radius of curvature of the wave front.

Equation (1.2) allows the light beam to be described as an electric field distribution in three dimensions, x, y, z, and also at any time, t.

The symbol $\mathbf{k}$ represents the propagation vector. In a particular direction it can be a scalar, and is then known as the propagation constant; it can be complex and is given (see Eq. (2.17)) by

$$\mathbf{k}^2(\mathbf{r}) = \omega^2 \mu \varepsilon(\mathbf{r})\left(1 - \frac{i\sigma(\mathbf{r})}{\omega\varepsilon}\right) \tag{1.3}$$

and where $\varepsilon(\mathbf{r})$ and $\sigma(\mathbf{r})$ are the permittivities and conductivities, respectively. The magnetic permeability μ will usually reduce to μ_0 for optical materials. This expression allows for the possibility of positional dependence of the refraction, absorption, gain or nonlinear response of a medium.

It may be required to modulate or attenuate the intensity, amplitude or phase of the light beam. A device may also be required to achieve the same result at all wavelengths, or selectively over a chosen band. It may be difficult to modulate only one of these properties at any one time and this can cause problems. Modulation can be achieved by various approaches – mechanical, optical, electrical or magnetic.

The term 'optical frequencies' covers a broad range, stretching from the far infrared (3×10^{12} Hz) to the ultraviolet (3×10^{15} Hz). Most of the time we will deal with the range 6×10^{14}–3×10^{13} or 0.5–10 μm in terms of wavelength. The telecommunications industry has a large interest in fast modulation so that the large bandwidth available at optical frequencies can be exploited to give a high information carrying capacity.

Very fast modulators and other devices can be limited by *electrical time constants* (Chapter 6), so the use of *optical control* (Chapter 4) operating at the speed of light becomes an interesting proposition. Control may be difficult to exercise in just one parameter. Some measurements may be unaffected, whereas others (such as in nonlinear optics) may be strongly affected. The response of detectors as a function of time and space may also lead to complications.

1.2.3 Guided waves

The mathematical description of electromagnetic waves in a dielectric waveguide is of importance in semiconductor diode lasers (Chapter 3 and Chapter 8), optical fibers (Chapter 9) and planar structures in 'integrated optics'. It is obtained by a solution of the wave equation, derived from Maxwell's equations (see Chapters 2 and 6), subject to the boundary conditions on $\mathbf{E}$ and $\mathbf{H}$ which specify that the tangential components of these fields are continuous on either side of a dielectric interface.

The wave equation may be written (see, for example, Yariv [1.1]):

$$\nabla^2 \mathbf{E}(\mathbf{r}) + k^2 n^2(\mathbf{r})\mathbf{E}(\mathbf{r}) = 0 \tag{1.4}$$

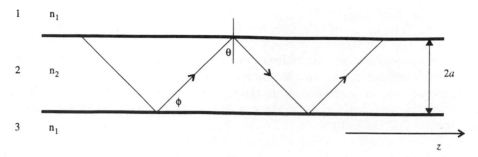

Figure 1.2 A slab waveguide.

where $k = \omega/c$, the propagation constant *in vacuo*, and $n(\mathbf{r})$ ($\equiv \varepsilon(\mathbf{r})/\varepsilon_0$) is the refractive index. The solution is of the form

$$\mathbf{E}(\mathbf{r}, t) = \mathbf{E}(\mathbf{r})e^{\,i[\omega t - \phi(\mathbf{r})]} \tag{1.5}$$

For waves with phase fronts normal to z, the waveguide axis, we can write $\phi(\mathbf{r}) = \beta z$, and (1.4) becomes

$$\left(\frac{\partial^2}{\partial x^2} + \frac{\partial^2}{dy^2}\right)\mathbf{E}(x, y) + [k^2 n^2(\mathbf{r}) - \beta^2]\mathbf{E}(x, y) = 0 \tag{1.6}$$

Although we mostly require to understand the *modes* resulting from the solutions of Eq. (1.6) for structures of comparable x and y dimensions (i.e. channel waveguides), the important features can be obtained from a planar model. This simplifies the mathematics by taking $\partial/\partial y = 0$, and is known as a slab waveguide. For circular-section optical fibers the formal solutions are even more complex mathematically but, again, the mode features remain similar. Figure 1.2 shows a simple planar waveguide index $n_2 > n_1$ and thickness $2a$.

For regions 1, 2 and 3 we have

1. $$\frac{\partial^2}{\partial x^2}\,E(x, y) + (k^2 n_1^2 - \beta^2)E(x, y) = 0 \tag{1.7a}$$

2. $$\frac{\partial^2}{\partial x^2}\,E(x, y) + (k^2 n_2^2 - \beta^2)E(x, y) = 0 \tag{1.7b}$$

3. $$\frac{\partial^2}{\partial x^2}\,E(x, y) + (k^2 n_1^2 - \beta^2)E(x, y) = 0 \tag{1.7c}$$

where $E(x, y)$ is a Cartesian component of $\mathbf{E}(x, y)$.

We require solutions to Eqs. (1.7), subject to the boundary conditions at the interfaces, i.e. that the tangential components of $\mathbf{E}$ and $\mathbf{H}$ are continuous. With $n_2 > n_1$, we note that, for frequency ω, if $\beta > kn_2$, it follows that $(1/E(x, y))(\partial^2 E(x, y)/\partial x^2) > 0$ everywhere, leading to a field that increases with x away from the guiding region, and which is therefore not physically real.

If $\beta < kn_2$, however, $(1/E(x, y))(\partial^2 E(x, y)/\partial x^2) < 0$, and the solution is sinusoidal in region 2. It remains exponential in regions 1 and 3 and decays with distance away from the guide. The values of β imposed by the boundary conditions are *discrete* in the propagation condition and the *modes* so defined are referred to as *confined* or *guided modes*, in close analogy to the modes of metallic waveguides in radar technology. Such discrete solutions are known as 'eigenvalues'; they are readily deduced by the condition of *transverse phase resonance* first used explicitly for dielectric slab waveguides by Tien and Ulrich and others [1.2, 1.1, 1.3].

We follow the path of one ray in Figure 1.2 and assume an observer moving along the *z*-axis who sees only the *transverse* motion, i.e. along *x*. The boundary conditions require that the phase shift for a round trip from $x = 0$ to $x = 2a$ and back to $x = 0$ must be a multiple of 2π. In terms of optical path length and phase changes on reflection, this becomes

$$4akn_2 \sin \phi - \delta_{21} - \delta_{23} = 2N\pi \quad (N = 0, 1, 2 \ldots) \tag{1.8}$$

where δ_{21} and δ_{23} are the phase changes on reflection from the interfaces 2–1 and 2–3, respectively (in our simple case these are equal). We define

$$p^2 = \beta^2 - n_1^2 k^2$$
$$q^2 = n_2^2 k^2 - \beta^2$$
$$r^2 = \beta^2 - n_1^2 k^2$$

For a ray whose electric vector is normal to the plane of incidence (i.e. containing the ray and the normal to the interface) Eq. (1.8) defines the Transverse Electric (TE) mode. The Fresnel relations give

$$\delta_{21} = 2 \tan^{-1} \left[\frac{(\cos^2 \phi - n_1^2/n_2^2)^{1/2}}{\sin \phi} \right]$$

$$= 2 \tan^{-1} \left[\left(\frac{\beta^2 - n_1^2 k^2}{n_2^2 k^2 - \beta^2} \right)^{1/2} \right]$$

where we have used $\beta = n_2 k \cos \phi$.

Using this result and that $\delta_{21} = \delta_{23}$, Eq. (1.8) yields

$$4aq - 2 \tan^{-1} \left(\frac{p}{q} \right) - 2 \tan^{-1} \left(\frac{r}{q} \right) = 2N\pi \quad (N = 0, 1, 2 \ldots)$$

which can be written as

$$\tan(2aq - N\pi) = \frac{(p + r)q}{q^2 - pr} \quad (N = 0, 1, 2, 3 \ldots) \tag{1.9}$$

This is the eigenvalue equation for TE modes in the dielectric slab waveguide.

Similarly, for the electric vector parallel to the plane of incidence, it can be shown that

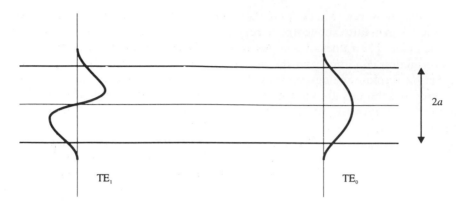

Figure 1.3 Lowest-order confined modes.

$$\tan(2aq - N\pi) = \frac{(n_1^2 p + n_1^2 r)n_2^2 q}{n_1^4 q^2 - n_2^4 pr} \quad (N = 0, 1, 2 \ldots) \tag{1.10}$$

This is the eigenvalue equation for Transverse Magnetic (TM) modes.

As the transverse dimension of the guide, $2a$, increases, the number of guided modes increases. At same value of $2a$ the mode TE_0 becomes confined. A further increase in $2a$ allows TE_1, as shown in Figure 1.3.

For a planar guide the TE_i and TM_i modes are described by a single integer, i, whereas for a channel (rectangular) or cylindrical (fiber) guide two indices are required. The lowest-order mode in a channel guide is known as the Transverse Electric Magnetic, TEM_{00} mode, and has no cutoff frequency. For a weakly guiding cylindrical fiber the equivalent mode is known as the LP_{01} mode (see Chapter 10).

1.3 Principles and properties of simple modulators

1.3.1 Continuous attenuation: the slowest modulator

The nonlinear optical properties of matter depend on various powers of the electric field amplitude E_0. The study of such power-dependent effects requires *continuous* control of irradiance (W/cm^2), loosely referred to as intensity and, often, a large dynamic range to bridge the gap between lasers and conventional light sources, or between Q-switched or mode-locked outputs and continuous-wave (cw) outputs. A device for achieving this control gives, for tutorial purposes, a good example of the problems and solution for a device with a distribution in space of intensity.

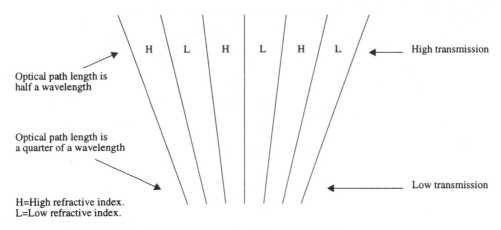

Figure 1.4 Wedge-shaped film providing continuous attenuation.

1.3.2 Variable attenuator for Gaussian laser beams: the use of spatial filtering

For studying effects using a laser beam it is often necessary to be able to attenuate the laser power whilst maintaining an output of constant Gaussian shape in space. An effective method for doing this was devised by Miller and Smith [1.4] in 1978. This method successfully combined the technologies of *spatial filtering* and *multilayer dielectrics*.

The control over the attenuation was achieved by using a 'wedged' thin-film layer (Figure 1.4). Interference effects are such that at one thickness along the wedge, maximum transmission will occur. At another point on the wedge the interference will cause minimum transmission. Between the two, the laser beam will be partially transmitted. So, just by moving the wedge across the beam, the attenuation is varied and the power of the beam is controlled.

The problem with using a 'wedge-shaped' film is that it introduces lateral disturbances into the spatial disribution of the intensity The original waveform may not be a good Gaussian shape, Figure 1.5a, and introducing the wedge makes this worse (Figure 1.5b). This can be overcome by using spatial filtering techniques. In the Fourier transform plane all the high-frequency spatial components that make up the rapidly changing disturbances on the laser beam profile are found on the outside of the pattern. These can be excluded by introducing an aperture in the Fourier transform plane and then the remaining optical information consists of low-frequency components, i.e. smoothly varying intensity profile. Combining the two technologies results in 'clean' Gaussian beams that can be controlled over several orders of magnitude of power (Figure 1.5c–g). We cite this example to emphasize the importance of achieving a known and simple beam profile for measurement of proper device operation.

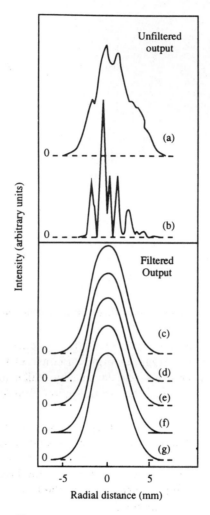

Figure 1.5 Beam intensity profile.

1.3.3 The simple chopper

A light beam is most simply modulated by a chopper blade, as shown in Figure 1.6(a), and is then focused onto a detector connected to an amplifier. This allows display of the detected signal. For an input beam of smaller cross-section than that of the chopper blade, the signal will have approximately a square wave format, as shown in Figure 1.6(b).

In practice all detectors have 'noise' sources which are introduced by effects such as thermal fluctuations – these noise sources are discussed in detail in Chapter 9.

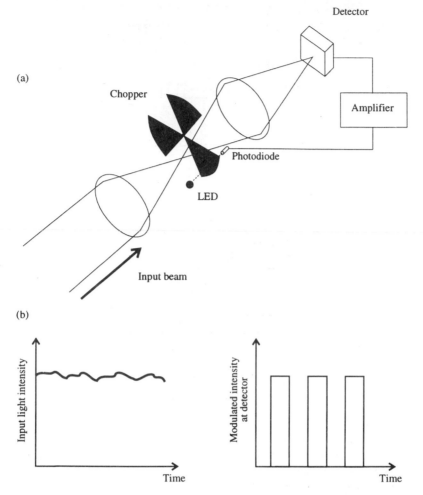

Figure 1.6 (a) Modulation of an optical beam with a simple chopper. (b) The intensities at input and output.

Noise is superposed onto the signal at all electrical frequencies. The use of a chopper limits the signal to a narrow band of 'wanted' frequencies centred on the chopping frequency. A frequency-selective amplifier can then preferentially present these signals. This enables, for example, a particular wanted signal to be differentiated from background, which does not pass through the chopper.

The noise, in either the detector or electronics, can also be reduced by restricting *response* to this narrow band of frequency. A separate electrical 'pick off' is used to synchronize the detection system to the chopper. This arrangement is known as the *phase-sensitive detector* (PSD, or *lock-in amplifier*) and is based on a switch operated by the reference input in phase with the signal (Figure 1.7a). This switch selects the

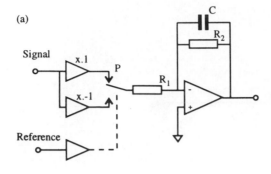

(a)

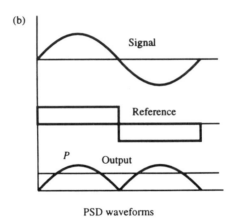

Phase-sensitive detector (PSD)

(b)

PSD waveforms

Figure 1.7 (a) Circuit of phase-sensitive detector or lock-in amplifier. (b) The signal (now depicted as a sine wave), reference and output of the phase-sensitive detector (the signal at the detector is always positive being an optical intensity, but the DC component is ignored by the amplifier giving a positive–negative AC component).

signal during the positive half-cycle of the reference and the inverted signal during the negative half-cycle, resulting in the waveform *P* at the switch output. This waveform is seen to be a full-wave rectified version of the input signal. The output stage is a low-pass filter which passes only the DC component, giving a final output that is proportional to the signal amplitude (Figure 1.7(b)).

Electrical noise is characterized by its power spectrum. This is defined as $N(f)$, where the power in a spectral interval δf centred on frequency f is $N(f)\delta f$. Typically, the power spectrum decreases from high values at low frequency ($N(f) \propto 1/f$), to a flat region, and eventually drops at high frequency where amplifier gains are reduced (see Figure 1.8).

A signal that is not at the reference frequency has a steadily changing phase difference ϕ. Since the output $\propto \cos(\phi)$, it oscillates at the difference in frequency

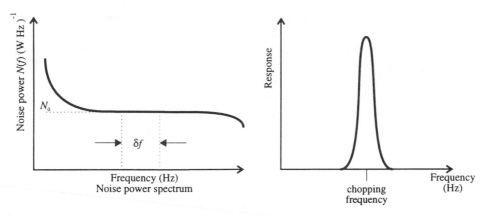

Figure 1.8 Noise performance of a PSD.

between the signal and reference (the *beat* frequency). Only noise components near to the reference frequency f_0 appear at the output, since the low-pass filter attenuates high beat frequencies. This can be quantified as follows.

If the filter transfer function is $A(f)$, where f is the *beat frequency*, the output noise power N_T is found by integrating over the frequency:

$$N_T = \int_{-\infty}^{\infty} N(f_0 + f)\,|\,A(f)\,|^2 \mathrm{d}f \tag{1.11}$$

Here the integral extends over negative and positive frequencies (since the beat frequency can have either sign) and, because we are integrating noise power, only the magnitude of the filter transfer function is important. Now $A(f) = 1(1 + 2\pi f\tau)$, where τ is the filter time constant, and we may assume that $N(f_0 + f)$ is constant over the narrow band of significant frequencies close to the reference frequency. With these approximations,

$$N_T = \int_{-\infty}^{\infty} \frac{N_0}{1 + (2\pi f\tau)^2}\, \mathrm{d}f = \frac{N_0}{2\pi\tau} \left[\arctan(x)\right]_{-\infty}^{\infty} = \frac{N_0}{2\tau}$$

where $N_0 = N(f_0)$. Note that the PSD shows the same noise power at its output as an ideal filter with bandwidth $1/(2\tau)$. This quantity, known as the *equivalent noise bandwidth*, can take very small values (e.g. 1–0.1 Hz), while f_0 may be of the order of 1 kHz. It would be very difficult indeed to achieve such narrow relative bandwidths with conventional filters because of the required stability in the centre filter frequency and in f_0.

An arrangement analogous to that shown above can also be used to provide 'spatial chopping'. This allows us to distinguish a point source of radiation from a uniform background by arranging that the chopper blade occupies just half of the optical aperture. A uniform input intensity then delivers a steady signal at the detector, whereas radiation from a point source is modulated. Modulation is generally used to distinguish wanted from unwanted radiation.

Spectral modulation can also be achieved using a PSD arrangement. In this case it is possible to have the chopper with blades arranged as cells filled with a gas such as CO_2. This gas will then only absorb radiation within a small band of frequencies and hence allows the detection of radiation from this particular spectral band. This type of arrangement has been used to measure the temperature of the atmosphere as a function of height from a satellite in the selective chopper radiometer [1.5]

1.4   General principles of display technologies

This section introduces the concept of system requirements and their impacts upon component devices. The following example underlies many of the devices, materials and effects described throughout the text.

The oldest and most abundant displays are cathode ray tube TVs (CRT TVs), with approximately 50 km^2 of display area worldwide in 1973; it is estimated that this will reach 2000 km^2 by 1999. However, CRTs are not ideally suited to numerous applications that require low power, small size or large-area displays. The largest demand is anticipated to be for small and intermediate displays with a seemingly unlimited demand for alphanumeric, digital or graphic displays linked to the expansion and improvement of computers of all sizes and their use in information and control.

1.4.1 Display classification

There are two important classes of display:

(i) Light-emitting, which are *active* displays, and include light-emitting diodes (LEDs), tungsten lamps, CRTs, DC or AC phosphors, and gas discharge plasma panels.
(ii) Light-modulating, which are *passive* displays and include liquid crystal displays (LCDs), and electrophoretic, electrochromic and ferroelectric devices.

Presently competing for the small and intermediate display markets are LECs, LCDs, gas discharge and DC electroluminescent phosphor lamps.

Displays can be grouped according to the number of characters per display element as shown in Table 1.2.

1.4.2 Comparison of types of display

Table 1.3 compares several types of display.

The important criteria, apart from those in Table 1.3, in choosing a display are the threshold voltage for switching, inherent memory, element size and range of colour. There are several different types of format for digital information. Characters can be preformed using indicator tubes, where a small number of preformed cathodes are placed behind each other in a gas discharge tube. Segmented formats such as those used for calculator displays are satisfactory for up to about 30 segments. A minimum of seven segments are needed for numerals and more for alphanumeric

Table 1.2 Display classification by number of characters per display element

	Number of characters per display unit				
	1–16 (alphanumeric)	56	256	1000	Bigger
Applications	Clocks	Minicomputers	Computers	Computers	Graphics editing
	Watches	Instrumentation	Control systems	Control systems	
	Calculators	Keyboards	Banking	Graphics editing	TV
	Instrumentation	Bar charts	Reservation systems	Videophones	Video camera monitors
	Indicators	Avoid radar	Patient monitors		
	Simple keyboard displays	Cockpit displays	Word processors	Word processors	HDTV

displays. The dot matrix format produces a more readable display with fewer errors; this is the only solution for big displays, but requires extensive circuitry. When deciding whether a display is suitable for a particular application one must also consider the ancillary electronics: it must be possible to design suitable driving and decoding circuitry. The circuitry must allow the functions shown in Figure 1.9 to be performed. There is a need for memory because neither the speed of delivery nor the form of the information is usually suitable for the actual display devices. The character generation

Table 1.3 Comparison of display systems

	Display type			
	LED	LCD	Gas discharge panel	DC electroluminescent EL panel
Appearance	Bright	Relatively dim	Bright	Bright
Semiconductor (S/C) compatible	Yes	Yes	No	No
Reliability	Very good	Good	Good	Satisfactory
Power consumption	High	Very low	High	Low/medium
Cost per element	High	Low	Low	Low
Switching speed	High	Very low	Sufficient	Sufficient
Comment	Poor in bright light	Poor in dim light		

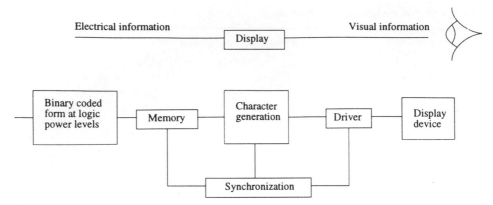

Figure 1.9 Schematic diagram showing how information is transformed in order to address a display device.

decodes the input and the driver provides either current or voltage at the right level. After selection of a particular 'picture element' (or *pixel*) of the display, light is either produced or modulated at the element. In the CRT both the process of selection and production or modulation are performed by the scanning electron beam.

1.4.3 Address methods

In this section we examine two addressing mechanisms: the crossbar addressing system, and point and line sequential addressing.

(i) *The crossbar address system.* In a simple display each element is separately connected and switched so that there is one connection for each element. This is obviously impractical for large displays as the backup circuitry would have to be extensive. To make the circuitry less complicated, a *crossbar address system* can be used. For an array of elements this reduces the electrodes needed from $(n \times m)$ to $(n + m)$. The system is based on a set of orthogonal electrodes with the elements at the electrode intersections. An element is switched on by the *coincident address* of two mutually perpendicular electrodes (Figure 1.10).

Since all the electrodes are connected via all other display elements, then unless there is a physical phenomenon that exibits an *input threshold*, all the unselected elements will be activated at half intensity. Obviously this is not desirable; the eye will not easily distinguish between the half and full intensity and so the display will be difficult to read.

In the absence of an in-built threshold the electronic circuitry has to be designed to provide that feature at each individual element. This is relatively complicated and expensive. Ideally, an in-built threshold is required that is greater than the current supplied by one electrode and less than the current from two electrodes. Then an address by two mutually perpendicular electrodes will switch on the element at that interconnection whilst leaving the other elements along the electrodes off.

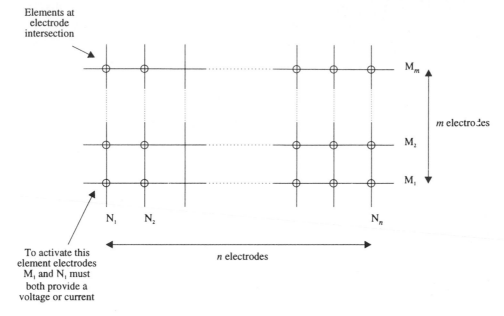

Figure 1.10 The crossbar address system.

(ii) *Point and line sequential addressing.* Individual elements can be addressed *point sequentially* or *line sequentially*. This imposes a time limit on addressing the complete array, i.e. it limits the *frame time*. Individual elements are only activated for a fraction of time as the scan proceeds across the whole display. This is expressed in terms of the duty factor F (Figure 1.11), where

$$F = \frac{\text{time element is on}}{\text{time between successive addresses}} = \frac{\tau}{T}$$

With the sequential system of addressing, each element is switched on in turn and then, if there is no memory to the system, will be off until the next scan. This requires that each element is bright enough and on long enough for the eye to see that it is there. In addition, the whole scan should be completed in less than the integration time of the eye to prevent the display appearing to flicker. The larger the number of elements, then the shorter the time for which current is supplied to the element and the brighter the light emitter needs to be to produce the same visual effect. The average intensity, I, of the element when it is on for a time τ is given by

$$I = I_0 \frac{1}{F}$$

where I_0 is the normal working light intensity.

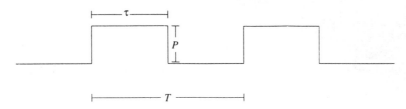

Figure 1.11 Graph showing the switching signal (e.g. voltage) at an element versus time.

Some elements have a memory function and will stay on until a second address pulse switches them off. This avoids the problem with the duty factor and the memory circuitry is simpler.

The response speed of the elements should be smaller or comparable to the address time. This is less important if the electro-optic effect has integration properties – then a number of successive address pulses will have an effect similar to that of a continuous pulse of the same length.

Aspects of matrix address cycles are expanded on in Chapter 5, with particular reference to liquid crystal devices. At the time of writing no display technology encompasses satisfactory properties regarding even the majority of requirements discussed in this section.

References

[1.1] A. Yariv, *Quantum Electronics*, 3rd edn (Wiley, New York, 1989).

[1.2] P.K. Thien and R. Ulrich, *J. Opt. Soc. Am.*, **60** 1325–37 (1970).

[1.3] M.J. Adams, *An Introduction to Optical Waveguides* (Wiley, New York, 1981).

[1.4] D.A.B. Miller and S.D. Smith, 'Variable attenuator for Gaussian laser beams', *Applied Optics*, **17** 3804–8 (1978).

[1.5] J.T. Houghton and S.D. Smith, 'Remote sounding of atmospheric temperature from satellites', *Proc. R. Soc. London* **A320**, 23–33 (1970).

2

The electronic and optical properties of device materials

2.1 Introduction

A treatise on optoelectronic devices needs to include the physical properties of the materials from which the devices are made. The properties of greatest importance here are those of electrons in crystalline solids, but the interaction between the electrons and the atoms of the crystal lattice is also significant; this implies that we must also understand the dynamics of the atoms themselves. The combination of these microscopic effects is responsible for determining the macroscopic optical and electrical properties of solids and in this chapter we give the necessary background on these topics. The electrical and optical properties of liquid crystals are rather different, and these are reviewed in Chapter 5. Crystalline linear and nonlinear optics are discussed in terms of macroscopic theory in Chapter 6.

The present chapter begins with a 'classical' treatment of the electrical conductivity of solids in which the electrons are regarded as charged particles obeying Newton's laws of motion and electrical resistance is assumed to arise from collisions with the vibrating lattice and/or impurities. Although very simple, this model is extremely useful conceptually. The theory is initially applied to 'stationary' or low-frequency situations but it can readily be extended to high frequencies where, with suitable approximations, a first treatment of optical properties is obtained. This is particularly useful for the subject of *dispersion* – the frequency dependence of the refractive index.

The classical theory does not give a method of predicting the natural frequencies of the oscillators which the model of the bound electrons requires; it is simply assumed to be some value 'ω_0'. To predict these frequencies for solids we need to use quantum mechanics, and this is discussed from Section 2.3 onwards.

2.1.1 Classical (Drude) DC conductivity

A theory was proposed in 1900 by P. Drude to explain thermal and electrical conduction in metals. In this model, the negatively charged outer *valence* electrons of the metal atoms are assumed to move freely around a static lattice of positive ions, thereby forming a 'gas' of *conduction* electrons which is then treated using the methods of kinetic theory. Thus the electrons are regarded as classical particles which have random thermal velocities in the absence of an applied field, so that the average velocity and hence the net current is zero. When an electric field is applied each electron is accelerated in the field direction until it undergoes a collision (considered by Drude to be a collision with the lattice), after which its velocity is randomized again.

Consider a uniform current I flowing through a slab of conducting material of length L and cross-sectional area A. From Ohm's law the resistance R is given by $R = V/I$, where V is the voltage difference between the ends of the conductor. R is a function of the sample dimensions and it is usual to remove this dependence by defining a *resistivity*, ρ, such that

$$\mathbf{E} = \rho\mathbf{j} \tag{2.1}$$

where $\mathbf{E}$ is the electric field at a point in the conductor, and $\mathbf{j}$ is the current density (the current passing through unit area perpendicular to the flow of charge). Both $\mathbf{E}$ (units V m^{-1}) and $\mathbf{j}$ (A m^{-2}) are vector quantities and ρ (Ωm) is, in general, a *tensor*. In the case of uniform current flow, E is related to the voltage drop V via $V = EL$, and j is just I/A, so that $\rho = RA/L$. It is sometimes more convenient to work with the reciprocal of the resistivity, which is known as the *conductivity*, σ, so that Eq. (2.1) becomes

$$\mathbf{j} = \sigma\mathbf{E} \tag{2.2}$$

Both conductivity and resistivity are *independent of sample dimensions*.

We now seek to write Eq. (2.2) in terms of microscopic quantities. We may write the current density in the form

$$\mathbf{j} = -\mathbf{n}e\langle\mathbf{v}\rangle \tag{2.3}$$

where $\langle\mathbf{v}\rangle$ is the average electronic velocity and n is the electron density. The negative sign arises because the charge on the electron is negative. In the absence of a field, $\mathbf{v}$ averages to zero, but when an electric field is applied the electrons acquire an average velocity in the direction opposite to the field, known as the *drift velocity*, $\mathbf{v}_D$. The effect of the field on a free electron is to produce an acceleration, $d\mathbf{v}/dt = -(e/m)\mathbf{E}$ (where m is the electronic mass), so we account for the constant nature of the resulting current by arbitrarily assuming that the electrons undergo frequent collisions that randomize the velocity. This is described by a *relaxation* or *scattering time*, τ, which is defined as the *mean free time between collisions* [2.1]. If we consider an individual electron whose velocity is $\mathbf{v}_0$ immediately after a collision, then after a further time t its velocity $\mathbf{v}$ will be $\mathbf{v}_0 - e\mathbf{E}t/m$ because of the effect of the electric field. Because $\mathbf{v}_0$ is random, it makes no contribution to the average electronic velocity $\langle \mathbf{v} \rangle$, which is therefore simply an average over the $-e\mathbf{E}t/m$ term, and we have

$$\langle \mathbf{v} \rangle = \mathbf{v}_D = -\frac{e\mathbf{E}\langle t \rangle}{m} = -\frac{e\mathbf{E}\tau}{m} \tag{2.4}$$

from the definition of the scattering time. Substituting this into Eq. (2.3) and comparing with (2.2) gives us, finally, the following expression for the conductivity:

$$\sigma = \frac{Ne^2\tau}{m} \tag{2.5}$$

The theory makes no prediction of the magnitude of the scattering time. Comparison with experimental measurements of conductivity gives values of τ which seem reasonable, and it is only when other phenomena are considered that the theory is seen to be deficient. Equation (2.4) is often written as $\langle v \rangle = \mu E$, where $\mu = (e\tau/m)$ is known as the *mobility*, i.e. the velocity per unit field (units m^2/volt-s or cm^2/volt-s).

The Hall effect

The optical and electrical properties of materials are strongly dependent on the *density of charge carriers* that they contain. The Hall effect, discovered in 1879 by E.H. Hall, is a means of determining both the *carrier density* and the *sign* of the charge carriers in a material. When a conductor carrying a current density $\mathbf{j}$ parallel to the x-axis is placed in a magnetic field $\mathbf{B}$ along the z-axis, then a potential difference appears across the conductor in the y-direction. The reason for this can be seen with reference to Figure 2.1. Carriers of charge q moving in a magnetic field experience a *Lorentz force*, $\mathbf{F}$, given by

$$\mathbf{F} = q(\mathbf{v} \times \mathbf{B}) \tag{2.6}$$

In Figure 2.1 both positively and negatively charged carriers experience a deflecting force in the y-direction, setting up a nonuniform charge distribution which results in an electric field, E_y. The magnitude and direction of E_y is such as to exactly balance the Lorentz force, so we have

$$-qE_y + qv_xB_z = 0 \tag{2.7}$$

We know from Eq. (2.3) that $j_x = nqv_x$, so the magnitude of the transverse electric field E_y is simply

$$E_y = \frac{j_x}{nq}B_z = R_H j_x B_z \tag{2.8}$$

where $R_H = 1/nq$ is known as the *Hall coefficient*. Clearly, measurements of the Hall effect result in a determination of both the charge density *and* the sign of the charge

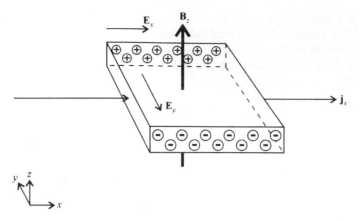

Figure 2.1 The Hall effect, illustrated for the case where the charge carriers are electrons. The sample width is a, the height b and the cross-sectional area is A. The magnetic field deflects the electrons (travelling in the x-direction) towards the foreground of the sample, setting up an electric field in the y-direction.

carriers. It turns out that in some metals the sign of the Hall coefficient is *positive*, a phenomenon that was only satisfactorily explained after the advent of quantum theory.

2.1.2 Optical properties: classical theory

The optical properties of materials are usually described in terms of various quantities derived from classical electromagnetic theory, and we set these out briefly here. We have already defined the conductivity, σ, in terms of the electric field $\mathbf{E}$ and the current density $\mathbf{j}$ in Eq. (2.2). We also recall here the relation between the electric field and the displacement $\mathbf{D}$:

$$\mathbf{D} = \varepsilon_0 \varepsilon_\mathrm{r} \mathbf{E} \tag{2.9}$$

and that between magnetic field $\mathbf{H}$ and flux density $\mathbf{B}$:

$$\mathbf{B} = \mu_0 \mu_\mathrm{r} \mathbf{H} \tag{2.10}$$

where ε_r is the relative permittivity of the medium, μ_r its relative permeability, and ε_0 and μ_0 are the permittivity and permeability of free space.

Maxwells equations are

$$\nabla \cdot \mathbf{D} = \rho \qquad \nabla \cdot \mathbf{B} = 0 \tag{2.11a}$$

and

$$\nabla \times \mathbf{E} = -\frac{\partial \mathbf{B}}{\partial t} \qquad \nabla \times \mathbf{H} = \mathbf{j} + \frac{\partial \mathbf{D}}{\partial t} \tag{2.11b}$$

where ρ is the electric charge density. In a region with no charge density, Eqs. (2.9–2.11) can be combined to give the well-known equation predicting the existence of electromagnetic waves:

$$\nabla^2 \mathbf{E} - \varepsilon_0 \varepsilon_r \mu_0 \mu_r \frac{\partial^2 \mathbf{E}}{\partial t^2} - \mu_0 \mu_r \sigma \frac{\partial \mathbf{E}}{\partial t} = 0 \qquad (2.12)$$

and a similar equation in $\mathbf{H}$.

The solution of Eq. (2.12) that is of greatest importance to us describes plane electromagnetic waves of angular frequency ω, in which $\mathbf{E}$ and $\mathbf{H}$ are perpendicular to each other and also to the direction of propagation. At a point at a distance z from a fixed origin measured along the direction of propagation the wave is described by its transverse component (see Chapter 1):

$$E = E_0 \exp i(\omega t - kz) \qquad (2.13)$$

where the propagation constant $k = 2\pi/\lambda = \omega/u$, u being the 'phase velocity' and λ the wavelength of the disturbance. More generally, we can replace the scalar quantity k by the vector $\mathbf{k}$ in the direction of propagation, then known as the *wavevector*, which is given by

$$\mathbf{k} = \frac{\omega}{u^2} \mathbf{u} \qquad \text{or} \qquad |k| = \left| \frac{\omega}{u} \right| \qquad (2.14)$$

where the vector $\mathbf{u}$ is in the propagation direction. Equation (2.13) may now be written

$$\mathbf{E} = \mathbf{E}_0 \exp i(\omega t - \mathbf{k} \cdot \mathbf{r}) \qquad (2.15)$$

A similar solution exists for the magnetic field such that

$$\mathbf{H} = \frac{1}{\omega \mu_r \mu_0} \mathbf{k} \times \mathbf{E} \qquad (2.16)$$

and the ratio $|\mathbf{E}|/|\mathbf{H}|$ is defined as the *wave impedance*, Z, of the medium.

The propagation constant k can be expressed in terms of the electric and magnetic constants of the material by substituting Eq. (2.13) into (2.12), giving

$$k^2 = \omega^2 \mu_r \mu_0 (\varepsilon_r \varepsilon_0 - i\sigma/\omega) \qquad (2.17)$$

To relate these quantities to measurements that may be made in optical experiments, we recall that for a non-absorbing medium the refractive index is defined as c/u, where c is the velocity of propagation in free space. For the absorbing case we define a *complex refractive index* $n - i\kappa$ such that

$$n - i\kappa = \frac{c}{u} = \frac{ck}{\omega} \qquad (2.18)$$

where $c = 1/\varepsilon_0 \mu_0$, and, using Eq. (2.17),

$$n^2 - \kappa^2 = \varepsilon_r \mu_r \qquad (2.19)$$

and

$$2n\kappa = \frac{\sigma\mu_r}{\varepsilon_0\omega} \tag{2.20}$$

At the frequencies of interest to us in this text, μ_r may usually be taken to be unity, in which case n and κ may be found for a given medium if the dielectric constant ε_r and the conductivity σ are known. In terms of n and κ, the solution (2.13) may be written

$$E = E_0 \exp i\left(\omega t - \frac{\omega}{c}(n - i\kappa)z\right)$$

$$= E_0 \exp\left(-\frac{\omega\kappa z}{c}\right) \exp i\omega\left(t - \frac{nz}{c}\right) \tag{2.21}$$

which describes a propagating wave of phase velocity c/n whose amplitude is attenuated proportionally to the *absorption index* κ.

Experimentally we measure the attenuation of the intensity, I, which is proportional to the square of the electric field. In a path of length z the ratio of the incident intensity I_0 to emergent intensity I is

$$\frac{I}{I_0} = \exp\left(-\frac{2\omega\kappa z}{c}\right) = \exp(-\alpha z) \tag{2.22}$$

where $\alpha = 2\omega\kappa/c$ is the *absorption coefficient*. This is normally measured in units of reciprocal length (m^{-1} or cm^{-1}), and clearly the intensity falls to $1/e$ of its initial value in a length α^{-1}.

The reflectivity R of a plane boundary between the medium and vacuum may also be expressed in terms of n and κ. For normal incidence, Fresnel's formula gives the reflectivity as

$$R = \frac{(n-1)^2 + \kappa^2}{(n-1)^2 + \kappa^2} \tag{2.23}$$

The complex refractive index can also be discussed, alternatively, in terms of a *complex dielectric constant* ε_c or a complex conductivity σ_c. Then k^2 in Eq. (2.17) can be written as either $\omega^2\mu_0\varepsilon_c\varepsilon_0$ or $i\omega\mu_0\sigma_c$, where

$$\varepsilon_c = \varepsilon_r - i\frac{\sigma}{\omega\varepsilon_0}, \qquad \sigma_c = -\sigma + i\omega\varepsilon_0\varepsilon_r \tag{2.24}$$

These quantities are important because from assumptions about the atomic structure of the medium it is possible to derive expressions for the polarization **P** (the electric dipole moment per unit volume):

$$\mathbf{P} = (\varepsilon_c - 1)\varepsilon_0\mathbf{E} \tag{2.25}$$

where $(\varepsilon_c - 1)$ is the *electrical susceptibility*, χ, or the current density in the presence of an applied field (via Eq. (2.2)).

Having established the relationships between the various optical and electrical constants, we now seek a model that will predict their actual values and frequency dependence. The simple classical model of Lorentz, developed further by Drude, provides us with a useful picture of the interaction between an optical field and a crystal, in which charged particles (electrons and ions) are bound by Hooke's law forces, to atoms for electrons, or to nearest neighbours for lattice ions (see below). The particles are assumed to be subject to a damping force proportional to velocity [2.2], so that when an external optical field of the form $E = E_0 \exp(i\omega t)$ is applied to the system, the equation of motion is that of a forced damped harmonic oscillator where the damping force is proportional to the velocity. In one dimension the equation of motion for a particle of mass m is the familiar forced oscillator equation:

$$m \frac{d^2 x}{dt^2} + mg \frac{dx}{dt} + m\omega_0^2 x = eE \tag{2.26}$$

where the second term, proportional to the velocity, describes the damping, and the third term represents the restoring force. g is a damping constant and is related to the carrier scattering time through $g = 1/\tau$. The restoring force has the form $F = -fx$ (Hooke's law) and is written above in terms of the resonant frequency of the system, $\omega_0 = (f/m)^{1/2}$. The particle oscillates and constitutes a changing electric dipole.

The steady state solution of Eq. (2.26) is

$$x = \frac{(e/m)E}{\omega_0^2 - \omega^2 + i\omega g} \tag{2.27}$$

which results in an induced polarization $P = Nex$, where N is the number of oscillating dipoles per unit volume:

$$\mathbf{P} = \frac{(Ne^2/m)\mathbf{E}}{\omega_0^2 - \omega^2 + i\omega g}. \tag{2.28}$$

Now, from Eqs. (2.17)–(2.20) and (2.25), we have

$$\varepsilon_c(\omega) = (n - i\kappa)^2 = 1 + \frac{\mathbf{P}}{\varepsilon_0 \mathbf{E}} \tag{2.29}$$

We can substitute for $\mathbf{P}/\mathbf{E}$ from Eq. (2.28) and we find that, separating real and imaginary parts, the real part of this expression describes the *dispersion*, i.e. the frequency dependence of *refraction*, given by

$$n^2 - \kappa^2 = 1 + \frac{(Ne^2/m\varepsilon_0)(\omega_0^2 - \omega^2)}{(\omega_0^2 - \omega^2)^2 + \omega^2 g^2} \tag{2.30}$$

and the imaginary part describes *absorption*:

$$2n\kappa = 1 + \frac{(Ne^2/m\varepsilon_0)\omega g}{(\omega_0^2 - \omega^2)^2 + \omega^2 g^2} \tag{2.31}$$

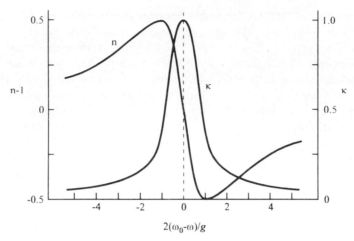

Figure 2.2 Variation of n and κ near a classical Lorentzian absorption line. Both $n-1$ and κ are in units of $Ne^2/2m\omega\varepsilon_0 g$ (after Houghton and Smith [2.3]).

which is clearly zero for the case where $g \to 0$, i.e. $\tau \to \infty$, which implies that there is no absorption or power dissipation in the absence of damping (collisions).

In many practical cases the dipole density N is very small so that the second term on the left-hand side of Eq. (2.30) is also small. Thus, with $n^2 \gg \kappa^2$ and $2n\kappa \approx 2\kappa$ we can obtain expressions for n and κ separately from the above equations. The result is the classic *Lorentz lineshape* for absorption and dispersion, illustrated in Figure 2.2.

There are several different mechanisms which cause dispersion and absorption in materials; we discuss two cases here.

Electronic case: free electron limit

Absorption by 'free' carriers is the dominant mechanism in metals, but it is also important in semiconductors, particularly for long wavelengths. If we take $\omega_0 = 0$ in Eq. (2.26) then there is no restoring force and the carriers are free. At optical frequencies $\omega^2 \gg g^2$ and for small absorption $n^2 \gg \kappa^2$. We therefore have, from Eq. (2.30),

$$n^2 \approx 1 - \frac{Ne^2}{m\varepsilon_0 \omega^2} \tag{2.32}$$

This reduction in the refractive index due to the presence of free carriers is important in nonlinear effects (cf. Section 2.2.1 and Chapter 4) and in semiconductor diode lasers where 'gain guiding' of the modes takes place (cf. Section 3.4.4).

The refractive index tends to unity as the frequency increases. We can include any residual contributions to the dielectric constant by defining $\varepsilon(\infty)$, the permittivity at energies far above any absorption edge, so that the refractive index is given by

$$n_{\text{total}}^2 \approx \left(\varepsilon(\infty) - \frac{Ne^2}{m\varepsilon_0 \omega^2} \right) \tag{2.33}$$

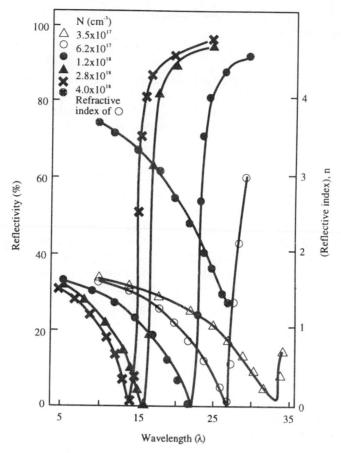

Figure 2.3 Reflectivity near the plasma edge in the n-doped semiconductor InSb (after Spitzer and Fan [2.4]).

The refractive index is therefore zero when ω is equal to the *plasma frequency* ω_p, given by

$$\omega_p^2 = \frac{Ne^2}{m\varepsilon_0\,\varepsilon(\infty)} \tag{2.34}$$

From Eq. (2.23) we see that the reflectance R tends to 1 at the plasma frequency. It turns out that R increases again very rapidly to lower frequencies, forming the so-called *plasma edge* (Figure 2.3).

Since the above expression is independent of τ (and hence of scattering mechanisms), measurement of the plasma edge provides a good method for determining the effective mass in situations where poor mobility (short τ) makes measurement of the cyclotron resonance frequency difficult (see below).

The free carrier *absorption* is given from Eq. (2.31), again assuming that $\omega^2 \gg g^2$:

$$\alpha = \frac{2\omega\kappa}{c} \approx \frac{Ne^2}{nc\omega^2\varepsilon_0}\frac{1}{\tau} = \frac{\sigma_0}{nc\varepsilon_0\omega^2\tau} \qquad (2.35)$$

This expression is clearly dependent on τ, which can have different frequency dependences for different scattering mechanisms. It therefore does not provide a suitable basis for the determination of effective mass parameters. Its importance is, in fact, somewhat negative in that free carrier absorption is the dominant loss mechanism in many optoelectronic devices.

Lattice absorption: phonons

A second process by which electromagnetic radiation interacts with a crystal involves motion of the constituent atoms in the lattice. The lattice has a set of discrete modes of vibration, known as *phonons*, which may have an associated oscillating electric dipole moment. Changes in the vibrational state of the crystal may be accompanied by changes in the dipole moment, allowing a coupling to take place between the atomic motions and the incident radiation field causing *absorption*. This interaction generally determines the optical properties of solids in the 10–100 μm wavelength range.

A simple one-dimensional model of a diatomic chain of atoms, with masses M and m, with a restoring force of the form $F = -fx$ between them, gives a dispersion curve of the form

$$\omega^2 = f\left(\frac{1}{m} + \frac{1}{M}\right) \pm \left\{\left(\frac{1}{m} + \frac{1}{M}\right)^2 - \frac{4\,\sin^2 qa}{Mm}\right\}^{1/2} \qquad (2.36)$$

where a is the lattice constant and q is the phonon wavevector. This is illustrated in Figure 2.4, where we see that the dispersion relation is split into an upper *optical* branch and a lower *acoustical* branch. The former is responsible for absorption in ionic crystals, since the mode of vibration sets up an oscillating dipole moment in the lattice. No such moment exists for the acoustical modes.

In order that a given mode can absorb, the conditions of conservation of energy and momentum must be satisfied. This means that different processes tend to give information about different parts of the phonon dispersion curves. For example, *Reststrahl* (residual ray) absorption, which occurs in ionic crystals only and is due to the creation of single phonons, is restricted to phonons of near-zero wavevector because the incoming photons have very low momentum. On the other hand, *multiphonon absorption* is biased toward high q values since it occurs when two or more phonons interact to produce an electric moment with which the radiation may couple. This process can occur in both ionic and homopolar crystals.

In the case of the ionic crystal, the simple theory predicts a line absorption at the optical phonon frequency corresponding to $q = 0$, i.e. at

$$\omega_{\text{max}} = \left\{2f\left(\frac{1}{m} + \frac{1}{M}\right)\right\}^{1/2} = \left\{\frac{2f}{\mu}\right\}^{1/2} \qquad (2.37)$$

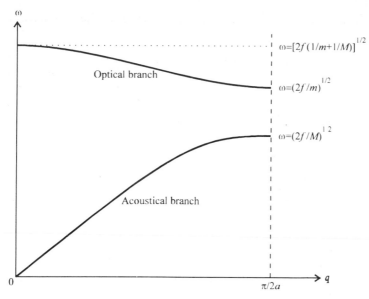

Figure 2.4 Dispersion curves for a diatomic chain.

where the reduced mass μ is given by

$$\frac{1}{\mu} = \frac{1}{m} + \frac{1}{M} \tag{2.38}$$

To deduce the dielectric constant, we can use the same model that we used for the electrons by using a (much larger) reduced mass μ and an effective charge e^*. Because electromagnetic waves are transverse waves, they only excite *transverse* optical modes in the crystal, so we write $\omega_{max} = \omega_T$. Ignoring any damping (i.e. $g = 0$), we have

$$\varepsilon(\omega) = \varepsilon(\infty) + \frac{(Ne^{*2}/\varepsilon_0\mu)}{\omega_T^2 - \omega^2} \tag{2.39}$$

where $\varepsilon(\infty)$ is a high-frequency dielectric constant that accounts for electronic effects. We may rewrite this in the form:

$$\varepsilon(\omega) = \varepsilon(\infty) + \frac{\varepsilon(0) - \varepsilon(\infty)}{1 - \omega^2/\omega_T^2} \tag{2.40}$$

where $\varepsilon(0)$ is the static dielectric constant obtained by putting $\omega = 0$ in Eq. (2.39). Equation (2.40) gives us the dielectric constant in terms of accessible parameters. The zero of $\varepsilon(\omega)$ defines the longitudinal mode frequency ω_L. Thus

$$0 = \varepsilon(\infty) + \frac{\varepsilon(0) - \varepsilon(\infty)}{1 - \omega_L^2/\omega_T^2} \tag{2.41}$$

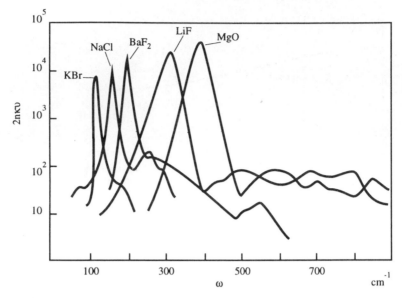

Figure 2.5 Absorption in some ionic crystals: LiF and MgO at 100 K, NaCl, KBr and BaF$_2$ at 300 K (after Price and Wilkinson [2.5]).

and this gives us the well-known Lyddane–Sachs–Teller relation:

$$\frac{\varepsilon(0)}{\varepsilon(\infty)} = \frac{\omega_L^2}{\omega_T^2} \tag{2.42}$$

It turns out that the longitudinal optical mode frequency is generally higher than the transverse mode frequency, so in the range $\omega_T < \omega < \omega_L$, $\varepsilon(\omega)$ is negative and therefore purely imaginary, i.e. $\varepsilon(\omega) = (n + i\kappa)^2 \to (i\kappa)^2$ and $n \to 0$. In this range the reflectivity R (given by Eq. (2.23)) tends to unity and one has total reflection. Historically, this was known as *Reststrahlen* (residual ray) reflection, from the early days of infrared technology. For ionic materials the effect is very strong, since the interaction between the optical modes of vibration and the electromagnetic field is strong. The absorption for some ionic crystals is illustrated in Figure 2.5. We see that it is dominated by a single, *strong* peak as predicted by the classical theory.

Each atom, molecule or crystal possesses a number of different resonances, each of which contributes to the optical properties. The resonances may be due to bound electrons, free electrons, vibrating atoms in a molecule or crystal, or rotating molecules, and each tends to dominate a different spectral region. This is illustrated in Figure 2.6 for a dipolar material, where a series of resonances in the polarization is clearly evident, with the lattice vibrations important in the far infrared.

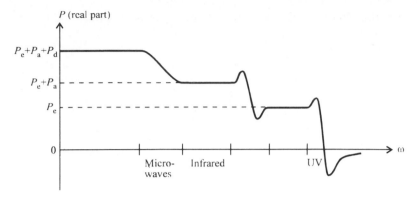

Figure 2.6 The real part of the polarization P as a function of frequency for a dipolar substance with a single atomic and electronic resonance frequency. P_e, P_a and P_d represent the electronic, atomic vibrational and dipolar rotational contributions to the total polarization P.

2.1.3 Optical properties: quantum mechanics

In order to predict the resonant frequencies of a system, quantum theory is required and this is the subject of the latter part of this chapter. We note here, however, that the quantum mechanical result for the polarization is not very different from that obtained using a classical oscillator model. The only substantial difference arises because the classical model allows only one resonant frequency, ω_0, per oscillator, whereas each atom, molecule or electron has a *characteristic set* of frequencies, ω_{0n}, corresponding to the radiation frequencies that it can emit or absorb. We can, however, proceed as before if we replace terms of the type

$$\frac{1}{\omega_0^2 - \omega^2 + i\omega g}$$

in Eq. (2.28), for example, with

$$\sum_n \frac{f_{n0}}{\omega_{0n}^2 + \omega^2 + i\omega g}$$

summing over all the frequencies ω_{0n}. The weighting function f_{n0} is known as the *oscillator strength*; this is a dimensionless quantity and is always less than unity. It turns out that for an electron in a free atom

$$\sum_n f_n = 1.$$

2.1.4 Free electron theory of metals

At the time that the classical theory described above was developed, it seemed reasonable to assume that the velocity distribution in the electron gas was described by Maxwell–Boltzmann statistics. However, the electronic contribution to the thermal heat capacity thereby deduced was very much larger than observed, and only with the advent of quantum theory was this discrepancy explained. Then it was recognized that, because of the *Pauli exclusion principle*, the electronic energies were described by the *Fermi–Dirac distribution*, rather than by Maxwell–Boltzmann statistics; Sommerfeld then modified the classical theory to take this into account [2.6]. In this section we review some of the fundamental concepts underlying Sommerfeld's model.

1-D particle in box

We seek to describe the ground state properties of N electrons confined to a box of side L. To do this we assume that the electrons are non-interacting, calculate the permitted energy levels for a single electron and then fill these up, taking the exclusion principle into account, until all N electrons have been accommodated. In one dimension the Schrödinger equation takes the form

$$\frac{\hbar^2}{2m} \frac{d^2\psi}{dx^2} + E\psi = 0 \tag{2.43}$$

where m is the free electron mass, with solutions

$$\psi = \frac{1}{\sqrt{L}} \exp ikx \tag{2.44}$$

and application of the *Born–von Karmann* boundary conditions gives $k_x = n_x \pi / L$, where n_x are integers. The resulting dispersion relation is parabolic:

$$E = \frac{\hbar^2 k_x^2}{2m} \tag{2.45}$$

The boundary conditions only permit certain discrete values of k_x. Applying this idea to three dimensions, we are immediately concerned with the actual crystal structure and symmetry, in terms of reciprocal lattice vectors $2\pi/\mathbf{a}$, $2\pi/\mathbf{b}$ and $2\pi/\mathbf{c}$. The allowed $\mathbf{k}$ values can be represented in extended $\mathbf{k}$-space (the extended zone scheme) or, equivalently, entirely within the first Brillouin zone (the reduced zone scheme), by adding the quantum number n, as indicated in Figure 2.7 and Eq. (2.46):

$$E(k_x, k_y, k_z, n) = \frac{\hbar^2}{2m} \left[\left(k_x + n_x \frac{2p}{a} \right)^2 + \left(k_y + n_y \frac{2p}{b} \right)^2 + \left(k_z + n_z \frac{2p}{c} \right)^2 \right] \tag{2.46}$$

where the quantum number n labels the first, second, third, etc., Brillouin zones in $\mathbf{k}$-space [2.7]. Note that the range $-\pi/a < k_x < \pi/a$ is the first Brillouin zone in the k_x direction. The full energy band $E = \hbar^2 k_x^2 / 2m$ can either be represented in extended k-space (extended zone) or reduced to the first Brillouin zone (reduced zone) by adding the quantum number n_x, as shown in Figure 2.7.

Vibrations of atoms in a 1-D lattice

The physics of this can readily be understood, with rather simpler mathematics, by referring back to the essentially similar case of lattice vibrations, Eq. (2.36). If we now

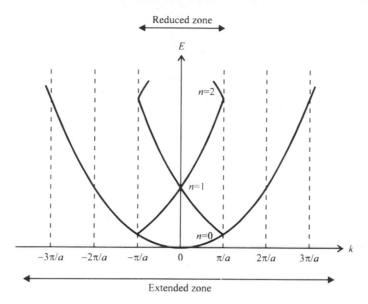

Figure 2.7 The quadratic electron dispersion relation in the 1-D particle in a box model.

consider a monatomic lattice of N atoms at a distance a apart, then the choice of standing wave solutions of the form $x_n = A \cos \omega t \cos kna$ and $x_n = A \cos \omega t \sin kna$ as trial solutions leads to the dispersion relation $\omega = \pm (4f/m)^{1/2} \sin(ka/2)$, which is illustrated in Figure 2.8.

We then wish to see what effect the imposition of appropriate boundary conditions has on this dispersion, and we specify these by requiring that the solutions be periodic over some sufficiently large distance in the chain. That is, we require $x_n = x_{n+N}$ (referred to as *periodic* boundary conditions), or $\cos kna = \cos k(n + N)a$, which means that k is restricted to the values $k = 2\pi l/Na$, where l has values $1, \ldots, \frac{1}{2}N$. This means that there are $\frac{1}{2}N$ cosine and $\frac{1}{2}N$ sine solutions and a total of N solutions equal to the total number of particles. The dispersion relation is seen to repeat itself outside the limits $k = \pm \pi/a$, which means that the larger values of k merely reproduce the motion described within this range. The range $-\pi/a < k < \pi/a$ is the same first Brillouin zone of this one-dimensional lattice. If we encounter larger values of k they can be handled by subtracting integral multiples of π/a which transform k back into the first zone; this is therefore the reduced zone scheme for our one-dimensional lattice. We thereby see that the allowed k-vectors, and hence the allowed frequencies, or energies, are a general consequence of confining wave motion within a periodic system.

Density of states
Although the wavevectors $\mathbf{k}_x$, $\mathbf{k}_y$ and $\mathbf{k}_z$ are discrete we are often interested in the number of states in a volume of k-space that is vast on the scale of π/L. Thus, to a

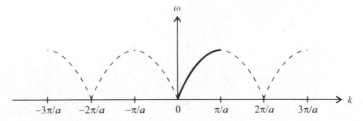

Figure 2.8 Dispersion relation for a monatomic linear chain.

good approximation, the number of states is just the volume of k-space divided by the k-space volume per state. The number of states per unit volume of k-space *per unit volume of real space* is known as the *density of (k) states*, $D(k)$; a knowledge of this function will be extremely important in our later consideration of the optical properties of materials.

We calculate the three-dimensional density of states by noting that any solution of Eq. (2.43) must satisfy the identity

$$\frac{L^2 k^2}{\pi^2} = n_x^2 + n_y^2 + n_z^2 \tag{2.47}$$

where $k_x = n_x \pi / L$ etc. for *fixed* boundary conditions. The states will be occupied in such a way as to minimize the total energy and thus they fill up the positive quadrant of a sphere of radius R_0, where $n_x^2 + n_y^2 + n_z^2 \leq R_0$ (see Figure 2.9). The number of states per unit volume between R and $R + dR$ in this quadrant (remembering the factor of $1/8$ for the positive quadrant) is

$$D(R)dR = 4\pi R^2 dR \left(\frac{1}{8}\right)\frac{1}{V} \tag{2.48}$$

i.e. the surface area of the shell multiplied by the thickness of the element. This can be expressed in terms of k, so the number of states in dk is given by:

$$D(k)dk = D(R)\frac{dR}{dk} dk$$

$$= \frac{1}{2\pi^2} 2k^2 dk = \frac{k^2}{\pi^2} dk \tag{2.49}$$

where the factor of 2 is due to the electron spin. This in turn, through the relation between E and k (Eq. (2.45)) can be expressed in terms of the *density of states per unit energy*, $D(E)$, such that the concentration of states between E and $E + dE$ is given by:

$$D(E)dE = D(k)\frac{dk}{dE} dE = \frac{1}{2\pi^2}\left(\frac{2m}{\hbar^2}\right)^{3/2} E^{1/2} dE \tag{2.50}$$

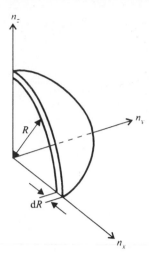

Figure 2.9 Distribution of states in '*n*-space'.

This is an important result: the *three-dimensional energy density of states*. It occurs in the description of all processes in which electron excitation occurs and determines such properties as the spectral distribution of emission from a light-emitting diode (LED). This result is required for the discussions in Chapter 3, particularly Sections 3.2.6 and 3.4.1.

Devices are, however, often in the form of very thin films or other *low-dimensional structures*, in which the density of states takes a different form. Using a similar argument to that presented above, we have, in *two dimensions*:

$$\frac{L^2 k^2}{2\pi^2} = n_x^2 + n_y^2 = R^2 \tag{2.51}$$

where R is now the radius of a disk in *n*-space, and

$$D(R)\mathrm{d}R = \frac{1}{4} \frac{2\pi R}{A} \, \mathrm{d}R \tag{2.52}$$

where A is the sample area. Also,

$$D(k)\mathrm{d}k = \frac{1}{\pi} \, k\mathrm{d}k \tag{2.53}$$

per unit area in real space. Thus

$$D(E)\mathrm{d}E = \frac{1}{\pi} \left(\frac{m}{\hbar^2}\right) \mathrm{d}E \tag{2.54}$$

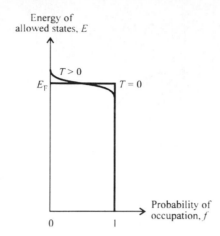

Figure 2.10 The Fermi–Dirac distribution.

and we see that in two-dimensional devices the density of states $D(E)$ is *constant* with respect to energy.

Similarly, in *one-dimension* we find that

$$D(E)dE = \frac{1}{2\pi}\left(\frac{2m}{\hbar^2}\right)^{1/2} E^{-1/2}dE. \tag{2.55}$$

The Fermi–Dirac distribution
At zero temperature, the electrons occupy the lowest energy states allowed to them by the exclusion principle. At finite temperatures this is no longer the case. Thermal energy allows some electrons to exist in higher energy states. In thermal equilibrium, the probability, f, that a state with energy E will be occupied is governed by Fermi–Dirac statistics:

$$f = 1\left/\left[1 + \exp\left(\frac{E - E_F}{kT}\right)\right]\right. \tag{2.56}$$

where E_F is known as the *Fermi energy* and is defined as the energy at which the probability of occupation is $1/2$. This function is shown in Figure 2.10 at zero and at finite temperatures.

2.1.5 Energy bands and energy gaps: semiconductors

Electrons in a periodic potential
In the presence of the perturbing periodic crystal potential, $V(x)$, the above simplicity is lost: this is the 'nearly-free electron model'. The Schrödinger equation in the x-direction becomes

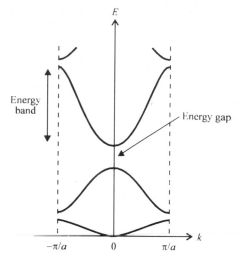

Figure 2.11 Energy bands in real semiconductors, characterized by gaps at the high symmetry points in the Brillouin zone.

$$-\frac{\hbar^2}{2m}\frac{d^2\psi}{dx^2} + V(x)\psi = E\psi \tag{2.57}$$

The main features of this model are as follows:

1. The electrons we are concerned with are confined within the lattice. Consequently, the wavefunction ψ is modulated at the period of the lattice. By symmetry arguments, the correct wavefunction can be shown to be:

$$\psi = \frac{1}{\sqrt{L}}\, u_k(x)\, \exp(ikx) \tag{2.58}$$

where $u_k(x)$ is a function with the periodicity of $V(x)$, and ψ is known as the Bloch function; it is a satisfactory solution to the Schrödinger equation.

2. Regions of energy appear in which there are no allowed solutions for $\psi_k(x)$; these are the so-called energy gaps (forbidden energy regions) that separate energy bands (allowed energy regions); see Figure 2.11. Note the analogy with Figures 2.4 and 2.8 for the case of vibrations of the lattice atoms where gaps also occur.

Effective mass
The motion of the electrons is, of course, affected by the periodic crystal potential. However, it is possible to construct a wavefunction for the electron so that its motion can be represented by that of a *particle with an effective mass, m**, which turns out to be somewhat smaller than the free mass. The actual magnitude of the effective mass is determined by the *E–k* relation, and its use allows a close analogy to be

drawn with the classical motion of a particle. It is then possible to use many of the classical results of Sections 2.1.1 and 2.1.2 in order to understand optical and electrical effects in semiconductors.

The wavevector, **k**, characteristic of the electron wavefunction, is closely related to the momentum of the particle as it moves through the crystal. Thus by considering the time rate of change of momentum, an analogy can be made with Newton's second law and so the effect of an external field can be expressed in terms of the **k**-vector.

Using the above description, the dispersion relation becomes

$$E = \frac{\hbar^2 k^2}{2m^*} \tag{2.59}$$

and we thus obtain, by differentiating twice,

$$\frac{1}{m^*} = \frac{1}{\hbar^2} \frac{\mathrm{d}^2 E}{\mathrm{d}k^2} \tag{2.60}$$

Therefore m^* is directly related to the band curvature.

Metals, insulators and semiconductors
The position of the Fermi level determines whether the solid is a metal, an insulator or a semiconductor. The highest band that is filled with electrons is called the valence band; the next highest band is called the conduction band; the energy gap is the valence/conduction band separation. For a metal, E_F is in the middle of the conduction band, and for an insulator, it is in the middle of the energy gap. Semiconductors are a class of crystals with small energy gaps. They are insulators at $T = 0$ K, but are conductors at finite temperatures due to either thermal excitation of electrons across the gap or the effect of 'doping' with impurity atoms.

2.1.6 Impurities: intrinsic and extrinsic conductivity

The key point about semiconductors from a device point of view is that their conductivity can be controlled. The basic method of control is to dope the host crystal with impurities either of valence greater than the host (*donors*, giving excess electrons so that the material is '*n*-type') or of valence less than the host (*acceptors*, giving excess holes and '*p*-type' material). The impurities are chosen to have very low binding energies so that the electrons (holes) are substantially ionized at room temperature to provide *n*-type (or *p*-type) conductivity (Figure 2.12). These shallow impurities can be treated by the simple Bohr theory of the hydrogen atom giving a set of discrete energy levels, E_n, but substituting the effective electron mass, m^*, for the usual mass of the electron:

$$E_n = \frac{m^*/m_0}{\varepsilon^2} \frac{13.6}{n_B^2} \text{ eV} \tag{2.61}$$

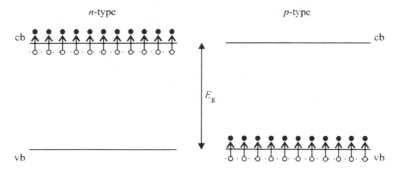

Figure 2.12 Doping in semiconductors. In *n*-type material, carriers are excited thermally from the donor states to the conduction band where they can contribute to the conductivity. In *p*-type material, electrons are effectively 'captured' by the acceptor states, leaving behind holes in the valence band which can also contribute to conduction.

where 13.6 eV is the binding energy of the hydrogen atom, ε is the dielectric constant and n_B is the principal (Bohr) quantum number. The binding energy is then

$$E_B = \frac{m^*/m_0}{\varepsilon^2} \times 13.6 \text{ eV} \tag{2.62}$$

For typical group IV semiconductors such as germanium and silicon, we have $m^*/m_0 \sim 0.2$–0.3, $\varepsilon \sim 12$–16, and for the III–V materials that are of particular importance to optoelectronics, such as GaAs, $m^*/m_0 \sim 0.01$–0.07, $\varepsilon \sim 12$–16. Thus E_B is typically a few meV (i.e. much less than E_g, and kT at room temperature).

The donor (acceptor) levels lie just below (above) the conduction (valence) band and carriers are therefore easily excited thermally into the neighbouring band where they are able to contribute to the overall conductivity (Figure 2.12). The total conductivity will then be $\sigma = \sigma_i + \sigma_e$, where the intrinsic conductivity, $\sigma_i = N_i e^2 \tau/m^*$, is that due to thermal excitation of electrons from the valence band to the conduction band, and the extrinsic conductivity, $\sigma_e = N_e e^2 \tau/m^*$, is that due to impurities. The intrinsic carrier concentration is then $N_i \propto \exp(-E_g/2kT)$. The extrinsic concentration is $N_e \propto \exp(-E_d/2kT)$ at low temperatures (i.e. where $kT < E_d$). At temperatures such that $kT > E_d$ (so that the impurities are all ionized) $N_e \rightarrow N_d$ (or N_A), the donor (or acceptor) concentration (see Figure 2.13 and Ref. [2.8]).

2.1.7 Interband optical transitions: excitons

Apart from thermal promotion of electrons and holes via interband (valence to conduction band) transitions responsible for intrinsic conductivity, it is possible to directly excite such transitions optically by subjecting the sample to radiation with $\hbar\omega > E_g$. If the lowest conduction band minimum is situated vertically above the valence band maximum (usually at $k = 0$), then the optical process is referred to as

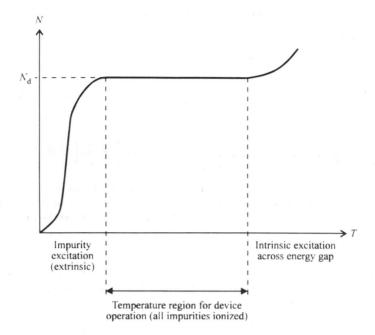

Figure 2.13 Variation in carrier density as a function of temperature in a typical semiconductor.

a *direct transition*. If the lowest conduction band minimum is situated at some other point in *k*-space than that of the valence band maximum, the process is referred to as an *indirect transition*. This requires the participation of optical phonons to satisfy momentum conservation and is therefore a weaker (higher-order) process. The above materials are referred to as *direct gap* and *indirect gap* semiconductors respectively (Figure 2.14).

The optical spectrum corresponding to the onset of interband electric dipole transitions is referred to as the *absorption edge*. The most widely used indirect gap materials are the group IV semiconductors Ge and Si. The best known direct gap materials are the III–V (InSb, GaAs, GaP, etc.), II–VI (HgTe, CdTe, ZnSe, etc.) and IV–VI (lead salts PbTe, PnSe and SnTe) and alloy combinations of all these. An important additional effect, which becomes progressively stronger as the energy gap becomes larger in a semiconductor, is the Coulomb interaction between the electrons and holes created by interband optical transitions. The electron is raised into an excited state but remains bound in a hydrogen-like orbit around the positive hole, creating a two-particle excitation called an *exciton*. The composite particle can then move bodily throughout the crystal. In an ideal crystal, therefore, the absorption spectrum should consist of a series of discrete lines corresponding to the ground and excited exciton states, followed at somewhat higher energies by the continuum associated with interband electronic transitions. The energy levels of the exciton

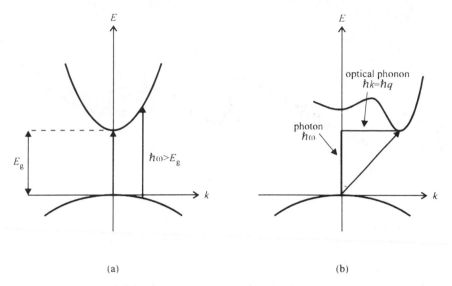

Figure 2.14 (a) Direct and (b) indirect gap semiconductors. Direct gap transitions involve only the absorption or emission of a single photon and are therefore very nearly 'vertical' transitions in *k*-space. On the other hand, indirect transitions have a 'two-stage' nature, involving the participation of both a photon and an optical phonon that carries sufficient momentum *hq* to allow the conduction band electron created in the absorption process to have a large nonzero wavevector.

spectrum measured from the beginning of the ionization continuum will be given by the Bohr formula:

$$E_{ex} = \frac{m_r}{\varepsilon^2} \frac{13.6}{n_B^2} \text{ eV} \tag{2.63}$$

where m_r is the reduced mass given by:

$$\frac{1}{m_r} = \frac{1}{m_e} + \frac{1}{m_h}$$

where m_e, m_h are the electron and hole effective masses. The exciton structures observed near the absorption edge of GaAs (weakly excitonic) and Cu_2O (strongly excitonic, large gap) are shown in Figure 2.15. The change in excitonic absorption on applying an electric field is used to make a modulator, as described in Section 3.5.3.

2.2 Examples of applications of the Drude model in the optical and transport properties of semiconductors

The Drude model, modified by the use of the effective mass, has been extremely helpful in interpreting optical and transport properties of semiconductors, and we

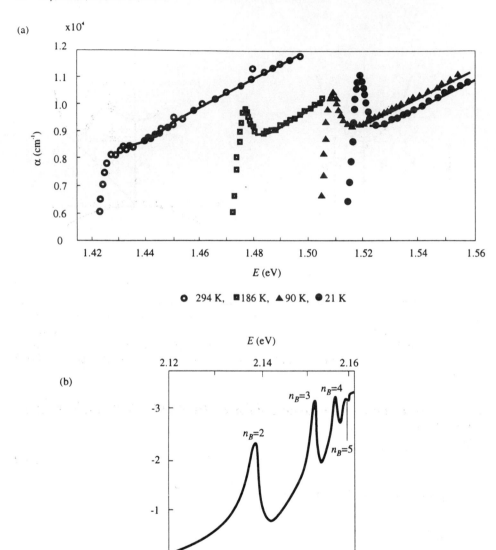

Figure 2.15 (a) Exciton structure at the band edge in bulk GaAs at various temperatures [2.9], and (b) in Cu_2O at 77 K [2.10].

give some examples here. In particular we give a simple classical treatment of photoconductivity, nonlinear refraction and magneto-optics.

2.2.1 Photoconductivity

When radiation excites interband (or impurity to band) transitions, electron–hole pairs are created, bringing about a change in conductivity. Normally a constant current bias is used and the voltage change across a load resistor is used as the output of this *photoconductive detector*. Many of the detectors discussed in Chapter 9 make use of this effect (Section 9.5.3).

The dark current conductivity is given by

$$\sigma_0 = e(n_0 \mu_e + p_0 \mu_h) \tag{2.64}$$

When the light is turned on, $\Delta n = \Delta p$, and

$$\sigma = \sigma_0 + e\Delta n(\mu_e + \mu_h) = \sigma_0 + e\Delta n \mu_h(1 + b) \tag{2.65}$$

where b is the mobility ratio μ_e/μ_h. Therefore the increase in the conductivity is

$$\Delta\sigma = e\Delta n \mu_h(1 + b) \tag{2.66}$$

Δn is determined by the rate of electron–hole creation balanced by recombination. The rate equation for a simple generation–recombination model is

$$\frac{dn}{dt} = G - \frac{\Delta n}{T_1} \tag{2.67}$$

where G is the generation rate per unit volume and T_1 is the carrier recombination time (i.e. *not* the carrier scattering time τ, typically $T_1 > 10^3\tau$). In the steady state $dn/dt = 0$, so $\Delta n = GT_1$.

For a slab of thickness d, the fractional power absorbed, $\Delta I/I = \alpha d$, where α is the absorption coefficient. Therefore, if there are $N(\omega)$ incident photons per second, the number absorbed per second is $= \alpha d N(\omega)$. If we assume that each photon creates one electron–hole pair, we have

$$G = \frac{\alpha d N(\omega)}{V} \tag{2.68}$$

But $N(\omega) = I(\omega)A/\hbar\omega$, where A is the area illuminated. So, finally, the excited electron density is

$$\Delta n = \frac{\alpha I(\omega)}{\hbar\omega} T_1 \tag{2.69}$$

and

$$\Delta\sigma = \frac{\alpha I(\omega) T_1 \mu_h(1 + b)e}{\hbar\omega} \tag{2.70}$$

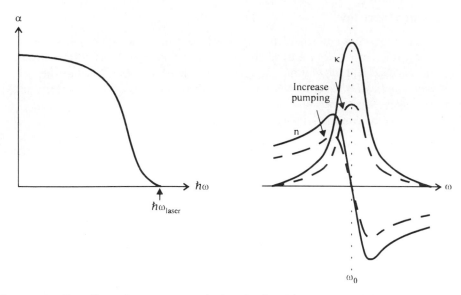

Figure 2.16 The effect of pumping into the band tail on absorption and dispersion.

2.2.2 Nonlinear refraction by interband optical pumping: a mechanism for optical bistability

In the study of semiconductor nonlinear optics it is found that illuminating the material at high intensities ('pumping') at energies corresponding to the tail of the absorption edge produces a strong intensity dependence in the refractive index. This is a direct result of photoexcited (valence to conduction band) carriers contributing to the refractive index, i.e.

$$n = n_0 + n_2 I \tag{2.71}$$

We can obtain an approximate expression for n_2 by using the Drude model.

Assume that each photon produces one electron–hole pair with a recombination time T_1. We once again use the rate equation (2.67), in a manner similar to that used in the photoconductivity discussion, where the generation rate per unit volume is just

$$G = \frac{\text{fractional power absorbed} \times N(\omega)}{V} \tag{2.72}$$

where $N(\omega)$ is the number of photons incident per second, so that

$$G = \frac{\alpha d N(\omega)}{V} = \frac{\alpha d}{V} \frac{IA}{\hbar\omega} = \frac{\alpha I}{\hbar\omega} = \frac{\Delta I}{\hbar\omega}. \tag{2.73}$$

In the steady state $dn/dt = 0$, so $\Delta n = \alpha I T_1 / \hbar\omega$. From the Drude theory (Eq. (2.33)) we have the refractive index in terms of the carrier density:

$$n^2 = \varepsilon_\infty - \frac{\Delta n e^2}{m^* \omega^2 \varepsilon_0}$$

giving, for small Δn,

$$n \approx \varepsilon_\infty^{1/2} + \tfrac{1}{2}\varepsilon_\infty^{-1/2}\,\frac{-\Delta n e^2}{m^* \varepsilon_0 \omega^2} \tag{2.74}$$

$$= n_0 - \left(\frac{e^2 \alpha T_1}{2n_0 m^* \hbar \omega^3 E_0}\right) I \tag{2.75}$$

or

$$n_2 = -\frac{e^2 \alpha T_1}{2n_0 m^* \hbar \omega^3 \varepsilon_0} \tag{2.76}$$

Clearly, n_2 makes a negative contribution to the refractive index and has a strong frequency dependence. This effect is responsible for part of the nonlinearity which has been utilized in optical bistability, optical logic and memory elements, and is described in Chapter 4. Figure 2.16 gives an analogy to explain the remaining interband blocking effect discussed in Section 4.2.4.

2.2.3 Classical magneto-optics of semiconductors

Cyclotron resonance and the Faraday effect
Magneto-optical experiments provide some of the most precise measurements of semiconductor band parameters. For example, where it is measurable (i.e. in high-quality, pure crystals) *cyclotron resonance* (CR) is the most accurate method for obtaining carrier effective masses. Furthermore, many materials become optically active when placed in a magnetic field and this phenomenon (known as the *Faraday effect*) has been utilized in modulator devices which act as *optical isolators* (see Section 7.4). Here we consider these magneto-optical effects from the point of view of the classical theory.

We can rewrite Eq. (2.26), the equation of motion for an electron, to take into account the presence of a magnetic field as follows:

$$m^* \frac{\mathrm{d}^2 \mathbf{r}}{\mathrm{d}t^2} + g m^* \frac{\mathrm{d}\mathbf{r}}{\mathrm{d}t} = e\mathbf{E} + e\,\frac{\mathrm{d}\mathbf{r}}{\mathrm{d}t} \times \mathbf{B} \tag{2.77}$$

where $\mathbf{E}$ is the electric component in the optical wave and $\mathbf{B}$ the externally imposed (static) magnetic component of the field. The term on the right-hand side is the Lorentz force, and the g-term is once again introduced to account for damping effects. We can write Eq. (2.77) for each component separately, putting B along z:

$$m^* \frac{\mathrm{d}^2 x}{\mathrm{d}t^2} + g m^* \frac{\mathrm{d}x}{\mathrm{d}t} - e\left(\frac{\mathrm{d}y}{\mathrm{d}t}\,B_z\right) = eE_x \tag{2.78}$$

$$m^* \frac{\mathrm{d}^2 y}{\mathrm{d}t^2} + g m^* \frac{\mathrm{d}y}{\mathrm{d}t} + e\left(\frac{\mathrm{d}x}{\mathrm{d}t}\,B_z\right) = eE_y \tag{2.79}$$

$$m^* \frac{d^2y}{dt^2} + gm^* \frac{dz}{dt} + 0 = eE_z \tag{2.80}$$

For circularly polarized light propagating along the z-direction, we can write the electric vector in the form

$$E = E_x \pm iE_y = E_0 (\cos \omega t \pm i \sin \omega t) = eE \exp(\pm i\omega t) \tag{2.81}$$

Thus, multiplying Eq. (2.79) by i and adding Eq. (2.78), we have for both polarized modes:

$$m^* \frac{d^2s}{dt^2} + gm^* \frac{ds}{dt} - iBe \frac{ds}{dt} = e(E_x \pm iE_y) = eE_0 \exp(\pm i\omega t) \tag{2.82}$$

where we define $s = s_0 \exp(\pm i\omega t)$ as a complex displacement. The solution to this is

$$s = \frac{(e/m^*)E_0 \exp(\pm i\omega t)}{(-\omega \pm eB/m^*) \pm ig} \frac{1}{\omega} \tag{2.83}$$

where eB/m^* is the cyclotron frequency, ω_c. Thus the external magnetic field has now 'turned on' a resonant frequency. Proceeding exactly as before for the Drude model, we now have an induced a polarization for the electrons, $P = nes$. Thus,

$$\varepsilon_\pm = (n_\pm + i\kappa_\pm)^2 = 1 + \frac{P}{E\varepsilon_0} = 1 + \frac{ne^2/m^*\omega\varepsilon_0}{(-\omega \pm \omega_c \pm ig)} \tag{2.84}$$

therefore

$$n_\pm^2 - \kappa_\pm^2 = 1 + \frac{(\omega \pm \omega_c)ne^2/m^*\omega\varepsilon_0}{(\omega \pm \omega_c)^2 + g^2} \tag{2.85}$$

and

$$2n_\pm\kappa_\pm = \frac{ne^2g/m^*\omega\varepsilon_0}{(\omega \pm \omega_c)^2 + g^2} \tag{2.86}$$

where the $+$ and $-$ signs refer to left and right senses of polarization. As we discussed earlier in the context of the Drude model, the imaginary part of the dielectric constant gives rise to absorption. For right-hand polarization we get *cyclotron resonance absorption* when

$$\omega = \omega_c = eB/m^* \tag{2.87}$$

ω_c is the frequency with which the electrons orbit about the magnetic field direction. To see a resonance, the electrons have to complete more than one orbit before they are scattered into another state by a collision. Thus we need $\omega_c \gg g$, i.e. $\omega_c\tau \gg 1$, to observe a well defined resonance. Clearly there is no *resonant* absorption for light polarized in the opposite sense.

For plane-polarized light with **k** parallel to **B**, the dispersive effects result in a rotation, θ, of the plane of polarization. This is known as *Faraday rotation*, and to understand this we note that the incoming plane-polarized light can be divided into two oppositely polarized circular components of equal magnitude. The refractive indices of the two polarizations are very slightly different so that they have different velocities as they travel through the crystal; this results in a phase shift given by

$$\delta = \frac{2\pi}{\lambda} d(n_+ - n_-) \tag{2.88}$$

where d is the sample width. After traversing the sample the resultant polarization has been rotated by $\theta = \delta/2$:

$$\theta = \frac{\pi d}{\lambda}(n_+ - n_-) = \frac{\omega d}{2c}(n_+ - n_-) \tag{2.89}$$

For *free carriers*, making the approximations $n^2 \gg \kappa^2$, $\omega^2 \gg g^2$, one finds that

$$n_+^2 - n_-^2 = \left(\frac{ne^2}{mE_0}\right)\left(\frac{2Be}{m\omega^3}\right) \tag{2.90}$$

so that

$$\theta = \frac{Bne^3}{2ncE_0 m^{*2}\omega^2} \tag{2.91}$$

This is a good nonresonant method for measuring m^*. It is independent of τ and therefore can be used in some situations where a cyclotron resonance measurement is not possible because the condition $\omega_c\tau \gg 1$ is not met.

Devices that take advantage of the above property to produce a rotation in the plane of polarization are known as *Faraday rotators*. These are used, in conjunction with a plane polarizer, to provide optical isolation in systems where back reflection from a resonator mirror is to be eliminated, such as in a laser amplifier chain. Alternatively, mode propagation in only one direction may be desirable, as in some ring laser cavities. This is achieved by designing the Faraday rotator so that it produces a rotation of 45° in the plane of polarization in one pass. Any beam that is reflected and traverses the rotator a second time will be subject to a 90° rotation in all, allowing it to be eliminated by a polarizer. The device is described in Section 7.4. Usually the Faraday effect of *bound* electrons is utilized.

2.3 Energy bands of real semiconductors

The models so far presented give useful simple explanations of many effects used in devices. We now discuss the quantum mechanics required for quantitative 'band structure engineering'.

2.3.1 Introduction: the nearly-free electron model

The physical model underlying most 'first-principles' energy band calculations is the so-called 'nearly-free electron' (NFE) model. We briefly outline this formalism here, making the simplification of one dimension.

The quantum mechanical model of the electron in a one-dimensional infinite potential well is taken as the starting point, as outlined in the previous section. The Schrödinger equation is therefore

$$H_0 \psi_k(x) = E_k \psi_k(x) \tag{2.92}$$

where H_0 is the Hamiltonian and E_k the energy eigenvalues. The solutions are in the form of the free-electron plane wavefunctions

$$\psi_k = \frac{1}{\sqrt{L}} \exp(ikx) \tag{2.93}$$

The *periodic* crystal potential is then treated as a perturbation and because it is periodic in space, it can be written as a Fourier series:

$$H'_j = \sum_j V_j \exp - i\left(\frac{2j\pi x}{a}\right) \tag{2.94}$$

where j is an integer. The V_j are simply coefficients in this expansion. The perturbed equation

$$H\psi'_{j,k} = (H_0 + H')\psi'_{j,k} = E'_{j,k}\psi'_{j,k} \tag{2.95}$$

is solved by perturbation theory, as described in many undergraduate textbooks on solid state physics [2.11]. Close to points in k-space where the unperturbed Hamiltonian gives bands of equal energy (*degenerate* bands), such as at the zone centre and zone boundaries in high symmetry directions (Figure 2.17), degenerate perturbation theory is used. In this case we can make a crude approximation including only the two energy bands adjacent to the degeneracy point in the interaction. This leads directly to the familiar quadratic expression for the energies of two (conduction-like and valence-like) energy bands separated by an energy gap:

$$\left(\frac{\hbar^2 k^2}{2m} - E'\right)\left(\frac{\hbar^2}{2m}\left(k - \frac{2\pi}{a}\right)^2 - E'\right) - V^2 = 0 \tag{2.96}$$

where V is the matrix element $\langle \psi_1 | H' | \psi_0 \rangle$ in our example for the $n = 0$ and $n = 1$ bands. At $k = \pi/a$ (i.e. the Brillouin zone boundary) this leads immediately to a simple expression for the energy:

$$E' = \frac{\hbar^2}{2m}\left(\frac{\pi}{a}\right)^2 \pm V \tag{2.97}$$

implying an energy gap of $\Delta E = 2\,V$. The difficulty with this approximation, however, lies in obtaining the functional form of the crystal potential in space; the actual magnitude of the V_j coefficients in Eq. (2.94) is not known.

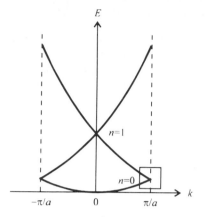

Figure 2.17 The two-band approximation. In this crude approximation, we consider only the interaction between two adjacent bands in regions close to degeneracy points. One such region is indicated by the box in the figure, the bands $n = 0$ and $n = 1$ are degenerate at $k = \pm \pi/a$ in the parabolic approximation which provides our basis. The interaction between these two bands is treated using degenerate perturbation theory.

Another problem with the NFE method is that an enormously large number of plane waves is needed in the perturbation expansion

$$\psi'_{jk} = \sum_j c_j \psi_{jk} \tag{2.98}$$

(where the c_js are simple coefficients) to accommodate the violently oscillating atomic potential in the region of the atom cores (note that only two coefficients were taken to obtain the simplified limit of Eq. (2.96)). In 'first-principles' calculations a more rapid convergence is achieved by putting in more realistic expressions for the atomic potential and electron wavefunctions at the start, including both the atomic and crystal symmetry.

A wide variety of different techniques have been used to obtain theoretical energy band structures of all commonly known semiconductors, including the so-called augmented plane wave, orthogonalized plane wave and pseudo-potential methods. The full body of this knowledge is then incorporated in the semi-empirical method most relevant to practical experiments, i.e. the $\mathbf{k} \cdot \mathbf{p}$ *perturbation theory* described in the next section. This has been called the single most important procedure for predicting and analyzing the experimental parameters determining the energy band structures of semiconductors. We have already seen that $\mathbf{k}$ is the wavevector associated with the plane wave part of the electron wavefunction, with a 'crystal momentum' $\hbar \mathbf{k}$ which is only equal to the free electron momentum in the limit where the crystal potential is constant. In $\mathbf{k} \cdot \mathbf{p}$ theory, $\mathbf{p}$ acts as the *momentum operator*, defined by $\mathbf{p} = -i\hbar\nabla$.

2.3.2 k·p perturbation theory

The **k·p** method is a semi-empirical perturbation technique for energy band calculations of real semiconductors. The known ordering of band states at particular symmetry points (most importantly the so-called Γ-point at $k = 0$) is taken from first-principles calculations. All known experimental information about band separations (energy gaps) is introduced empirically and the number of independent unknown elements (i.e. energy band, or effective mass, parameters) is reduced to a minimum by group theoretical symmetry considerations. The theory, taken to the order $\mathbf{k}^2$, defines the so-called *effective mass approximation*. The resulting band structure is then obtained in terms of a fixed set of band parameters to be determined by direct comparison with experimental results. No attempt is made to give an absolute calculation of the parameters; it is only necessary to know the symmetry of the problem and the energy gaps.

We begin with the electronic functions in the Bloch form:

$$\psi_{j,k}(\mathbf{r}) = u_{j,k}(\mathbf{r}) \exp i\mathbf{k} \cdot \mathbf{r} \tag{2.99}$$

for a particular band j. In general the parity of the $u_{j,k}$ (i.e. the way the function transforms under the action of the symmetry operator) is undefined, but at $k = 0$ (the Γ-point) the $u_{j,0}$ functions have the symmetry of the parent atomic states. For most of the semiconductors we are concerned with (i.e. group IV elements such as Ge, Si, α-Sn, group III–V elements GaAs and InSb, and group II–VI elements HgTe, CdTe, ZnSe), the valence states transform like atomic p-functions (symmetry X, Y, Z) and the conduction states like atomic s-functions (symmetry S); see Figure 2.18. In the presence of spin–orbit coupling, which splits the valence states, one has to consider the spin component (spin up or down) and the valence p-states have mixed-spin character.

The perturbed function $\psi_{j,k}(\mathbf{r})$ is then expanded in terms of the complete set of Γ-point functions whose symmetry is known (e.g. for Ge the valence band functions transform like atomic 4p-functions and the conduction band like atomic 5s-functions):

$$u_{j,k}(\mathbf{r}) = \sum_i A_{ji}(\mathbf{k}) u_{j,0}(\mathbf{r}) \tag{2.100}$$

where the summation is over all band symmetries at $k = 0$. It is then necessary to solve the perturbed problem:

$$H\psi_{j,k} = \left(\frac{\mathbf{p}^2}{2m} + V(\mathbf{r}) \right) \psi_{j,k} = E_{j,k} \psi_{j,k} \tag{2.101}$$

where now the momentum operator $\mathbf{p} = -i\hbar\nabla$ operates on a product function.

Taking one dimension for simplicity we can operate on $\psi = ue^{ikx}$:

$$\frac{\partial}{\partial x} ue^{ikx} = ikue^{ikx} + e^{ikx} \frac{\partial u}{\partial x}$$

$$\frac{\partial^2}{\partial x^2} (ue^{ikx}) = ik\left(uike^{ikx} + e^{ikx} \frac{\partial u}{\partial x} \right) + ike^{ikx} \frac{\partial u}{\partial x} + e^{ikx} \frac{\partial^2 u}{\partial x^2}$$

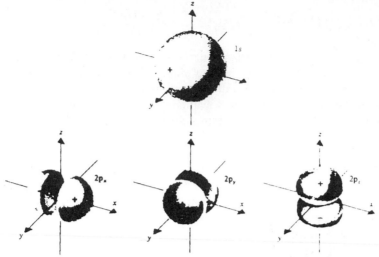

Figure 2.18 Atomic orbitals of the hydrogen atom, showing the characteristic symmetry of the s, p_x, p_y and p_z orbitals. The lobes are derived from $|\psi|^2$, but the sign of ψ is also shown.

The term e^{ikx} cancels in the Schrödinger equation, giving the following equation in terms of the Bloch functions:

$$-\frac{\hbar^2}{2m}\left(-k^2 u_j + 2ik\frac{\partial u_j}{\partial x} + \frac{\partial^2 u_j}{\partial x^2}\right) + V u_j = E u_j \tag{2.102}$$

At the Γ-point we assume the energies of the band extrema (including E_v and E_c) are known from first-principles calculations and from experiments; i.e. we know solutions to:

$$H_0 u_{j,0} = \left(-\frac{\hbar^2}{2m}\frac{\partial^2}{\partial r^2} + V\right) u_{j,0} = E_j(0) u_{j,0} \tag{2.103}$$

The general equation is then:

$$\left(H_0 + \frac{\hbar^2 k^2}{2m} + \frac{\hbar}{m}\mathbf{k}\cdot\mathbf{p}\right) u_{j,k} = E_j(k) u_{j,k} \tag{2.104}$$

The first term gives the Γ-point band extrema (assumed known at $k = 0$); the second term is the free electron energy; and the third term ($\mathbf{k}\cdot\mathbf{p}$) is the effective mass term and is treated by perturbation theory. The latter arises directly from the periodic crystal potential through the choice of the Bloch form for the electronic wavefunction.

We then expand $u_{j,k}(\mathbf{r})$ in terms of the complete set of unperturbed functions (Eq. (2.99)) and perform second-order perturbation theory. We assume for simplicity that the axes are coincident with the crystal axes. Then, to second order (i.e. k^2), we have for a single band j (e.g. the conduction band):

$$E_j(k) = E_j(0) + \frac{\hbar^2 k^2}{2m} + \frac{\hbar}{m}\,\mathbf{k}\cdot\mathbf{p}_{jj} + \frac{\hbar^2}{m^2}\sum_{i\neq j}\frac{|\mathbf{k}\cdot\mathbf{p}_{ji}|^2}{E_j(0) - E_i(0)} \qquad (2.105)$$

The first term, as before, gives the band energies at $k = 0$ (e.g. $E_c(0)$ for the conduction band) and the second is the free electron energy. The third term is the first-order perturbation which vanishes for band extrema at the Γ-point, and the fourth term defines the *effective mass* for the energy evaluated to order k^2. This expansion is valid for small k (i.e. $k \ll \pi/a$). The momentum matrix elements are

$$p_{ji} = \int\limits_{\text{unit cell}} u_{j0}^* p u_{i0}\,\mathrm{d}^3 r \qquad (2.106)$$

where the Bloch functions are normalized over the unit cell and not over all space. The matrix elements are constant band parameters to be determined by experiment; we therefore only need to know the symmetry of functions $u_{j,0}(\mathbf{r})$ and we are not concerned with the absolute value of the p_{ji}. Thus we have

$$E_j(k) \equiv E_k(0) + \frac{\hbar^2 k^2}{2m_j} \qquad (2.107)$$

defining the effective mass for the jth band:

$$\frac{1}{m_j} = \frac{1}{m} + \frac{2}{m^2}\sum_{i=j}\frac{|p_{ji}|^2}{E_j(0) - E_i(0)} \qquad (2.108)$$

In some experimental situations, particularly for narrow-gap semiconductors such as InSb ($E_g \sim 0.2$ eV), it is possible to make the extreme simplification of considering only the valence and conduction bands. In this *two-band model* for the conduction band we take only the valence band in the summation:

$$E = E_c(0) + \frac{\hbar^2 k^2}{2m} + \frac{|\hbar k p_{cv}/m|^2}{E_g} \qquad (2.109)$$

where the energy gap $E_g = E_c(0) - E_v(0)$. Thus, we can write

$$E = E_c(0) + \frac{\hbar^2 k^2}{2m}\left(1 + \frac{|2p_{cv}^2/m|}{E_g}\right) = E_c(0) + \frac{\hbar^2 k^2}{2m_c} \qquad (2.110)$$

where m_c/m is the effective mass of the conduction band, given by:

$$\frac{m}{m_c} = \left(1 + \frac{|2p_{cv}^2/m|}{E_g}\right) \qquad (2.111)$$

In general, the free mass energy term is negligible for semiconductors and the effective mass is substantially smaller than m. The resulting (first-order) perturbed functions are

$$u_{v,k} = u_{v,0} + c_v u_{c,0} = u_{v,0} + \frac{\hbar}{m}\frac{k p_{cv}}{E_g} u_{c,0} \qquad (2.112a)$$

$$u_{c,k} = u_{c,0} + c_c u_{v,0} = u_{c,0} + \frac{\hbar}{m}\frac{k p_{cv}}{E_g} u_{v,0} \qquad (2.112b)$$

The previous treatment has been particularly simple because we have been considering a single, s-like, conduction band. The p-like valence band is more complicated because it consists of three states, initially degenerate at $k = 0$, which transform as X, Y and Z. We have to go more carefully over the procedure leading to Eq. (2.105) and apply degenerate perturbation theory for the three $u_{v,0}(\mathbf{r})$ functions; i.e. treat the valence states together. This then gives a (3×3) generalized secular equation for the triply degenerate valence band. If we then add spin–orbit coupling, H_{SO}, it is necessary to include the spin components of the wavefunctions, $u_{v,0}(\mathbf{r})$. The main result of the inclusion of H_{SO} is that the correct valence band functions which diagonalize the Hamiltonian have mixed-spin character (in atomic terms this is referred to as the $|J, m_J\rangle$ representation), and the threefold degeneracy is partially raised. Thus, one has the well-known *light-* and *heavy-hole* valence bands (degenerate at $k = 0$):

$$E(k) = Ak^2 \pm (B^2 k^4 + C^2(k_x^2 k_y^2 + k_y^2 k_z^2 + k_x^2 k_z^2)) \qquad (2.113)$$

and spin–orbit split-off band

$$E(k) = -\Delta + Ak^2 \qquad (2.114)$$

The cubic symmetry of the crystal has made it possible to express the valence band dispersion in terms of just three independent effective mass parameters A, B and C. The overall picture for the conduction band and valence bands near $k = 0$ is shown in Figure 2.19 (taking the top of the valence band as the zero of energy).

Equation (2.111) shows that the effective mass in the conduction band is related directly to the energy gap; the *interband momentum matrix element*, $\mathbf{p}_{cv}$, *is found to be fairly constant for all semiconductors considered.* Thus we find that for InSb, $m_c \sim 0.015 m_0 (E_g \sim 0.2\ \text{eV})$ and for GaAs $m_c \sim 0.067 m_0 (E_g \sim 1.5\ \text{eV})$. The light-hole valence band is close to a mirror image of the conduction band, but a little

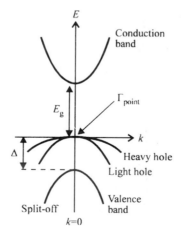

Figure 2.19 Energy bands of a direct gap III–V semiconductor near the zone centre.

(typically $\sim 20\%$) heavier, and the heavy-hole band has a substantially larger mass, typically $m_{hh} \sim 0.4m_0$.

The description given above applies to direct gap semiconductors. For indirect gap semiconductors like Ge and Si, the valence band near $k = 0$ is accurately described by Eq. (2.113), but the lowest conduction band minima form ellipsoids of revolution about the $\langle 111 \rangle$ and $\langle 100 \rangle$ directions, respectively, at the *Brillouin zone boundary*. Another class of technically important semiconductors comprises the lead salts and their alloys, particularly for infrared diode lasers. The band structure of these materials is significantly different from that described above since they have both valence band maxima and conduction band minima at the so-called L-point (in the $\langle 111 \rangle$ direction) of the Brillouin zone. Thus both conduction and valence bands are anisotropic ellipsoids of revolution, defined by longitudinal and transverse effective masses.

Energy gaps and effective masses for some technically important group IV, III–V and lead salt semiconductors are shown in Table 2.1.

2.4 Interband optical absorption

A very large number of semiconductor optical devices, including most diode light sources and detectors, are based upon electric dipole interband optical transitions. It is appropriate to consider the subject here, utilizing the simple band model developed in the previous section. For convenience, we consider the valence to conduction band absorption process, taking the simple two-band model in the first instance, as shown in Figure 2.20.

2.4.1 Interband transitions

We have interband absorption for incident photons of energy, $\hbar\omega$, if

$$\hbar\omega = E_c - E_v = \frac{\hbar k^2}{2}\left(\frac{1}{m_c} + \frac{1}{m_v}\right) + E_g = \frac{\hbar k^2}{2m_r} + E_g \qquad (2.115)$$

where m_r is called the *reduced mass*. From time-independent perturbation theory for electric dipole transitions, the optical transition rate for photons of energy ω between single states, c and v, is given by the Fermi golden rule [2.13]:

$$W_{cv}^{opt} = \frac{2\pi}{\hbar}|H_{cv}^{opt}|^2\delta(E_c - E_v - \hbar\omega) \qquad (2.116)$$

where H_{cv}^{opt} is the optical transition probability given by

$$H_{cv}^{opt} = \boldsymbol{\mu}\cdot\mathbf{E} = \tfrac{1}{2}erE_0[\exp i(\omega t - k_{opt}x) + \text{complex conjugate}] \qquad (2.117)$$

for an optical field $\mathbf{E}$ and electron dipole moment $\boldsymbol{\mu}$.

The total number of transitions per second, N_t, in a crystal of volume V is obtained by integrating over the density of states:

$$N_t = V\int_0^\infty D(k)W_{cv}^{opt}dk \qquad (2.118)$$

Table 2.1 Energy gaps, effective masses and position of band extrema for some technically important semiconductors [2.12]

	InSb	InAs	GaAs	Ge	Si	GaP	PbTe
Minimum energy gap at 300 K (eV)	0.18	0.36	1.43	0.67	1.106	2.24	0.30
k at conduction band minimum	0	0	0	At zone edge in (1,1,1)	Near zone edge in (1,0,0)	Near zone edge in (1,0,0)	At zone edge in (1,1,1)
m_e^*/m	0.015	0.024	0.072	m_l^*/m 1.6 m_t^*/m 0.082	0.97 0.19	1.5 0.25	0.2 0.03
k at valence band maximum	0	0	0	0	0	0	At zone edge in (1,1,1)
m_h^*/m	0.6 0.015	0.41 0.04	0.6 0.15	0.3 0.04	0.5 0.16	0.5 0.15	m_l^*/m 0.25 m_t^*/m 0.03

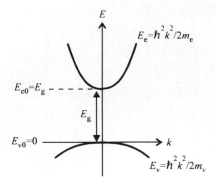

Figure 2.20 The simple two-band model.

with, from Eq. (2.49),

$$D(k) = \frac{2}{(2\pi)^3}\, 4\pi k^2 dk \tag{2.119}$$

giving

$$N_t = \frac{2V}{\pi\hbar}\int_0^\infty |H_{cv}^{opt}(k)|^2 \delta\!\left(\frac{\hbar^2 k^2}{2m_r} + E_g - \hbar\omega\right) k^2 dk \tag{2.120}$$

We change the variable to

$$X = \left(\frac{\hbar^2 k^2}{2m_r} + E_g - \hbar\omega\right) \tag{2.121}$$

giving

$$dX = \frac{\hbar^2}{m_r}\, k\, dk$$

and

$$k = (X + \hbar\omega - E_g)^{-1/2}\left(\frac{2m_r}{\hbar^2}\right)^{1/2} \tag{2.122}$$

Thus

$$N_t = \frac{2V}{\pi\hbar}\int_0^\infty |H_{cv}^{opt}(k)|^2\, \frac{m_r}{\hbar^2}\left(\frac{2m_r}{\hbar^2}\,(X + \hbar\omega - E_g)\right)^{1/2}\delta(X)\, dX \tag{2.123}$$

$$= \frac{V}{\pi} |H_{cv}^{opt}(k)|^2\, \frac{(2m_r)^{3/2}}{\hbar^4}\,(\hbar\omega - E_g)^{1/2} \tag{2.124}$$

since the only nonzero value of the delta function is $\delta(X = 0) = 1$, i.e. when

$$X = \frac{\hbar^2 k^2}{2m_r} + E_g - \hbar\omega = 0. \tag{2.125}$$

Lambert's law is an expression for the absorption coefficient obtained by relating the reduction in intensity dI to the thickness of the absorber dx:

$$dI = \alpha I_0 dx \tag{2.126}$$

so that

$$\alpha = \frac{dI/dx}{I_0} = \frac{\text{power absorbed/unit volume}}{\text{incident power/unit area}} \tag{2.127}$$

Therefore

$$\alpha = \frac{N_t \hbar \omega / V}{\varepsilon_0 n E_0^2 c / 2} \tag{2.128}$$

where n is the refractive index. In our one-dimensional example of the preceding section (so $r \to x$), we replace $H_{cv}^{opt}(k)$ by $eE_0 x_{cv}/2$, where $x_{cv} = \langle \psi_{c,k} | x | \psi_{vk} \rangle$ and $\psi_{j,k} \sim u_{j,0} \exp(i\,kx)$, giving

$$\alpha(\omega) = \frac{\omega e^2 x_{cv}^2 (2m_r)^{3/2}}{2\pi\varepsilon_0 n c \hbar^3} (\hbar\omega - E_g)^{1/2} \tag{2.129}$$

This function is illustrated in Figure 2.21. For typical semiconductors, this 'absorption edge' is dominated by the term $(\hbar\omega - E_g)^{1/2}$ and rapidly reaches values of $\alpha \sim 5 \times 10^3$ cm^{-1}.

The absorption spectrum measured at the onset of interband (i.e. valence to conduction band) transitions is called the *absorption edge* of a semiconductor. Measurement of this feature with optical (or magneto-optical) spectroscopy provides an extremely accurate method of determining the energy gap. The form of Eq. (2.129) recurs in the discussion of *emission* with respect to semiconductor lasers in Section 3.2.3.

2.4.2 Selection rules for electric dipole transitions

The transition rate, from the initial valence state v to the final conduction band state c, is given by Eq. (2.124):

$$N_t = \frac{V}{\pi} |H_{cv}^{opt}|^2 \frac{(2m_r)^{3/2}}{\hbar^4} (\hbar\omega - E_g)^{1/2}$$

with $H_{cv}^{opt} = eE_0 r_{cv}/2$; we have retained the r operator for generality.

In order to establish selection rules we write this in terms of band structure momentum matrix elements. Noting that $p = mv \approx mr\omega$, we have

$$H_{cv}^{opt} = \frac{eE_0}{m\omega} p_{cv} = \text{constant} \times (\mathbf{p}\cdot\hat{\varepsilon})_{cv} \tag{2.130}$$

where $\hat{\varepsilon}$ is the unit vector defining the radiation polarization. Now we also have

$$(\mathbf{p}\cdot\hat{\varepsilon})_{fi} = \langle u_c e^{ik'\cdot r} | \mathbf{p}\cdot\hat{\varepsilon} | u_v e^{ik\cdot r} \rangle$$

$$= \int (u_c^* e^{-ik'\cdot r} e^{ik\cdot r} | \mathbf{p}\cdot\hat{\varepsilon} | u_v + u_c^* u_v e^{-ik'\cdot r} | \mathbf{p}\cdot\hat{\varepsilon} | e^{ik\cdot r}) d^3 r \tag{2.131}$$

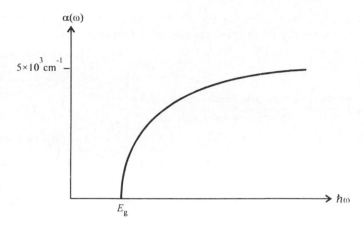

Figure 2.21 Theoretical optical absorption close to the band edge.

Remember that u_j is normalized over a unit cell. We make the approximation that $e^{i\mathbf{k}\cdot\mathbf{r}}$ is slowly varying over a unit cell and hence separate the expression into one integral for the u_j over the unit cell and the other for $e^{i\mathbf{k}\cdot\mathbf{r}}$ over the whole crystal. Then

$$\langle \mathbf{p}\cdot\hat{\boldsymbol{\varepsilon}} \rangle_{\text{fi}} = (\mathbf{p}\cdot\hat{\boldsymbol{\varepsilon}})_{\text{cv}}\delta_{kk'} + \delta_{\text{cv}} \int e^{i\mathbf{k}\cdot\mathbf{r}} \left(\frac{\hbar}{i} \nabla E \right) e^{i\mathbf{k}\cdot\mathbf{r}} d^3 r \tag{2.132}$$

$$= (\mathbf{p}\cdot\hat{\boldsymbol{\varepsilon}})_{\text{cv}}\delta_{kk'} + \hbar k \delta_{\text{cv}}\delta_{k'k}$$

The momentum matrix element can be shown by symmetry to be zero between functions of the same parity. The first term in Eq. (2.132) is therefore associated with strong *allowed* transitions between bands of different parity, that is with *interband transitions*. The presence of the delta function, $\delta_{kk'}$, shows that these are only allowed for $\Delta k = 0$ and the transitions are 'vertical' in k-space. The second term is associated with the free electron momentum, which is negligibly small, but allows the presence of transitions within a band which are known as *intraband transitions*. With the participation of optical phonons, terms such as this form the basis for the quantum-mechanical description of free-carrier absorption, where the momentum conservation condition is satisfied by the phonon momentum.

Direct transitions
Thus the optical selection rules for direct ($\Delta k = 0$) allowed transitions are determined by considering the function

$$N_{\text{t}} \propto |(\mathbf{p}\cdot\hat{\boldsymbol{\varepsilon}})_{\text{cv}}|^2 \tag{2.133}$$

For example, it is possible to preferentially populate one or other conduction band spin state by pumping with circularly polarized light. This depends on the details of the band structure, and is therefore different for simple cubic and wurzite materials and also for the lead salts.

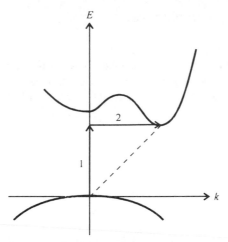

Figure 2.22 The 'two-step' nature of indirect transitions.

Direct interband *absorption* is the basis of a large class of photoconductive or photovoltaic detectors (see Chapter 9). On the other hand, for light-emitting devices we are concerned with the reverse process, i.e. interband *emission,* in direct gap materials (see Chapter 3).

Indirect transitions
As with free carrier absorption, indirect transitions require the participation of optical phonons to conserve momentum. Therefore the process can be formulated in two steps (Figure 2.22), it is a higher-order (weaker) one, with an absorption edge threshold that has a more gradual dependence on frequency than for direct transitions. This process applies in Si and Ge.

Forbidden transitions
Intra-valence band transitions are forbidden at $k = 0$ since all the bands have the same parity. However, at $k \neq 0$, first-order functions resulting from $H_{\mathbf{k} \cdot \mathbf{p}}$ give nonzero interband terms resulting from the interaction of the conduction and valence bands. The conduction band function is therefore mixed with the valence band function away from the Γ point:

$$u_{v,k} = u_{v,0} + c_v u_{c,0} = u_{v,0} + \frac{\hbar}{m} \frac{k}{E_g} p_{cv} u_{c,0} \tag{2.134}$$

and the transition probability $(\mathbf{p} \cdot \hat{\varepsilon})_{vv'}$ for $v' \neq v$ is nonzero away from $k = 0$, since there are now cross-terms between bands of opposite parity. Strong intra-valence band absorption is observed in heavily doped p-type materials and is the basis of the fast photon drag and hot hole detectors used for transient laser spectroscopy [3.14]. These are described in Chapter 9.

2.5 Electron–hole recombination

Excited electrons and holes must at some point recombine and return to the original equilibrium distribution that was disturbed. To do this they must dispose of their excess energy. There are various processes by way of which this can occur, and these can be divided into three main categories: (1) direct (radiative) recombination; (2) Auger (three-body) recombination; and (3) Shockley–Read recombination through traps.

2.5.1 Radiative recombination

It is possible to describe the rate of radiative recombination without treating indirect and direct transitions separately, since the rate depends only on the total absorption due to band-to-band transitions. The emission rate at a certain frequency can be found from the product of the probability that a photon will be absorbed per unit time and the density of photons $\rho(v)$ found from Planck's radiation law, i.e.

$$R(v)\mathrm{d}v = P(v)\cdot\rho(v)\mathrm{d}v \tag{2.135}$$

Now Planck's law gives

$$\rho(v)\mathrm{d}v = \frac{8\pi h n^3 v^3}{c} \left[\exp(hv/kT) - 1\right]^{-1}\mathrm{d}v \tag{2.136}$$

for emission of radiation from a solid of refractive index n, and we can relate the probability $P(v)$ to the mean lifetime of photons $\tau(v)$:

$$P(v) = \frac{1}{\tau(v)} = \alpha(v)\cdot v_{\mathrm{g}} \tag{2.137}$$

where v_{g} is the group velocity of photons through the medium and $\alpha(v)$ is the absorption coefficient. These can then be substituted into the equation for the emission rate, giving

$$R(v)\mathrm{d}v = \alpha(v)8\pi v^2 n^2 c^{-2}\left[\exp(hv/kT) - 1\right]^{-1}\mathrm{d}v \tag{2.138}$$

This expression relates the absorption to the expected emission. To obtain the total recombination rate, R, this expression must be integrated over all values of v. The following expression is obtained:

$$R = \frac{8\pi n^2 (kT)^3}{c^2 h^3} \int_0^\infty \frac{\alpha(v)u^2}{\mathrm{e}^u - 1} \, \mathrm{d}u \tag{2.139}$$

where $u = hv/kT$. This is valid for transitions between any set of states that are in thermal equilibrium and limits the application to a fairly narrow range of frequencies close to the fundamental absorption edge. As the frequency increases outside that range the factor e^u grows rapidly.

It is now possible to calculate the minority-carrier lifetime (i.e. the lifetime of the less prevalent carrier type). We define the quantities involved as follows: *n* is the carrier density in the upper state, p is the carrier density in the lower state and P_{cv} is the probability that one carrier per cm^3 in the upper state will make a radiative transition to one vacancy per cm^3 in the lower state. The radiative recombination rate is then given by:

$$R = np P_{cv} \qquad (2.140)$$

For an intrinsic material where the number of holes is equal to the number of electrons, this then becomes $R = n_i^2 P_{cv}$. This is true for thermal equilibrium in nondegenerate situations. If deviations from thermal equilibrium are considered, then the total radiative recombination rate R_c is given by

$$R_c = np(R/n_i^2) \qquad (2.141)$$

A deviation from the equilibrium will produce changes in the concentrations of the carriers, which can be expressed as:

$$n = n_0 + \Delta n \qquad (2.142)$$

$$p = p_0 + \Delta p \qquad (2.143)$$

where p_0 and n_0 give the carrier numbers for the intrinsic case. Substituting these into the expression for the total recombination rate gives

$$R_c = R + \Delta R = (n_0 + \Delta n)(p_0 + \Delta p)R/n_0 p_0 \qquad (2.144)$$

and neglecting the term in $\Delta n \Delta p$ we have

$$\Delta R = R(n_0 \Delta p + p_0 \Delta n)/n_0 p_0 \qquad (2.145)$$

For an n-type material $n_0 \Delta p \gg p_0 \Delta n$, so the expression can be simplified to

$$\Delta R = R \frac{\Delta p}{p_0} \equiv \frac{\Delta p}{\tau_p}, \qquad (2.146)$$

expressed in terms of the minority-carrier lifetime $\tau_p = p_0/R$. Following the same procedure for the carrier lifetime in p-type material gives $\tau_n = n_0/R$. For an intrinsic semiconductor $n_0 = p_0 = n_i$, so that $\tau_i = n_i/2R$.

Radiative efficiency
In addition to the radiative recombination that occurs there are also nonradiative processes through which carriers may lose energy (see Figure 2.23). These often occur via intermediate states and the effect is clearly to reduce the total number of carriers taking part in the radiative recombination. The effective carrier lifetime in the upper state, τ_{eff}, is then given by

$$\frac{1}{\tau_{eff}} = \frac{1}{\tau} + \frac{1}{\tau'} \qquad (2.147)$$

where τ' is the nonradiative lifetime. The radiative efficiency is then defined by

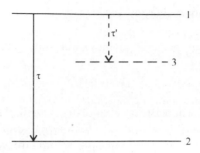

Figure 2.23 Simple three-level system involving one radiative recombination path (1 → 2) and one nonradiative path (1 → 3).

$$\eta = \frac{R_c}{R_T} = \frac{1}{1 + \tau/\tau'} \tag{2.148}$$

2.5.2 Auger recombination

This is a concept of a three-body interaction, of earlier general application in wider context which has subsequently been applied to solid state transitions. Auger recombination is a three-particle process in which an electron in the conduction band recombines with a hole in the valence band, and the recombination energy is transferred by electron–electron interaction to another electron in the conduction band, which is then promoted to a higher excited state (known as an Auger state). The main effect of nonradiative recombination is that it can substantially reduce the minority-carrier lifetimes. In some semiconductors this is the dominant recombination process and is observed by the absence of emitted photons. These loss processes include surface recombination as well as the Auger effect.

There are many different types of Auger recombination. One example is the *chcc transition*, which involves the conduction band–heavy hole and conduction band–conduction band transitions. The 'activation' energy for this transition is given by

$$E_A = \frac{m_e E_g}{m_e + m_h} \tag{2.149}$$

where m_e and m_h are the electron and hole effective masses. The lifetime can be written as $1/\tau = Cn_0^2$, where

$$C \propto \frac{E_A^{1/2}}{[(kT)^{1/2}\exp(E_A/kT)]} \tag{2.150}$$

The exponential factor dominates, so that the Auger recombination increases with increasing temperature and with decreasing energy gap, E_g. Device design can minimize unwanted effects of this recombination as described in Section 9.5.

2.5.3 Shockley–Read recombination through traps

For direct gap semiconductors, the excess carrier lifetimes permitted by the combination of radiative and Auger recombination are likely to be quite small (in the range 10^{-10}–10^{-7}s depending on the equilibrium concentration of electron and hole densities). However, the mutual satisfaction of energy and momentum conservation required for Auger recombination is much more difficult to accomplish in an indirect gap semiconductor, and in addition the radiative recombination is very slow, corresponding to the inefficient indirect absorption process. Thus, extremely long excess carrier lifetimes might be expected if they were determined purely by *band-to-band* processes.

In practice, lifetimes of 10^{-7}–10^{-3}s are typical because the recombination is dominated by two-step processes via traps or flaws. For direct gap semiconductors these processes are less important since the direct recombination processes are so efficient. The flaw states that facilitate electron–hole recombination are usually not the ones which are major contributors to the carrier density at equilibrium but are typically close to the middle of the bandgap.

A derivation of the expressions for lifetime will not be given here (cf. Ref. [2.15]). When the flaw density is small enough to keep Δn and Δp essentially the same, then the minority hole lifetime equals the minority electron lifetime, and has the form

$$\tau = \frac{\tau_0(n_0 + p_0) + \tau_\infty \Delta n}{(n_0 + \Delta p + \Delta n)} \tag{2.151}$$

where τ_0 and τ_∞ are the limiting values of lifetime for very small and very large disturbances from equilibrium.

2.6 Low-dimensional systems: quantum wells

2.6.1 Introduction

So far in this chapter we have been concerned almost entirely with the bulk properties of crystalline solids. Such properties can normally be specified in terms of coefficients that are independent of the shape and size of a particular specimen. Thus, for example, the electrical conductivity and resistivity, or the optical absorption coefficient and refractive index are properties only of the material and not of the sample dimensions. When one dimension or more of a solid is reduced sufficiently, the properties are no longer given by these bulk coefficients. The sample is then described as a *low-dimensional system* (LDS), and such systems form the basis of many optoelectronic devices. Low-dimensional systems are classified according to the number of dimensions that are large. Thin films and heterojunctions are *two*-dimensional since only the film thickness is reduced; fine wires are one-dimensional since only one dimension, the length, is large; and 'dots' are zero-dimensional.

Departures from bulk behaviour occur when the size of the sample becomes comparable to the electron wavelength of the important excitations in the solid – a phenomenon sometimes described as the quantum size effect. We illustrate this here by considering the most important example in optoelectronics of electrons confined to a semiconductor film or junction, known as the two-dimensional electron gas (2-DEG).

In passing, we note that we have already discussed the creation of a kind of confinement in bulk systems by the application of an external magnetic field in the z-direction, B_z. The classical discussion showed that under these circumstances the electrons undergo circular (cyclotron) motion in the x–y plane but the motion in the z-direction parallel to the magnetic field is unaffected. In quantum-mechanical terms, k_x and k_y are no longer good quantum numbers and we are left with the quantum number k_z. From the point of view of the density of states we have a one-dimensional system, with motion along k_z. Thus the k_z motion is unchanged, with energy given by the usual quadratic dispersion:

$$E_z = \frac{\hbar^2 k_z^2}{2m^*} \tag{2.152}$$

The motion in the x,y plane is given by the solution to a harmonic oscillator equation, where the resonant frequency ω_c is the cyclotron frequency. This gives rise to quantized energy states in the k_x–k_y plane

$$E_\perp = (l + \tfrac{1}{2})\hbar\omega_c \tag{2.153}$$

where l is the magnetic orbital ('Landau') quantum number. The original three-dimensional energy band is coalesced into one-dimensional subbands called Landau subbands:

$$E = \frac{\hbar^2 k_z^2}{2m^*} + (l + \tfrac{1}{2})\hbar\omega_c + \text{spin term} \tag{2.154}$$

where we have included a spin term that has not been considered so far.

2.6.2 Quantum wells

By 'sandwiching' a thin layer of one semiconductor between two layers of another semiconductor of greater bandgap, it is possible to create a discontinuity in the conduction and valence band edges so that the carriers are restricted in 1-D (say z), but are free within the junction plane (x–y). This is known as a 'quantum well'. See Ref. [2.16]. In the z-direction, then, the energy dispersion is no longer parabolic but is discrete, as it is in the case of the 'particle in a box' problem referred to earlier. This type of structure is illustrated in Figure 2.24.

We begin, however, by assuming that the potential steps defining the quantum well between the layers of each material are infinite, so Schrödinger's equation for the z-direction has the simple form

(a)

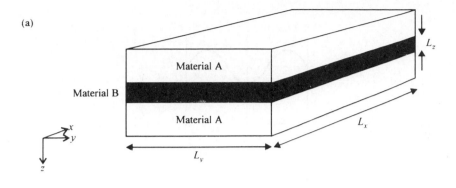

(b)

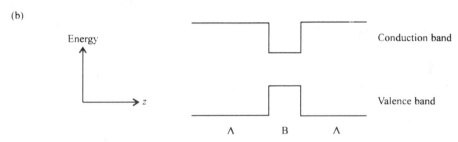

Figure 2.24 (a) A simple two-dimensional quantum well structure, where L_z is of the order of the electron wavelength and (b) a diagrammatic representation of the resulting energy band structure, illustrating the discontinuities in both valence and conduction bands (note that this is only one possible configuration).

$$-\frac{\hbar^2}{2m}\frac{d^2\psi}{dz^2} = E\psi \tag{2.155}$$

inside the well, with solutions $\psi = A \sin(kz)$ and eigenvalues $E = \hbar^2 k_z^2 / 2m^*$. Imposing the boundary condition that $\psi = 0$ at $z = 0$ and $z = L$ results in the familiar quantization of the wavevector:

$$k_z = \frac{n_z \pi}{L_z} \tag{2.156}$$

giving

$$E_{n_z} = \frac{\hbar^2 \pi^2 n_z^2}{2m^* L_z^2} \tag{2.157}$$

and

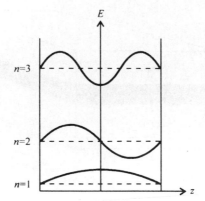

Figure 2.25 Infinitely deep quantum well energy levels and resulting wavefunctions.

$$\psi_{n_z} = A \sin\left(\frac{n_z \pi}{L_z} z\right) \tag{2.158}$$

where $n_z = 1, 2, 3$, etc. Motion is essentially unrestricted in the k_x, k_y plane, so the normal quasi-continuum in energy is obtained in these directions. Thus, if we have, say, $L_x = L_y \sim 10^5 L_z$, then the total energy is

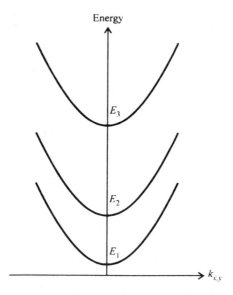

Figure 2.26 The quadratic dispersion of each subband with respect to k_x and k_y.

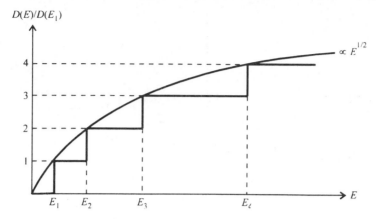

Figure 2.27 Two-dimensional density of states as a function of energy. The 3-D DOS is also shown for comparison.

$$E = \frac{\hbar^2 \pi^2}{2m^*} \left(\frac{n_x^2 + n_y^2}{L^2} + \frac{n_z^2}{L_z^2} \right) \tag{2.159}$$

where the first term in the parentheses represents the quasi-continuum, and the second term the widely spaced quantum well subbands. For example, if we have $L_x = L_y = 1$ mm, $L_z = 3$ nm, $m^* \sim m_0$, then we find that the energy spacing of the states in the x- and y-directions is of the order of 4×10^{-10} meV, whereas the separation of the $n_z = 1$ and $n_z = 2$ levels is about 120 meV (Figure 2.25). Thus we can put approximately:

$$E = \frac{\hbar^2 (k_x^2 + k_y^2)}{2m^*} + \frac{\hbar^2 \pi^2}{2m^*} \frac{n_z^2}{L_z^2} \tag{2.160}$$

so if we plot total energy E against k_x or k_y, then we obtain a set of subbands as shown in Figure 2.26. In the sample discussed above, if $E_F < 3\hbar^2\pi^2/2m_0 L_z^2 = 120$ meV and $kT \ll 120$ meV, then we can safely assume that all electrons are in the lowest ($n_z = 1$) subband.

The density of states is localized in k_z, but in the k_x, k_y plane it is just given by the 2-D expression (including spin) $D(E) = m^*/\pi\hbar^2$ and is independent of E and n_z. For the lowest subband $E_1 = \hbar^2\pi^2/2mL_z^2$ and if we put $D(E_1) = 1$, then for the next subband ($n_z = 2$) we have $E_2 = 4E_1$ and $D(E_2) = 2$. For $n_z = 3$, we have $E_3 = 9E_1$, $D(E_3) = 3$ and so on (Figure 2.27).

For the 2-DEG the sheet carrier concentration n_S (i.e. the number of carriers per unit area in the 2-DEG) is given very simply in terms of the Fermi energy:

$$n_S = \int_0^{E_F} \frac{m^*}{\pi\hbar^2} \, dE = \frac{m^*}{\pi\hbar^2} E_F \tag{2.161}$$

if only the first subband is occupied. We can also write this in terms of the Fermi wavevector k_F:

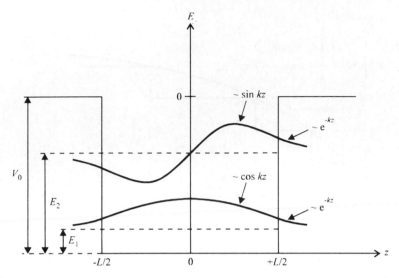

Figure 2.28 The lowest two energy levels in a finite quantum well, together with the associated wavefunctions [2.16].

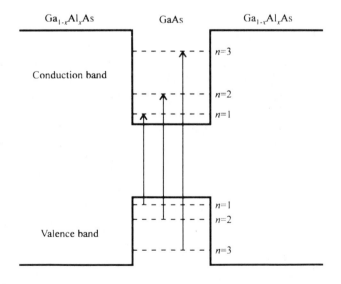

Figure 2.29 Parity-allowed transitions in $Ga_{1-x}Al_xAs/GaAs$ quantum wells.

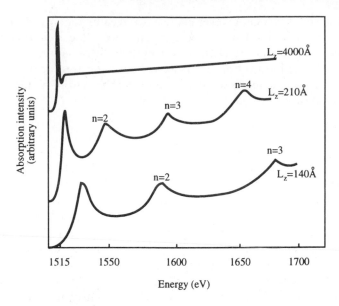

Figure 2.30 Absorption as a function of photon energy in quantum wells of thickness 140, 210 and 4000 Å. The absorption curve for the 4000 Å layer displays clear three-dimensional behaviour [2.17].

$$n_S = \frac{m^*}{\pi \hbar^2} \left(\frac{\hbar^2 k_F^2}{2m^*} \right) \tag{2.162}$$

giving

$$k_F = (2\pi n_S)^{1/2} \tag{2.163}$$

If we take a typical example of a 2-DEG, with $n_S = 10^{12}$ cm^{-2} and $m^* \sim m_0$, this gives $E_F = 2.3$ meV ($k_F = 2.5 \times 10^6$ cm^{-1}). The resulting Fermi temperature ($E_F = kT_F$) is $T_F \approx 25$ K, and the sample is therefore degenerate at liquid helium temperature.

The preceding picture is modified a little in real low-dimensional structures. The most common of these is the Ga$_{1-x}$Al$_x$As/GaAs quantum well, usually grown by molecular beam epitaxy (MBE), where a confining potential is obtained by 'sandwiching' a thin layer of GaAs (material B in Figure 2.24) between two much thicker layers of Ga$_{1-x}$Al$_x$As (which forms the barrier material, material A). The thickness of the GaAs quantum well layer is usually less than 150 Å, whereas the barrier material is at least 500 Å thick and often more. This creates a *finite* confining potential for both holes and electrons in the direction perpendicular to the plane of the layers (Figure 2.24). The finite nature of the potential means that there is some penetration of the carrier wavefunction into the Ga$_{1-x}$Al$_x$As layer, which alters it slightly, and the resulting functions are illustrated for one band in Figure 2.28, together with the associated energy eigenvalues.

   The density of states in a GaAs layer may be deduced by looking at the absorption of electromagnetic radiation that is associated with electronic excitation between the conduction and valence band. Since the wavelength of the radiation is long compared to the width of the well, transitions only occur between states for which the spatial variation of the wavefunction is similar, leading to the selection rule $\Delta n = 0$. The absorption should occur at frequencies ω_n, given by

$$\hbar\omega_n = E_{cn} - E_{vn} \tag{2.164}$$

where $E_{cn} - E_{vn}$ is the energy difference between the nth bound states in the conduction and valence bands. These transitions are illustrated in Figure 2.29. The measured absorption spectra for GaAs layers of thickness 140, 210 and 4000 Å are shown in Figure 2.30; the expected step structure is clearly visible for the two thinner layers. The onset of absorption is marked by peaks at energies just below the predicted values. This results from the existence of a bound state between the created electron and hole, the difference being the binding energy of the electron–hole pair – the two-dimensional exciton. An exciton peak is clearly visible on the absorption curve for the 4000 Å GaAs layer, but the smooth step-like structure has disappeared, indicating that the well is by now so wide that the density of states has become characteristic of 3-D behaviour.

   The confining effect of the quantum well potential on the 2-D exciton is quite pronounced in thin films, producing strong enhancement of the exciton binding energy and the oscillator strength. This means that excitons become clearly observable even at room temperature; in recent years this property has been utilized in the optical processing device known as the SEED (self-electro-optic effect device) described in Section 3.5.3.

References

[2.1] The concept is not a trivial one! For a full discussion see B.I. Bleaney and B. Bleaney, *Electricity and Magnetism*, 3rd edn. (Oxford University Press, 1976), p. 65.

[2.2] A simple consideration of the nature of the damping forces in the Drude model (i.e. collisions) leads to a damping term of this form. See N.W. Ashcroft and N.D. Mermin, *Solid State Physics* (Holt, Reinhardt and Winston, 1976), pp. 9–11.

[2.3] J.T. Houghton and S.D. Smith, *Infrared Physics* (Oxford University Press, 1966).

[2.4] W.G. Spitzer and H.Y. Fan, *Phys. Rev.* **106**, 882 (1957).

[2.5] W.C. Price and G.R. Wilkinson, *Final Report*, US Army Contract DA-91-591-EUC-2127, OI-26489-B (1964).

[2.6] For a fuller discussion of the various statistics see, for example, R.A. Smith, *Wave Mechanics of Crystalline Solids* (Chapman and Hall, London, 1969), pp. 35–49.

[2.7] See Ref. [2], pp. 162–66, for a discussion of the concept of Brillouin zones.

[2.8] For a full treatment see, R.A. Smith, *Semiconductors* (Cambridge University Press, 1979), pp. 77–96.

[2.9] M. Sturge, *Phys. Rev.* **127**, 768 (1962).

[2.10] Baumeister, *Phys. Rev.* **121**, 359 (1961).

[2.11] O. Madelung, *Introduction to Solid State Theory* (Springer-Verlag, Berlin, 1981), p. 50.

[2.12] R.J. Elliot and A.F. Gibson, *Solid State Physics* (Macmillan, 1974).

[2.13] For a discussion of the Golden Rule, see E. Merzbacher, *Quantum Mechanics* (Wiley, New York, 1970), pp. 475–81.

[2.14] See R.A. Smith, *Semiconductors* (Cambridge University Press, 1979) for a brief outline of photon drag detectors.

[2.15] J.S. Blakemore, *Semiconductor Statistics* (Pergamon Press, Oxford, 1962).

[2.16] C. Weisbuch and B. Vinter, *Quantum Semiconductor Structures* (Academic Press, London, 1991).

[2.17] R. Dingle, W. Wiegmann and C.H. Henry, *Phys. Rev. Lett.* **33**, 827 (1974).

3

---

pn-junction-based devices

3.1 Introduction

The simple *pn* junction is the basis for an enormous variety of optoelectronic devices. In this chapter we introduce the basic principles of the *pn* junction and the light-emitting devices based upon them, namely light-emitting diodes (LEDs) and semiconductor diode lasers (SDLs). The technology of the *pn* junction has evolved into more complicated structures which have yielded increasingly efficient light-emitting devices, and its properties can also be used in detectors and modulators. In general the devices have applications in displays, telecommunications, sensors, control devices (TV controllers), compact disc players and in laser printers. More recently, derivatives of the *pn* junction have further expanded this range of applications to include self-electro-optic effect devices (SEEDs), giving modulators and optical logic telements (described later in this chapter) as well as a new series of sensitive detectors based upon carrier exclusion and extraction (described in Chapter 9).

A wide range of materials has been developed in the constant quest for shorter emission wavelengths and also for applications requiring specific wavelengths. Only in direct gap semiconductors is the radiative recombination process efficient enough for the manufacture of practical light-emitting devices, and clearly it must also be possible to dope the material both *n*- and *p*-type to a fairly high density ($\sim 10^{18}$ cm^{-3}) in order to fabricate junctions. These requirements rule out many materials and thus the field has been dominated to date by III–V materials such as GaAs and associated ternary materials $Ga_{1-y}Al_yAs$ and $GaAs_{1-x}P_x$. However, there has been considerable interest in II–VI materials, and *pn* junctions have recently been fabricated in ZnSe-based material systems allowing the production of blue-green and blue LEDs and lasers in the laboratory

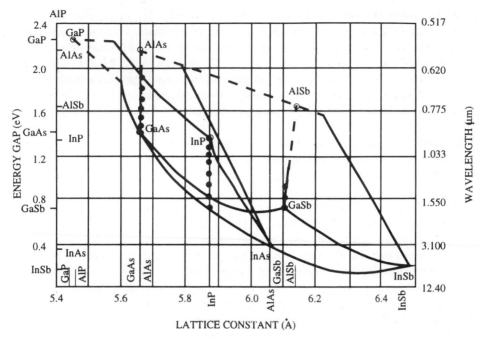

Figure 3.1 Energy bandgap and lattice constants for various III–V semiconductors. Lines joining the binary materials indicate ternary materials with corresponding emission wavelength as shown on the right. Dashed lines indicate indirect gap materials which are unsuitable for laser operation [3.1].

3.2 *pn* junctions

A *pn* junction is formed in semiconductor material at the boundary between two regions, one of which has an excess of mobile negative charge carriers (called *n*-type material) and the other an excess of positive charge carries (*p*-type material). Such a junction has properties that are used as the basis of many electronic and optoelectronic devices and, in particular, of *light-emitting diodes* (LEDs) and *semiconductor diode lasers* (SDLs), where the recombination of electrons and holes occurs in a light-emitting process. In these devices, the wavelength of the emitted radiation is dependent on the materials via the width of the forbidden energy state region known as the *bandgap*. The spectral regions covered by the binary III–V compounds and associated ternary alloys are shown in Figure 3.1.

3.2.1 The *pn* junction in equilibrium: zero bias

For good electrical conductivity, practical devices are made from single crystals of semiconductor material so that defects and impurities are kept to a minimum. These are either slices cut from much larger single crystals, into which dopants are introduced from the surface by some means (e.g. diffusion) to form the *pn* junction, or they are grown *epitaxially* (literally layer-by-layer on an atomic scale, with the atoms in each new layer aligned with those in the material below) on a single-crystal substrate, with appropriate dopants added during the growth of each layer. Either way, the pn junction is never entirely abrupt but extends over a finite distance, depending on the growth technique involved. In the following treatment, however, we can consider the junction to be an abrupt one to a good approximation, and we do so for mathematical convenience.

Although real junctions are made from single crystals, it is useful first to imagine that they are formed by bringing together two separate pieces of crystal, one doped *n*-type by the addition of N_D donor atoms per unit volume, and the other doped *p*-type by the addition of N_A acceptors, as in Figure 3.2(a,b). Each piece of material is electrically neutral, the charge of the excess free carriers in each case being balanced at every point by the residual charge on the ionized impurities from which they arise. When they are brought together, electrons diffuse from the *n*-type side, where they are abundant (they are the *majority carriers*), to the *p*-type side, where they are very few (they are the *minority carriers*), and a momentary electron *diffusion current*, j_e^{diff}, arises. Diffusion occurs in all materal systems where there are differences in concentrations of mobile molecules, atoms or particles, and is here simply a consequence of the imbalance between the electron number densities in the two types of material. A similar 'hole' diffusion current, j_h^{diff}, occurs in the opposite direction, from *p* to *n*. However, the flow of carriers produces an immediate loss of charge neutrality, the *n*-type side becoming positive because of the net charge on the ionized donors left behind, and the *p*-type becoming negative. A potential gradient is established across the junction, which quickly forms a barrier to any further charge movement and reduces the current to zero. The magnitude of the resulting potential barrier (V_B in Figure 3.2c) is such as to produce drift currents (j_e^{drift} and j_h^{drift}), which exactly balance the diffusion currents due to the concentration gradients. We thus observe that for each carrier type there are two contributions to the current flow due to diffusion and drift, which, in the absence of an applied voltage, balance in dynamic equilibrium. Clearly, we may change this equilibrium situation by the application of an external field, which, if it is of the correct polarity, will allow steady diffusion currents to occur once again, resulting in *carrier injection* across the junction.

Returning to the zero applied field case, we see that because of the 'built-in' equilibrium field a region must exist on either side of the junction in which there are few mobile carriers since they will be swept away by the field. This region, called the *depletion layer*, has a finite space charge caused by the presence of the ionized impurities (Figure 3.2f).

To obtain expressions for the 'built-in' voltage and the depletion layer width, we equate to zero both the electron and hole currents, j_e and j_h, respectively. For the electrons we have

$$j_e = j_e^{\text{drift}} + j_e^{\text{diff}} \qquad (3.1)$$

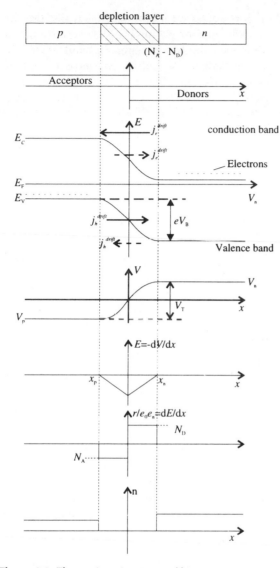

depletion layer

(a) The pn junction.

(b)The acceptor and donor concentration at a step junction

(c) The energy band structure that raises through the charge redistribution.

(d) The potential variation brought about by the charge redistribution indicating the magnitude of the potential barrier. Clearly $V_T=V_B$ with no applied field.

(f) The space charge in the depletion region.

(g) The carrier density across the junction.

Figure 3.2 The *pn* junction in equilibrium.

This gives

$$j_e = -n_e(x)e\mu_e \frac{dV(x)}{dx} + eD_e \frac{dn(x)}{dx} = 0 \qquad (3.2)$$

i.e.

$$\mu_e \frac{dV}{dx} = D_e \frac{1}{n_e} \frac{dn_e}{dx} \qquad (3.3)$$

where $V(x)$ is the electrical potential, μ_e is the electron mobility, $n_e(x)$ is the electron density and D_e is the electron diffusion coefficient. Integrating Eq. (3.3) between points x_n, x_p far enough from the junction that its influence is negligible, and where the carrier densities are n_n and n_p and the potentials are V_n and V_p, we have

$$\int_{x_p}^{x_n} \mu_e \frac{dV}{dx} \, dx = \int_{v_p}^{v_n} D_e \frac{1}{n} \frac{dn}{dx} \, dx \tag{3.4}$$

or

$$V_n - V_p = \frac{D_e}{\mu_e} \ln\left(\frac{n_n}{n_p}\right) \tag{3.5}$$

Assuming that all the impurities are ionized (a good approximation at room temperature), we know that

$$n_n = N_D \quad \text{and} \quad n_p = \frac{n_i^2}{N_A} \tag{3.6}$$

so that

$$V_n - V_p = \frac{D_e}{\mu_e} \ln\left(\frac{N_A N_D}{n_i^2}\right) = V_B \tag{3.7}$$

where V_B is the *barrier potential*, the potential difference between the two sides, which prevents the diffusion of electrons away from the regions of highest carrier density. n_i is the *instrinsic carrier density*, given by $n_i^2 = np$ in intrinsic material (this is elaborated upon below). A similar expression can of course be obtained by considering the holes, from which we have

$$V_B = \frac{D_h}{\mu_h} \ln\left(\frac{N_D N_A}{n_i^2}\right) \tag{3.8}$$

and thus

$$\frac{D_e}{\mu_e} = \frac{D_h}{\mu_h} \tag{3.9}$$

Here we introduce general expressions for the carrier densities n and p in terms of band and Fermi energies, which are true for both intrinsic and extrinsic cases:

$$n = A_c \exp\{-(E_c - E_F)/kT\},$$
$$A_c = 2(2\pi m_e kT)^{3/2}/h^3 \tag{3.10}$$

and

$$p = A_v \exp\{-(E_F - E_v)/kT\},$$
$$A_v = 2(2\pi m_h kT)^{3/2}/h^3 \tag{3.11}$$

where m_e and m_h are the electron and hole masses, respectively. Together, these give the product np in terms of the bandgap, E_g:

$$n_i^2 = np = A_cA_v \exp\{-(E_c - E_v)/kT\} = A_cA_v \exp\{-(E_g)/kT\} \qquad (3.12)$$

Clearly this a strong function of both temperature and bandgap, and it turns out that the product is three times greater in intrinsic Ge ($E_g \sim 0.67$ eV at room temperature) than in Si ($E_g \sim 1.15$ eV), and undoped GaAs ($E_g \sim 1.42$ eV) is almost insulating.

In equilibrium, the Fermi energy is a constant right through the system. Observing Figure 3.2(c), we can write V_B in the form

$$V_B = \{(E_c - E_F)\}_{p-side} - \{(E_c - E_F)\}_{n-side} \qquad (3.13)$$

and using Eqs. (3.6), (3.10) and (3.11) we find

$$-V_B = \frac{kT}{e} \ln\left(\frac{N_D N_A}{n_i^2}\right) \qquad (3.14)$$

We therefore have the important result that

$$\frac{kT}{e} = \frac{D_{e,h}}{\mu_{e,h}} \qquad (3.15)$$

This is the Einstein relation, which holds for both electrons and holes.

Clearly the magnitude of V_B depends on the concentrations of the majority carriers on both sides of the junction. Because of the n_i^2 factor, V_B is much larger in Si than Ge and is even larger in GaAs.

The depletion layer in equilibrium and with applied bias
The potential, electric field and depletion layer width are obtained by solution of Poisson's equation. In the depleted parts of the *p*-region, the only charges are due to the negatively charged acceptor ions, assumed to be evenly distributed in space. Thus we have

$$\frac{d^2V}{dx^2} = \frac{dE}{dx} = \frac{\rho}{\varepsilon_r\varepsilon_0} = -\frac{eN_A}{\varepsilon_r\varepsilon_0} \qquad (3.16)$$

where ε_0 is the permittivity of free space and ε_r is the relative permittivity of the material. The solution of this, with boundary conditions $E = 0$ at $x = x_p$, the depletion layer boundary, gives

$$E = eN_A(x_p - x)/\varepsilon_r\varepsilon_0 \qquad (3.17)$$

Similarly, in the n-type material,

$$E = eN_D(x - x_n)/\varepsilon_r\varepsilon_0 \qquad (3.18)$$

where x_n is the depletion layer boundary in the *n*-region (Figure 3.2e). The maximum field is obtained at the boundary between the *p*- and *n*-regions, and is given by

$$E_{max} = eN_A x_p / \varepsilon_r \varepsilon_0 = eN_D x_n / \varepsilon_r \varepsilon_0 \qquad (3.19)$$

Thus we have

$$N_A |x_p| = N_D |x_n| \qquad (3.20)$$

and we see that the depletion layer thickness will be greater where the doping is lighter.

Integration of Eqs. (3.17) and (3.18) yields the potential drop across the depletion layers in the *n*- and *p*-regions, respectively. The boundary condition here is that the total voltage across the junction from x_p to x_n is $V_T = V_B + V_A$, where V_B is the 'built-in' voltage described above, and V_A is the externally applied voltage. V_A is clearly zero in the absence of an applied field, but if $V_A \neq 0$ then the depletion layer width, $w = |x_n| + |x_p|$, is altered. We find that V_T is given by

$$V_T = \frac{e}{2\varepsilon_r \varepsilon_0} (N_D x_n^2 + N_A x_p^2) \qquad (3.21)$$

so that if there is an imbalance in the doping, more of the applied voltage appears across the lightly doped side. If one side is degenerate (i.e. very heavily doped so that the Fermi energy lies in the conduction or valence band), then the voltage dropped across it may be neglected altogether. Application of the condition that the net positive charge on the *n*-side equals the net negative charge on the *p*-side yields expressions for $|x_n|$ and $|x_p|$ separately and for the total depletion layer width, w, each of which is found to be proportional to $V_B^{1/2}$.

The carrier concentrations on either side of the junction are related by a simple Boltzmann factor:

$$\frac{n_n}{n_p} = \exp\left(-\frac{eV_B}{kT} \right) \qquad (3.22)$$

Minority carrier injection
When the *p*-side of the *pn* junction is made more positive (*forward biased*) by the application of an external potential V_A, the potential barrier to the flow majority carriers is reduced, and thus they flow across the junction to become minority carriers on the other side (Figure 3.3a). This flow of carriers constitutes a measurable current.

Charge neutrality is maintained outside the depletion regions so that majority and minority carriers increase and decrease in equal numbers. This results in a large proportional change in minority carrier density, whereas the majority carrier density is almost unchanged and in most cases can be assumed to be its equilibrium value. Denoting equilibrium values by a zero subscript, we have approximately that

$$\frac{p_n}{p_{p0}} = \exp\left(-\frac{eV_T}{kT} \right) = \frac{n_p}{n_{n0}} \qquad (3.23)$$

Rewriting Eq. (3.22) with appropriate subscripts we also have

$$\frac{p_{n0}}{p_{p0}} = \exp\left(-\frac{eV_B}{kT} \right) = \frac{n_{p0}}{n_{n0}} \qquad (3.24)$$

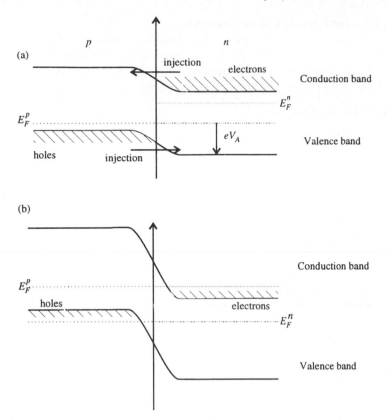

Figure 3.3 Energy bands of a pn junction diode (a) forward and (b) reverse biased.

The applied voltage $V_A = V_T - V_B$, so combining Eqs. (3.23) and (3.24), we obtain

$$\frac{p_n}{p_{p0}} = \exp\left(-\frac{eV_A}{kT}\right) = \frac{n_p}{n_{p0}} \tag{3.25}$$

which relates the perturbed and equilibrium values of carrier density through the exponential of the externally applied voltage. From Figure 3.3(a) we see that for forward biasing V_A is negative so that p_n is exponentially *increasing*. Clearly the opposite is the case for reverse biasing (Figure 3.3b)

It is said that in the forward bias case there is an *injection* of excess minority carriers, i.e. electrons move into the *p*-side and holes move into the *n*-side. This has the effect of modifying the width of the depletion region, and so now we have

$$w \propto (V_B - V_A)^{1/2} \tag{3.26}$$

Unlike all other types of laser it is this *transport* of electrons and of holes into the same spatial region which provides the 'pumping' or 'population inversion' mechanism in SDLs.

3.2.2 The *I–V* characteristics of *pn* junctions

We wish to be able to take the results of the previous section and transform them into expressions for current flow. To do this the *continuity equation* [3.2] is used to describe the excess minority carrier density Δn as a function of distance from the junctions. The solution for electrons turns out to be

$$\Delta n_p(x) = \Delta n_p(0) \exp(-x/L_e) \tag{3.27}$$

where L_e is the electronic diffusion length, with a similar expression for holes. From Eq. (3.25) we know that

$$\Delta n_p = n_{po}\left(\exp\left(\frac{e|V_A|}{kT}\right) - 1 \right) \tag{3.28}$$

The current density of the diffusing electrons is given by:

$$|j_e| = eD\left|\frac{d(\Delta n)}{dx}\right| \tag{3.29}$$

Substituting (3.28) into (3.29) an expression for the electron current density in the material is obtained. A similar expression is found for the current carried by the holes:

$$|j_e| = eD_e n_{po} \left(\exp\left(\frac{e|V_A|}{kT}\right) - 1 \right)\frac{1}{L_e} \tag{3.30}$$

$$|j_h| = eD_h p_{no} \left(\exp\left(\frac{e|V_A|}{kT}\right) - 1 \right)\frac{1}{L_h} \tag{3.31}$$

where L_h is the diffusion length for holes. The total current density, j, across the junction will be the sum of the current density carried by the electrons and that carried by the holes:

$$j = j_0\left(\exp\left(\frac{q|V_A|}{nkT}\right) - 1 \right) \tag{3.32}$$

where

$$j_0 = \{(D_e n_{po}/L_e) + (D_h p_{no}/L_h)\} \tag{3.33}$$

i.e. the current is determined by *diffusion effects* and not *drift*. For a diode of area A the total current will be Aj, so defining $I_0 = Aj0$, we have finally that

$$I = I_0\left(\exp\left(\frac{q|V_A|}{nkT}\right) - 1 \right) \tag{3.34}$$

where we have added in an empirical ideality factor, n, which has values between 1 and 2. A perfect junction has no recombination in the depletion region and $n = 1$, but for strong recombination in the depletion region n is nearer to 2. Because the movement of the electrons and holes is *governed by diffusion*, the response of a junction to an applied voltage is fairly slow. The sign of the applied voltage, V_A, is conventionally

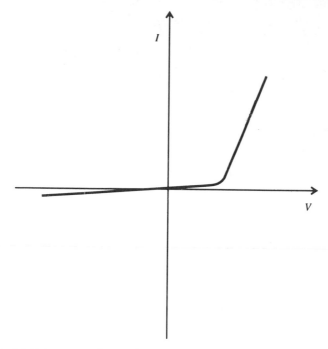

Figure 3.4 Typical *I–V* (current–voltage) characteristics of a *pn* junction diode.

positive for a forward bias, and typical current–voltage (*I–V*) characteristics for a *pn* junction diode are shown in Figure 3.4.

3.2.3 The quantum physics of LEDs and SDLs

The physics of interband optical transitions in semiconductors have already been covered extensively in Section 2.4, but the important results are repeated here for convenience. We know that if electrons are present in the conduction band, and if there are 'holes' (unoccupied states) in the valence band, then the electrons can lose their excess energy by 'dropping into' the holes, with a photon being emitted in the process. The converse process, involving the promotion of an electron from the valence band to the conduction band by *absorption* of a photon, also occurs. In the interband absorption process, the absorption coefficient $\alpha(\omega)$ depends on the incident photon energies, the bandgap, E_g, and the carrier populations in the valence and conduction bands.

If a radiation field exists in a semiconductor then both processes occur and a dynamic equilibrium is set up between absorption and emission. Because relaxation processes (the ways in which carriers can lose excess energy) tend to be much quicker within a band ('intraband') rather than between bands ('interband'), an electron

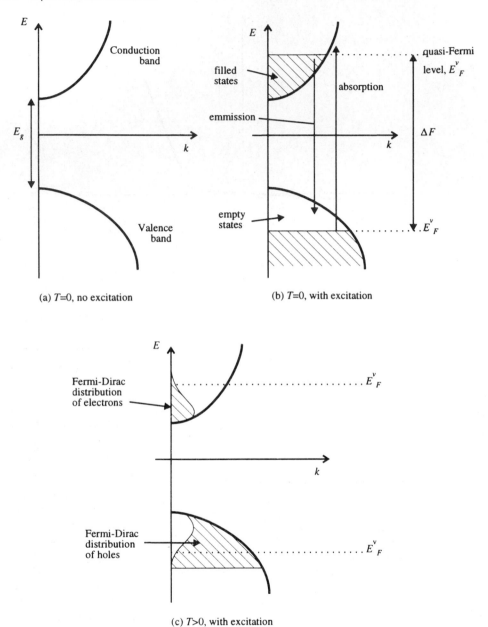

Figure 3.5 Carrier distribution in bands for (a) $T = 0$ with no excitation, (b) $T = 0$ with excitation, and (c) $T > 0$ with excitation.

population tends to build up at the bottom of the conduction band and a hole population at the top of the valence band. These populations are described by

'quasi-Fermi levels' E_F^v and E_F^c, as illustrated in Figure 3.5(b) (the prefix 'quasi-' simply means that E_F^v and E_F^c describe the equilibrium within each band, and not within the material as a whole). Clearly the quasi-Fermi level positions are dependent on the intensity of the exciting radiation or the injection current in a diode laser.

In these circumstances, we are able to identify three different absorption regimes:

$$\alpha(\omega) = 0 \qquad\qquad \text{where } \hbar\omega < E_g \qquad\qquad (3.35)$$

$$\alpha(\omega) = K(\hbar\omega - E_g)^{1/2} \qquad \text{where } \hbar\omega > E_{fc} - E_{fv} \qquad (3.36)$$

$$\alpha(\omega) = -K(\hbar\omega - E_g)^{1/2} \quad \text{where } E_g < \hbar\omega < E_{fc} - E_{fv} \qquad (3.37)$$

where K is a function of frequency and also includes terms that are dependent on the material (see Eq. (2.139)). It has a value of approximately 6×10^3 cm^{-1} (eV)$^{-1/2}$ in GaAs. E_g is the bandgap energy and $E_F^c - E_F^v$ is the quasi-Fermi level separation, ΔF. The negative sign in Eq. (3.37) indicates that photons are emitted rather than absorbed; this is *stimulated emission*, characteristic of lasers.

Comparison of Eqs. (3.36) and (3.37) shows that to be able to obtain *strong emission* it is necessary to have *strong absorption*, and hence that direct gap semiconductors are more effective that indirect. This is most easily understood by considering the case of a semiconductor at zero temperature, as in Figure 3.5. Figure 3.5(a) shows the band structure at $T = 0$ in the absence of any radiation and (b) illustrates the quasi-equilibria that are established in each band when above-bandgap excitation is present. In the latter case it can be seen that the only transitions with energies in the range $E_g < \hbar\omega < \Delta F$ occur between a filled conduction band state and an empty valence band state, i.e. only stimulated emission can occur, and no absorption. A transition from the valence band to the conduction band requires an energy greater than ΔF, when only absorption occurs. At finite temperatures, the statistically determined state filling means that both sets of transitions can occur at the same time but the original process still predominates. So for $E_g < \hbar\omega < \Delta F$ the rate of stimulated emission is greater than the rate of absorption and optical amplification occurs, equivalent to the 'classical' laser situation of population inversion. There are more electrons in the conduction band that satisfy the requirements for a downward transition to occur than those in the valence band for which an upward transition is possible.

In solid state lasers several methods are used to excite the semiconductor and produce population inversion. For example, electron beam excitation may be used, where an intense electron beam creates electron–hole pairs through electron–electron collisions. Optical pumpng is another possibility, in which a laser beam with photon energies greater than the bandgap creates the electron–hole pairs via interband absorption. However, the most efficient pumping method is by current injection across a *pn* junction, in which electrons and holes are injected from the appropriate sides of a forward-biased *pn* junction into the recombination region.

3.2.4 The Einstein relations

We wish to quantify the relationships between absorption, stimulated emission and spontaneous emission, and we use Einstein's treatment [3.3] adapted to energy bands

rather than discrete atomic states. First, we consider the transition between state $|1\rangle$ (energy E_1) in the valence band to state $|2\rangle$ (energy E_2) in the conduction band by absorption of a photon of energy $\hbar\omega_{21} = E_2 - E_1 = E_{21}$. The rate at which photons are absorbed may be written as:

$$r_{12} = B_{12}f_1(1 - f_2)P(\hbar\omega_{21}) \tag{3.38}$$

where B_{12} is the probability that the transition can occur, f_1 is the probability that state $|1\rangle$ is occupied, $(1 - f_2)$ is the probability that state $|2\rangle$ is empty, and $P(\hbar\omega)$ is the density of photons of energy E_{21}, given by the Planck radiation law. The terms f_1 and f_2 are given by the Fermi–Dirac function

$$f_1 = \frac{1}{\exp[(E_1 - E_F)/kT] + 1} \tag{3.39}$$

$$f_2 = \frac{1}{\exp[(E_2 - E_F)/kT] + 1} \tag{3.40}$$

where E_F is the Fermi level.

In addition to their being absorbed, photons can also stimulate the emission of a similar photon by the transition of an electron from state $|2\rangle$ to state $|1\rangle$. The rate of stimulated transitions is expressed by

$$r_{21}(\text{stim}) = B_{21}f_2(1 - f_1)P(\hbar\omega_{21}) \tag{3.41}$$

where B_{21} is the transition probability, f_2 is the probability that state $|2\rangle$ is occupied and $(1 - f_1)$ is the probability that state $|1\rangle$ is empty.

In addition to the stimulated transition, an electron may spontaneously return to state $|1\rangle$ without interaction with the radiation field $P(\hbar\omega_{12})$, the rate for such *spontaneous* transitions being

$$r_{21}(\text{spont}) = A_{21}f_2(1 - f_1) \tag{3.42}$$

By considering the thermal equilibrium between the levels, in which the absorption balances the total emission, we may deduce the following *Einstein relations* between the coefficients A_{21}, B_{21} and B_{12}:

$$A_{21} = (8\pi n^3 E_{21}^2/h^3 c^3)B_{21} \tag{3.43}$$

and

$$B_{21} = B_{12}, \tag{3.44}$$

showing that absorption and spontaneous and stimulated emission are all related.

Clearly, the above description of transitions between two discrete states is not appropriate in semiconductors, where we are interested in transitions between bands of states. As we shall see later, we need to take into account the actual band structure by including the density of states in the calculation, and it is this, together with the quasi-Fermi functions, which determine the spectral emission width of an LED and the gain lineshape of an SDL.

3.2.5 Band theory and transition probabilities

Here we are interested in obtaining physical expressions describing the transitions induced by the interaction between a semiconductor and a radiation field. Using the band theory and the transition probabilities given in Chapter 2, we can write the transition rate for an electron going from state a to state b by using Fermi's Golden Rule (Eq. (2.126)):

$$W_{cv}^{opt} = \frac{2\pi}{h}(H_{cv}^{opt})^2 \delta(E_g - E_c - \hbar\omega) \tag{3.45}$$

where H_{cv}^{opt} is the Hamiltonian matrix element for the interaction, $\langle \psi_c | H^{opt} | \psi_v \rangle$ and H^{opt} is the Hamiltonian operator. Another way of writing this is in terms of the transition probability B_{12} [3.4]:

$$B_{12} = \frac{\pi}{2h}|H_{cv}^{opt}|^2 \tag{3.46}$$

and we therefore need to deduce the form of the operator H^{opt}. Using the classical expression for the Lorentz force on an electron from interaction with electric (E) and magnetic (B) fields and writing it in the well-known Hamiltonian form:

$$H = (1/2m)(\mathbf{p} - e\mathbf{A})^2 + V(\mathbf{r}) \tag{3.47}$$

where $\mathbf{p}$ is the momentum operator, $\mathbf{A}$ is the vector potential and $V(\mathbf{r})$ is the electric potential, we find that

$$H^{opt} = -\frac{e}{m}\mathbf{A}\cdot\mathbf{p} \tag{3.48}$$

If the radiation field is a simple harmonic wave polarized in the x-direction then we find that the interaction Hamiltonian has the form

$$H^{opt} = \frac{eE_x}{mi\omega}\frac{i\omega x m}{} = \frac{exE_x}{2} \tag{3.49}$$

where ω is the forced harmonic oscillation frequency, x is the matrix element $\langle u_{vk} | x | u_{ck} \rangle$, and the u_{jk} are valence and conduction band Bloch functions, as discussed in Chapter 2. Transitions described by a Hamiltonian of this form are known as electric dipole transitions.

3.2.6 Absorption and emission in semiconductors

The expressions derived for absorption and emission in Section 3.2.4 involved two discrete levels only, and these need to be adapted for use in semiconductors where we are dealing with bands of states. In particular, we need to take into consideration the density of states per unit energy, $D(E)$, in both the conduction and valence bands. In the parabolic approximation, this has the form

$$D_c(E - E_c) = \frac{1}{2\pi^2}\left(\frac{2m_e}{\hbar^2}\right)^{3/2}(E - E_c)^{1/2} \qquad (3.50)$$

in the conduction band, with E_c the energy of the conduction band edge and m_e the electronic effective mass. A similar expression is obtained for the valence band.

The density of states function weights the probability functions f_1 and f_2 in Section 3.1.4. Thus for transitions between the valence and conduction bands, for example, the probability of occupancy of state $|1\rangle$, f_1, is replaced by the density of occupied states in the valence band, $D_v(E_v - E)f_v$. Similarly, the conduction band empty-state probability $(1 - f_2)$ becomes the density of unoccupied states $D_c(E - E_c)(1 - f_c)$.

In addition, the Fermi functions (3.39) and (3.40) are modified to reflect the band structure, now being described in terms of the quasi-Fermi levels that we discussed earlier. Thus we have

$$f_c = \frac{1}{\exp[(E_1 - E_F^c)/kT] + 1} \qquad (3.51)$$

and

$$f_v = \frac{1}{\exp[(E_1 - E_F^v)/kT] + 1} \qquad (3.52)$$

Now that we are dealing with absorption and emission between bands of states rather than discrete states, the various quantities calculated must involve transitions between all states separated by $\hbar\omega$ and we therefore need to integrate over all these states. Needless to say, the expressions begin to become quite complicated and so we omit discussion of them here. It is clear, however, that with the above modifications we can still apply the ideas of Section 3.1.4 to the more complex case of semiconductor materials.

3.3 Light-emitting diodes (LEDs)

We know, from the preceding sections, that a non-equilibrium carrier distribution can be created in a semiconductor by carrier injection across a *pn* junction, and that these injected carriers can lose their energy through radiative recombination, producing photons. This is the working principle of the *light-emitting diode*, or LED. The output from LEDs can be in the infrared or the visible range of the electromagnetic spectrum; this versatility in a wide variety of applications has led to enormous demand. LEDs share the important characteristics of other semiconductor devices of small size and weight, high mechanical stability, reliability, long life and low price. In addition, they have low operating voltages, medium current, high speed and, most importantly, they are efficient emitters of radiation over a very narrow spectral bandwidth. These properties make the *visible* LED useful in display technologies – an application with which we are all familiar – whereas the *infrared* (IR) LED finds its most important application in optical communication, where the emission wavelengths of, for example, InGaAsP lasers are well suited to available optical fiber technologies.

3.3.1 Material requirements

The choice of material in the manufacture of LEDs is limited by a series of requirements that are met by only a limited number of semiconductors. These are as follows:

1. the semiconductor bandgap must correspond to the desired photon energy;
2. a recombination mechanism resulting in effective radiative emission must exist;
3. it must be possible to dope the material both *n*- and *p*-type effectively, i.e. to a high density; and
4. the resulting *pn* junction must have excellent material properties so that, if condition (3) is fulfilled, there is effective current injection. This usually means that LEDs are made from single crystals.

The first requirement can be stated more plainly in terms of the required emission wavelength, λ, which is related to the bandgap energy by

$$E_g(\text{in eV}) = \frac{1240}{\lambda(\text{in nm})} \tag{3.53}$$

The wavelength sensitivity of the human eye is illustrated in Figure 3.6; we see that visible LEDs will have wavelengths in the range 400–700 nm, corresponding to photon energies of 3.2–1.8 eV.

The requirement that an efficient radiative mechanism should exist essentially rules out indirect gap materials, although sometimes the presence of shallow, impurity-related localized centres can considerably increase their efficiency. However, the common elemental semiconductors silicon and germanium are ruled out as LED materials.

The most important group of materals generally satisfying all the above conditions are the binary III–V compounds and their associated ternary and quaternary alloys. These are semiconductor *compounds* of the group III elements Al, Ga and In with the group V elements P, As and Sb, which crystallize in the zincblende structure and cover the range of bandgaps from 0.18 eV (InSb) to 2.26 eV (GaP), as illustrated in Figure 3.1. The only compound of the group III elements with nitrogen of current interest is GaN, which has the wurtzite structure and a very high bandgap of 3.4 eV. Also apparent in Figure 3.1 is the nature of the bandgap in each material. Excluding the nitrides, GaAs is the direct gap binary material with the highest bandgap energy, corresponding to an emission wavelength of about 860 nm, i.e. in the near IR. GaP, although indirect, is one of those materials whose radiative efficiency can be greatly enhanced by doping with, for example, nitrogen, to produce 'isoelectronic' centres. This results in reasonably efficient operation at an energy some 50 meV below the bandgap, producing green light.

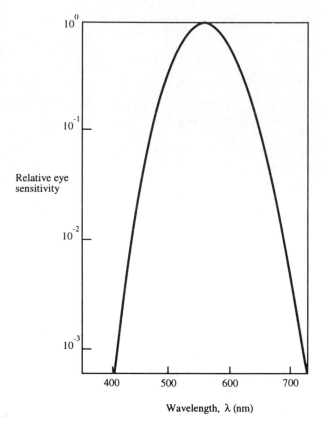

Figure 3.6 Sensitivity of the human eye as a function of wavelength [3.5].

The zincblende binaries can all be doped *n*- or *p*-type. Group III elements such as S, Se and Te are used to produce *n*-type doping and typical donor energies are below 100 meV. Thus all the donors are ionized at room temperature. Similarly, the group II elements Be, Mg, Zn and Cd form shallow acceptors. Of the wurtzite III–V compounds, only GaN has been doped successfully, and this only *n*-type.

Single crystals are usually required for the fabrication of LEDs and methods of growth are discussed in Appendix A. Bulk crystals of GaAs, GaP, InP, InAs, InSb and GaSb are readily available in both *n*- and *p*-type.

An extremely useful feature of III–V semiconductor materials is that, because of their similar properties, they can be combined to form mixed 'ternary' crystals which have properties intermediate between those of the binaries from which they are formed. This means that semiconductors can be 'tailored' for specific applications. The most obvious example is the tuning of the bandgap available in a system like GaAs–InAs, as illustrated in Figure 3.1. By varying the In concentration, x, in the ternary $In_xGa_{1-x}As$, the bandgap can be tuned continuously over the wavelength range 0.87–3.1 μm and the lattice constant changes from 5.65 to 6.06 Å.

The situation is slightly more complicated in a material like $GaAs_{1-y}P_y$, where continuous tuning of the direct bandgap is available only up to P contents of $y = 0.45$ ($E_g = 1.99$ eV), above which the bandgap becomes indirect. Thus in the range $0.45 < y \leqslant 1.0$ the bandgap has more of a 'GaP' character and, in a similar way to the binary material, the radiative efficiency can be improved by the addition of, for example, nitrogen. The tunability of the lattice constant means that special care has to be taken when growing epitaxial layers of this material.

Another very important ternary material is $Ga_{1-x}Al_xAs$, which is direct bandgap in the range $0 \leqslant y \leqslant 0.44$ and has the important property that the lattice constant varies only slightly with x. This allows easier epitaxial growth on GaAs substrates and very much easier growth of 'heterostructures' – structures comprising layers of more than one composition.

Also important in the manufacture of IR LEDs is the *quaternary* $In_{1-x}Ga_xAs_{1-y}P_y$ system. Quaternary systems allow an additional degree of freedom so that, for example, the bandgap can be tuned while the lattice parameter is kept constant. This allows lattice matching to an InP substrate (i.e. growth of an epilayer at the InP lattice constant) whilst tuning the bandgap over the range 0.7–1.35 eV.

Other semiconductor systems we have yet to mention include IV–IV and II–VI materials. SiC is an example of a IV–IV semiconductor upon which commercially available blue LEDs are based. These are quite inefficient and the high growth temperatures (2000 °C) required during growth make manufacture very difficult. II–VI materials such as ZnS_xSe_{1-x} and $ZnCd_ySe_{1-y}$ have been studied extensively of late. These tend to have bandgaps greater than those of the III–Vs, making them of considerable interest in the production of blue and blue-green devices, but they have suffered from the commensurate increases in self-compensation, making them difficult to dope *p*-type. However, these problems have recently been overcome in the above systems and efficient pn junctions have been fabricated. Already demonstrated in research, it seems that it will only be a matter of time before LEDs (and lasers) based on these materials become available commercially.

3.3.2 Types of LED: materials and structure

Commercial LEDs fall into two broad classes: visible and infrared. The infrared class of LEDs is dominated by $Ga_{1-x}Al_xAs$ devices, whilst visible LEDs are manufactured in the GaP and $GaAs_{1-y}P_y$ material systems in addition to $Ga_{1-x}Al_xAs$.

GaAs and $Ga_{1-x}Al_xAs$
GaAs is a direct bandgap semiconductor that has a bandgap of 1.44 eV, corresponding to a wavelength of about 860 nm. The addition of aluminium, giving $Ga_{1-x}Al_xAs$, increases the bandgap, as shown in Figure 3.1, pushing the emission wavelength closer to the visible as shown in Figure 3.7. Both of these materials can easily be doped both *p*- and *n*-type by the addition of Si, a group IV material, by careful control of the melt temperature.

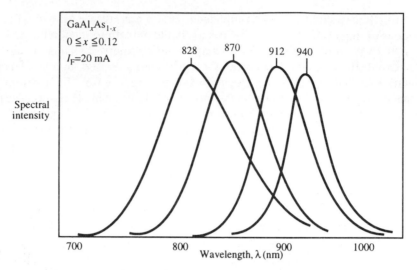

Figure 3.7 Emission spectra of $Ga_{1-x}Al_xAs:Si$ diodes with various aluminium contents [3.5].

In these devices the main luminescent transition is between the conduction band and a complex acceptor structure formed about 100 meV above the valence band edge. Thus the emission energy lies somewhat below the bandgap energy so that reabsorption of photons is low and quantum efficiencies are consequently quite high ($\sim 10\%$).

Typical device structures for these diode types are illustrated in Figure 3.8. Both

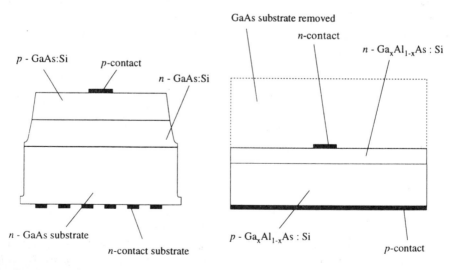

Figure 3.8 Typical infrared LED device structures. On the left is a GaAs:Si diode and on the right a $Ga_{1-x}Al_xAs:Si$ device [3.5].

structures are grown epitaxially on GaAs substrates from a single melt which produces both *n*- and *p*-type material. The process is relatively straightforward for the GaAs diode, but less so for the $Ga_{1-x}Al_xAs$ device where the material growth can present problems. In addition, the GaAs substrate is highly absorbing at $Ga_{1-x}Al_xAs$ emission wavelengths and so has to be removed completely.

The emission closer to the visible broadens due to the presence of an aluminium content gradient in the active layer. Moreover, the efficiency decreases steadily as Al is added. Typical device characteristics are given in Table 3.1.

Table 3.1 Typical GaAs and GaAlAs diode data

Parameter	GaAs : Si	$Ga_{1-x}Al_xAs$: Si
Maximum mean forward current (mA)	150	100
Output power at $I_f = 100$ mA (mW)	12–18	16–26
Output power at $I_f = 1.5$ A, pulsed (mW)	110–160	210–350
Peak wavelength (nm)	940	880
Forward voltage at $I_f = 100$ mA (V)	1.25	1.40
Forward voltage at $I_f = 1.5$ A, pulsed (V)	1.9	3.0

Despite the decreasing efficiency of $Ga_{1-x}Al_xAs$ devices with increased Al content, this method has been used to produce red LEDs, and a typical device structure is illustrated in Figure 3.9. It comprises a heterojunction with a *p*-type (Zn-doped) $Ga_{0.6}Al_{0.4}As$ active region and an *n*-type (Te-doped) $Ga_{0.3}Al_{0.7}As$ 'electron-injector' layer which is transparent to the emitted radiation at 650 nm. Such devices have a *p*-contact which covers the full area of the device because the substrate absorbs the emitted radiation anyway.

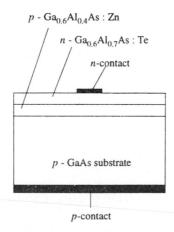

p - $Ga_{0.6}Al_{0.4}As$: Zn

n - $Ga_{0.6}Al_{0.7}As$: Te

n-contact

p - GaAs substrate

p-contact

Figure 3.9 A typical GaAlAs red LED device structure (after Ref. [3.5]).

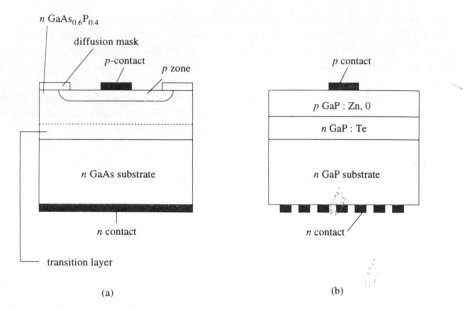

Figure 3.10 Red LED device structures: (a) GaAsP and (b) GaP:Zn,O (after Ref. [3.5]).

$Ga_{1-x}Al_xAs$ red LEDs are extremely bright and efficient, and have become very popular as a result. This improved performance arises because they are produced by liquid phase epitaxy (LPE) so that the material quality is high, the injection efficiency at the heterojunction is nearly 100%, the radiative efficiency of the p-type material exceeds that of the n-type, and there is little reabsorption in the n-type layer. They are thus superseding, to some extent, the longer-established GaP and $GaAs_{1-y}P_y$ technologies in the red, but we nonetheless discuss them here, because they are still produced in large quantities.

3.3.3 $GaAs_{1-y}P_y$ and GaP devices

A typical $GaAs_{1-y}P_y$ LED has a phosphorus content of about 40% and an emission wavelength of 660 nm. The $GaAs_{0.6}P_{0.4}$ layer is deposited on a GaAs substrate by vapour phase epitaxy (VPE) with a transition layer of graded composition in between. A typical structure is illustrated in Figure 3.10(a). The n-region is usually Te-doped and the p-region is formed in this structure by diffusing Zn into a masked-off region. The contact to the p-region is small to allow light to pass largely unobstructed from the surface. No light is emitted from the bottom of the device because of absorption by the GaAs, so the n-contact can cover the whole device area.

Typically, the diffusion layer is 5 μm thick in order to optimize the balance between the effects of surface recombination of the carriers and reabsorption, and the overall epilayer thickness is about 50 nm.

$GaAs_{1-y}P_y$ diodes are also used to produce orange and yellow emission, by increasing the phosphorus content to 60% and 80%, respectively. These structures tend to have lattice constants that are closer to that of GaP than to GaAs and so are grown on GaP substrates, but are otherwise similar to the above. In these devices, nitrogen is added to aid recombination by forming 'isoelectronic centres'.

We have already mentioned that GaP LEDs are green-emitting devices, but by doping the *p*-type layer with both Zn and O, an emission wavelength of 690 nm is obtained by recombination at impurity centres. The typical device structure shown in Figure 3.10(b) has the advantage that the emitted radiation has an energy well below the GaP bandgap and so is not reabsorbed in either the epilayers or in the GaP substrate. Thus the *n*-contact covers only a part of the surface, leaving the main part reflecting so as to increase the output intensity through the opposite face.

3.4 Semiconductor diode lasers (SDLs)

3.4.1 Population inversion and stimulated emission

The optical output of a light-emitting diode occurs over a spectral range (typically 50–800 nm) that is broad compared with that from a laser, and the phase of the emitted photons is random. In an LED, no special physical structures are incorporated into the device to prevent optical loss. Diodes, on the other hand, have a coherent, directional output that occurs over a much narrower bandwidth. Light is confined in an optical cavity where a coherent electromagnetic field builds up due to optical feedback. The feedback can be provided simply by the cleaved ends of the crystal

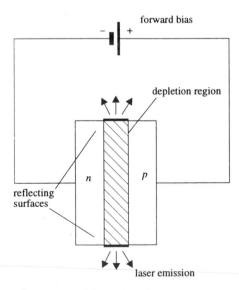

Figure 3.11 A simple diode laser.

which reflect about 30% of the optical power, or the reflectance can be increased by depositing gold or dielectric mirrors on the ends of the *pn* junction structure (Figure 3.11). The simplest diode lasers are *homostructure pn* junctions with cleaved ends at 90° to the current flow that provide optical feedback. More advanced *heterostructures* use carrier and current confinement to reduce threshold currents and also to increase device lifetimes.

A well-known quantum-mechanical principle due to Dirac states the following: 'In the interaction between a photon and an electron, the photon can be absorbed and the electron excited to a higher energy state, or, if the electron is already in an excited state, the emission of a second photon can be stimulated (with simultaneous relaxation of the electron) which shares the same characteristics of phase, direction and wavelength, as the incident photon'. In the absence of an excitation source, the majority of electrons in a semiconductor material are in the unexcited or ground (valence) state so that any emitted photons are usually reabsorbed. However, in diode lasers an external 'pumping' source of energy is used to excite a large number of electrons into the conduction band, creating a situation in which the stimulation of a second photon is more likely than absorption of the first: this is known as *population inversion*. The carrier populations in the diode are, of course, directly related to the *currents* flowing across the pn junction, and the inversion is caused by this *transport*. Recombination of this type is known as *stimulated* emission, which is related to the strength of the electromagnetic field, unlike *spontaneous* emission, which is independent of the field. Because of the population inversion the photons are not strongly reabsorbed by ground state electrons and amplification of the optical field occurs; this is the basis of laser action.

In order to consider this in detail, we need to return to the simple two-band analysis of Section 3.2.4 and its extension to transitions between energy bands.

From Eqs. (3.41) and (3.42), we know that the ratio of stimulated to spontaneous transitions within a two-level system is simply

$$\frac{r_{21}(\text{stim})}{r_{21}(\text{spont})} = \frac{B_{21}}{A_{21}} P(E_{21}) \tag{3.54}$$

where

$$P(E_{21}) = P(\hbar\omega_{21}) = \frac{8\pi n^3 E_{21}^2}{h^3 c^3} \left\{ \exp\left(\frac{E_{21}}{kT}\right) - 1 \right\}^{-1} \tag{3.55}$$

and $E_{21} = \hbar\omega_{21} = E_2 - E_1$.

Using Eq. (3.43), we can rewrite Eq. (3.54) more simply as

$$\frac{r_{21}(\text{stim})}{r_{21}(\text{spont})} = \left\{ \exp\left(\frac{E_{21}}{kT}\right) - 1 \right\}^{-1} \tag{3.56}$$

Thus, broadly speaking, the ratio of stimulated emission to spontaneous emission increases in a two-level system as the energy level spacing becomes small compared to *kT*. This has the effect of increasing the probability of occupation of the upper state, f_2, and decreasing that of the lower state, f_1. However, the occupancy of each level in such a system can never exceed one-half of the total population, and in the limit where $E_{21} \ll kT$ the probabilities f_1 and f_2 are equal.

To achieve a situation where net stimulated emission results from interaction with the optical field, rather than absorption, we require that

$$r_{21}(\text{stim})_{\text{net}} = r_{21} - r_{12} > 0, \tag{3.57}$$

that is,

$$r_{21}(\text{stim})_{\text{net}} = B_{12}P(E_{21})\{f_2(1 - f_1) - f_1(1 - f_2)\} \tag{3.58}$$

$$= B_{12}P(E_{21})(f_2 - f_1)$$

Thus we require that $f_2 > f_1$, *which cannot be achieved in a two-level system.*

However, we can achieve this condition if the system has more than two levels, or if we inject additional carriers into the system from an external source, as in a semiconductor laser. We have already seen that such current injection produces a non-equilibrium carrier distribution in the energy bands of the semiconductor, described by quasi-Fermi levels E_F^c and E_F^v and accompanying Fermi functions. Thus the condition for net stimulated emission now becomes

$$1 + \exp\left(\frac{E_v - E_F^v}{kT}\right) > 1 + \exp\left(\frac{E_c - E_F^c}{kT}\right) \tag{3.59}$$

i.e.,

$$E_F^c - E_F^v > E_c - E_v$$

which is known as the *Bernard–Durrafourg inversion condition*. In addition, we also wish to obtain $r_{21}(\text{stim}) > r_{21}(\text{spont})$ so that we have *coherent* amplification of the beam. This means that

$$P(E)B_{21} > A_{21} \tag{3.60}$$

from Eq. (3.54), i.e. a high photon density is required.

We wish to obtain the behaviour of the spontaneous and stimulated emission under various equilibrium and non-equilibrium conditions. To do this we write the above expressions in terms of the macroscopic quantity α, the absorption coefficient, measured in cm^{-1}. For a two-level system the stimulated emission rate is related to α via

$$r_{21}(\text{stim}) = -\alpha P(E)c/n \tag{3.61}$$

where c is the velocity of light in a vacuum and n is the refractive index of the medium. However, we have $r_{21}(\text{stim})$ from Eq. (3.41) and we obtain

$$\alpha = -\frac{\text{n}}{c} B_{12}(f_2 - f_1) \tag{3.62}$$

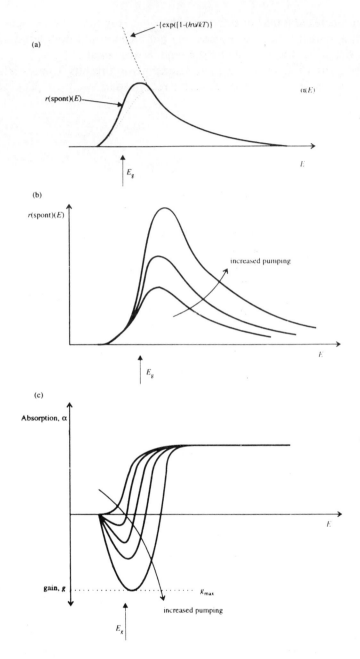

Figure 3.12 Absorption and emission at energies close to the band edge of a semiconductor. (a) Energy dependence of the spontaneous emission and the absorption in a thermal equilibrium situation, i.e. without any carrier injection. (b) and (c), behaviour of the spontaneous emission and the absorption/gain function as the carrier injection is increased.

At this point we note that under conditions where stimulated emission dominates, the absorption becomes negative and is then referred to as the *gain*, *g*. Thus $g = -\alpha$.

In a semiconductor material the expressions become much more complex because the band structure has to be taken into account. In this case we obtain

$$r_{21}(\text{stim}) = B_{12}P(E)[f_c(E_c) - f_v(E_v)]D_c(E_c)D_v(E_v) \qquad (3.63)$$

and

$$\alpha(E) = \int \frac{\text{n}}{c} B_{12}[f_c(E_c) - f_v(E_v)]D_c(E_c)D_v(E_v)\delta[(E_c - E_v) - E]dE \qquad (3.64)$$

where the δ-function ensures conservation of energy.

The spontaneous emission rate is given by

$$r_{21}(\text{spont}) = -\frac{8\pi\text{n}^3 E_{21}^2}{h^3 c^2 n}\left\{1 - \exp\left[\frac{E - (E_F^c - E_F^v)}{kT}\right]\right\} \cdot \alpha \qquad (3.65)$$

These functions are plotted in Figure 3.12(a–c), where we can see the behaviour of the spontaneous emission and the absorption/gain for various pumping levels close to the band edge. The most important aspect of these functions is illustrated in Figure 3.12(c), where we see that as pumping is increased the energy region in which net gain is obtained gets wider and the peak ($g_{\max}$), corresponding to the energy of laser operation, moves to higher energy.

The simple homojunction laser

We are now in a position to bring together the various aspects of *pn* junction technology and our knowledge of stimulated emission processes in semiconductors to form a simple laser. It is made entirely of a single semiconductor compound, usually GaAs, with different parts of the device having different dopings, and the junction is formed between *n*- and *p*-type layers of the same material. For this reason the device is known as a *homojunction* laser and its band structure under forward-bias conditions is illustrated in Figure 3.13. Under conditions of high injection, the carrier distributions in a small volume at the centre of the *pn* junction depletion region satisfy the conditions discussed in the previous section, and stimulated emission dominates. On either side of this region, recombination still occurs but it is dominated by spontaneous emission.

To take advantage of the stimulated emission to produce laser *oscillation*, feedback has to be introduced in the system. This is provided in our simple laser (Figure 3.11) by the two end facets of the semiconductor chip (typically only 500 μm long because of the high gain) which are cleaved to reflect some of the light back along the axis of the laser and define a *laser cavity* (in more sophisticated devices facets often have coatings to increase or decrease their reflectivity, as appropriate). The reflected light then stimulates the emission of more photons along the laser cavity, so that if the round-trip gain exceeds the total loss in the system, laser oscillation results.

Several aspects of diode laser operation will be discussed in the following sections. First, we examine how the stimulated emission and optical feedback in a diode laser

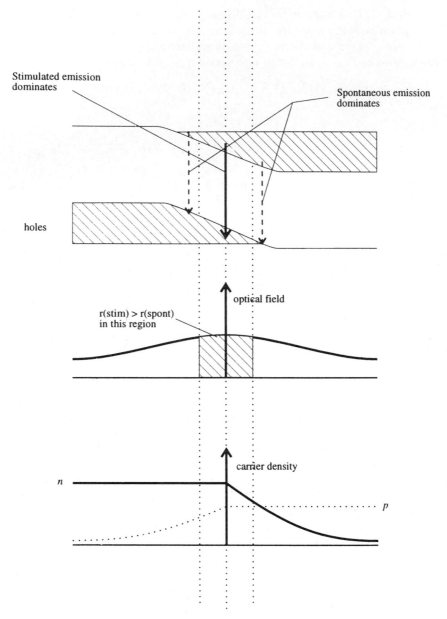

Figure 3.13 The optical field and carrier density distributions in the vicinity of the pn junction in a homojunction laser. The top diagram indicates the regions in which stimulated emission and spontaneous emission dominate.

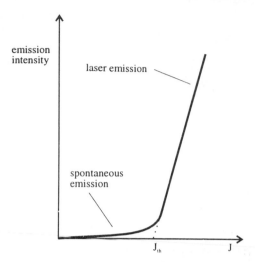

Figure 3.14 Emission versus current density for a typical diode laser. The threshold current density J_{th} is indicated. The much steeper slope in the region of laser operation indicates the greatly increased efficiency.

produces an output that is very different from that of light-emitting diodes. Because stimulated emission is proportional to the optical field intensity, wavelengths at which the emission is greatest tend to be amplified at the expense of wavelengths at which the emission is weaker, so that the gain spectrum is much narrower than the LED emission bandwidth. In addition, the presence of reflecting facets results in an emission spectrum that is *discrete*, rather than continuous. The set of allowed emission wavelengths are known as *longitudinal modes*, and these are the subject of the next section.

Laser action will only occur in our device if the injection current exceeds a value that is sufficient to bring about a population inversion. This value is known as the *threshold current*, below which the diode acts very much like an LED and above which the output efficiency very rapidly increases, as in Figure 3.14. The threshold current is a very important figure of merit for diode lasers, since lower-threshold devices dissipate less energy and tend to have higher efficiencies. The threshold current is measured in terms of the current density, J, which is the current per unit junction area, and is nominally deduced by extrapolating the linear part of the emission curve back to the zero-output line, as shown in Figure 3.14. Since the threshold current is so important, we will spend some time obtaining a theoretical value for it in Section 3.4.4.

3.4.2 Mode properties of SDLs

In this section we consider the effect of the structure of the laser itself on the output from the device. As we shall see, this structure imposes restrictions on the spectral and spatial distribution of the laser oscillation so that the output comprises one or more of a set of discrete *modes*.

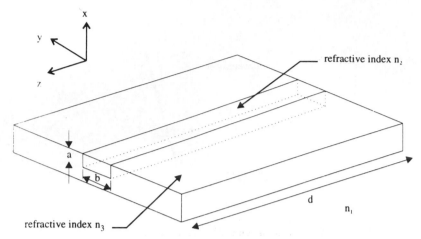

Figure 3.15 The waveguide defined by the semiconductor laser stripe and heterostructure. The refractive index of the surrounding material is n_1.

The *pn* junction, together with the variations on heterostructure, can be well represented by a rectangular dielectric waveguide as depicted in Figure 3.15. Within this the gain of the laser excites a series of modes.

The ends of the guide, acting as partial mirrors, cause the guide to act similarly to more familiar laser resonators and define longitudinal modes. These have a frequency separation

$$\Delta v = \frac{c}{2nd} \tag{3.66}$$

or, in terms of wavenumbers, expressing the denominator in cm,

$$\Delta \omega = \frac{1}{2nd} \tag{3.67}$$

For a typical length of $d = 400$ μm and n = 3.5 we thus have $\Delta \omega \sim 3$ cm^{-1}. For a GaAs laser at 850 nm wavelength the mode separation is ~ 0.5 nm. This implies that with a gain width of 50 nm, as many as 100 longitudinal modes can be excited. The spectral output structure of a diode laser as the current and hence the gain is increased is shown in Figure 3.16. In general, the spectral width of the output narrows compared with the spontaneous emission, separates into modes and, eventually, one or a few of these modes dominate the emission.

The rectangular structure also imposes, through boundary conditions, *guided wave modes*. The theory is similar to that described in Section 1.2.3, i.e. TE and TM modes. These are of the form, for TE modes,

$$E_y = A \cos(qx) + B \sin(qx) \tag{3.68}$$

within the guided region, where $q = (n_2^2 k^2 - \beta^2)^{1/2}$ and β, the propagation constant, is obtained from $n_{\text{eff}} = \beta/k$ and $k = 2\pi/\lambda = \omega\sqrt{\varepsilon_0 \mu_0}$. Above and below the slab, theory gives

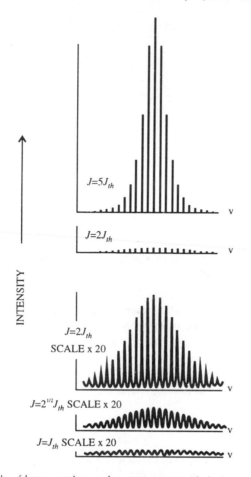

Figure 3.16 The growth of laser modes as the pump current is increased [3.7].

$$E_y = A \exp - p_1 x \tag{3.69}$$

and

$$E_y = [A \cos(qa) - B \sin(qa)]\exp p_3(x + a) \tag{3.70}$$

where $p_{1,3} = (\beta^2 - n_{1,3}^2 \kappa^2)^{1/2}$. Solving also for the H_z field one obtains an eigenvalue equation for the TE modes:

$$\tan(qd) = \frac{q(p_1 + p_3)}{q^2 - p_1 p_3} \tag{3.71}$$

which can be solved numerically. This calculation then defines the output frequencies due to guided, or transverse, modes of the diode laser.

The TE_0 mode is that of the lowest β. Typical spatial profiles of TE modes are given in Figure 3.17. Similar results follow for TM modes. The mode structure has

(a) Fundamental mode   (b) 2nd order mode

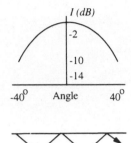
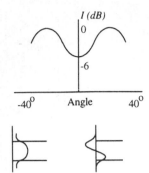

(c) Ray model of waveguide

(d) Energy model of waveguide in fundamental (left) and 2nd order modes (right)

Figure 3.17 The spatial profiles of various transverse output modes of a diode laser. (a) E^y_{11}, (b) E^y_{21}, (c) E^y_{12}, (d) E^y_{31}, (e) E^y_{22} and E^y_{41} [3.6].

consequences for the spatial form of the output of diode lasers as well as the frequencies.

The near field (i.e. at the exit facet of the laser) generally corresponds to a mode of the passive resonator structure. The wavefront is generally plane at the exit and, assuming a properly designed single transverse mode waveguide, is uniphase. The divergence is diffraction limited and can be roughly estimated from the formula for diffraction through a slit – assuming the correct effective mode width, d. Thus the half-angle, θ_D, is given by

$$\theta_D = \lambda/d. \tag{3.72}$$

Usually this gives a highly elliptical far-field because of the very different transverse mode dimensions, as illustrated in Figure 3.18.

3.4.3 Double-heterostructure lasers: optical and carrier confinement

Early homojunction diode lasers had very high threshold currents, resulting in inefficient operation and short device lifetimes. This meant that only pulsed mode operation was possible (i.e. the lasers were modulated at frequencies of 10^2–10^3 Hz with a low duty cycle ratio) and then only at cryogenic operating temperatures. The reasons for this inefficient operation can be understood from Figure 3.13. In a simple pn junction, carriers rapidly leak away from the junction region so that they are unable to contribute to the population inversion and, when they recombine, the emitted photons do not contribute to laser oscillation. The resulting optical field also spreads out far beyond the junction region, so that only a small part of it coincides spatially with the gain volume close to the middle of the depletion region.

Because of this highly inefficient operation, homojunction lasers were replaced by devices known as *heterostructure* (or *heterojunction*) *lasers*. In these devices, layers of

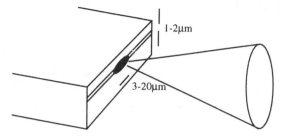

Figure 3.18 The far-field pattern of a typical diode laser.

different material composition lie next to the active region where, because of their different bandgaps and refractive indices, they tend to confine both carriers and the optical field. This increases the efficiency of the devices and allows operation at room temperature with much lower threshold currents. Early heterostructure devices were of the *single-heterostructure* (SH) type, where a layer with a different bandgap lies on one side of the active layer only. Thus in a typical device the active layer, usually GaAs, would be sandwiched between a layer of GaAs and a layer of $Ga_{1-x}Al_xAs$ and this results in a device which shows pulsed operation at room temperature with high peak powers. However, cw operation is still not possible with these devices and they have largely been superseded by double-heterostructure devices.

Double-heterostructure (DH) lasers use two heterojunctions for improved confinement. Thus the active layers (usually GaAs) are sandwiched between two layers of different material (typically $Ga_{1-x}Al_xAs$), as shown in Figure 3.19. These outer layers act as a waveguide in a similar way to the cladding of optical fibers, because they have a smaller refractive index than the active layer. The refractive index and the thickness of the active layer determine what fraction, Γ, of the optical field is confined, and this in turn determines the efficiency of the device. Room-temperature thresholds of the order of 500 A cm^{-2} are easily achievable in this way, with cw operation, and double-heterostructure lasers have therefore become the prevalent type for most applications. A number of different classes of DH laser are in use at the present time, but we put aside discussion of these for the moment in order to discuss further factors that affect the laser threshold current density.

3.4.4 The gain–current relation and threshold current density

We wish to obtain an analytical expression for the gain–current relation and a theoretical value for the injection current at which laser action begins. To do this we first need to think about the *lasing condition*, that is, the condition for obtaining overall gain in the laser cavity. This condition is clearly that the gain generated in the optical medium by stimulated emission must exceed the total optical losses within the cavity, including light emitted from the ends and light scattered out of the cavity. It can be shown [3.8] that this condition is satisfied when

$$\Gamma g = \alpha_i + \frac{1}{2L} \ln \frac{1}{R_1 R_2} \tag{3.73}$$

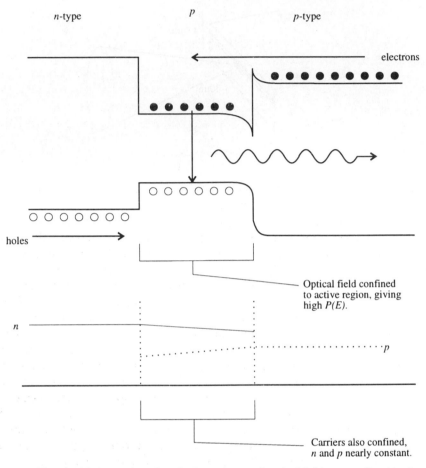

n-type   *p*   *p*-type

electrons

holes

Optical field confined
to active region, giving
high *P(E)*.

n

p

Carriers also confined,
n and *p* nearly constant.

Figure 3.19 The double heterojunction. Both carriers and optical field are confined in the active GaAs region, with a consequent increase in efficiency.

where g is the 'local gain' in the medium, Γ is the fraction of the optical power coupled into the gain region, as previously described, α_i are the scattering losses, L is the length of the cavity and R_1 and R_2 are the facet reflectivities. The last term on the right-hand side therefore represents the end losses, and both terms on the right taken together constitute the 'modal gain', G. For simplicity, the facet reflectivities are usually taken to be equal, so Eq. (3.73) reduces to

$$\Gamma g = \alpha_i + L^{-1} \ln R^{-1}. \tag{3.74}$$

In order to derive an expression for the gain–current relation and threshold current density we adopt the following approach. First, we need to relate the injection current going into the recombination region to the amount of light leaving it. This is in principle quite difficult, since we have two emission processes to consider, but we make the simplifying assumption that, below threshold, the recombination is entirely spontaneous. This is not unreasonable because the optical field $P(E)$ is comparatively small below threshold, and so $\int r(\text{stim})dE \ll \int r(\text{spont})dE$. In addition, we assume that the material is ideal in the sense that no nonradiative recombination takes place. Thus we can obtain the total radiative recombination rate in terms of what we will call a spontaneous recombination current, J_{spont}, which is a function of the separation of the quasi-Fermi levels. In addition, we shall write down empirically obtained expressions for the gain as a function of spontaneous recombination current and then combine the above with Eq. (3.74) to give us an expression for the gain–current relation and the current density at threshold.

Given the above conditions, the total radiative recombination rate, R, is given by

$$R = \int_0^\infty r(\text{spont})dE \tag{3.75}$$

where $r(\text{spont})$ is given by Eq. (3.65). If current I is flowing through a device of unity quantum efficiency with area A and thickness d, then we have that

$$R = I/|e|ad \tag{3.76}$$

where e is the electronic charge. Thus the spontaneous recombination current density is

$$J_{\text{spont}}(\text{A cm}^{-2}) = I/a = |e|Rd \tag{3.77}$$

For the purposes of comparison, J_{spont} is often quoted for a device of thickness $d = 1$ μm; this is known as the *nominal* current density, $J_{\text{nom}}(\text{A cm}^{-2}\ \mu\text{m}^{-1})$, the current required to maintain the actual recombination rate in a device of thickness 1 μm. For non-unity quantum efficiency we have that

$$\eta = \frac{Rad}{I/|e|} \tag{3.78}$$

and then the actual current density in the active layer (for an active mode propagating entirely within the active layer) is simply

$$J(\text{A cm}^{-2}) = J_{\text{nom}}d/\eta. \tag{3.79}$$

We also need to relate J_{nom} to the maximum in the gain coefficient, g_{max}, which moves to higher energy as pumping is increased, as we have already seen. In $Ga_{1-x}Al_xAs/GaAs$ double-heterostructure lasers it is found experimentally that this relation has the form

$$g_{\text{max}} = \beta(J_{\text{nom}} - J_0) \tag{3.80}$$

where J_0 is the current density at which g_{max} is zero, and β is an experimentally determined constant, as in Figure 3.20. The physical significance of J_0 is that it is the current at which neither absorption nor gain occurs so that the medium is transparent to photons of one particular above-bandgap energy. This is therefore known as the transparency current.

Now we could combine the above with the Eqs. (3.63)–(3.65) to produce a highly complicated expression for the nominal current density which includes its dependence

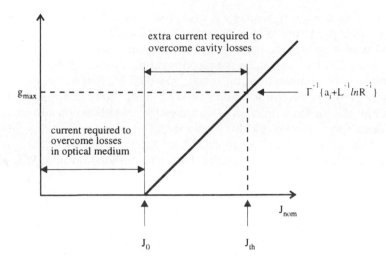

Figure 3.20 The value of maximum gain as a function of nominal current density. J_0 is the transparency current and Jth the threshold current.

on quasi-Fermi level separation and temperature, but it is more helpful, particularly in the interpretation of experimental data, to obtain an expression for the threshold current density, J_{th}. Remembering that $J_{th} = J_{nom}(g_{max})d/\eta$, substitution of Eq. (3.80) into Eq. (3.74) yields

$$J_{th}(\text{A cm}^{-2}) = \frac{J_0 d}{\eta} + \frac{d}{\eta \Gamma \beta}\left\{\alpha_i + \frac{1}{L}\ln\frac{1}{R}\right\} \tag{3.81}$$

Here, the first term on the right-hand side represents that part of the threshold current required to invert the carrier population (which we could call *material losses*) and the second term is the current required to overcome *cavity losses*.

3.4.5 Waveguiding and laser heterostructures

Equation (3.81) shows how important it is for the optical field to be coupled effectively to the gain region, keeping the confinement factor Γ as high as possible, since this results in a reduction in threshold current density. A further reduction in threshold current is obtained by reducing the width of the laser active region. In the earliest lasers, active regions of the order of 50 or 100 μm wide were common, but nowadays this figure has been reduced to 1–10 μm in so-called *stripe geometry* lasers. The stripe geometry results in improved performance in other ways, often resulting in single longitudinal and transverse mode operation making the device easier to use in applications such as compact disc players and optical communications.

Restriction of the gain region to a narrow stripe is achieved in two ways, both of which involve different types of laser structure, as illustrated in Figure 3.21. In *gain-guided* structures, the current flow is restricted to a narrow stripe by the formation of high resistivity regions on the top of the device. These regions, formed either by

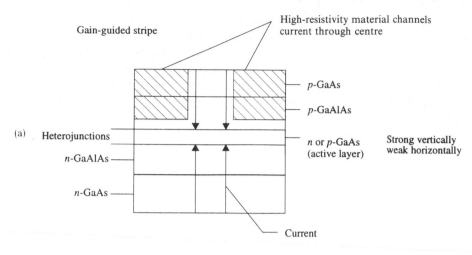

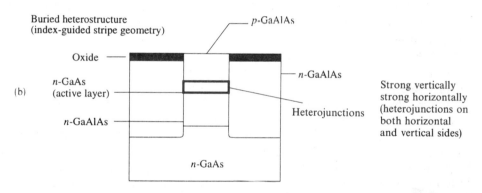

Figure 3.21 (a) Gain-guided and (b) index-guided stripe geometries [3.1].

growth of an oxide layer on the surface or by proton bombardment, confine the current and hence the population inversion to the stripe region, thus limiting the active laser volume. The gain region also has a very slightly different refractive index from that of the surrounding material (the population inversion changes the dielectric properties of the material), so that there is a weak waveguiding effect on the optical mode.

The simplicity of gain-guided lasers makes them easier to produce, but the weakness of the optical confinement increases the threshold current and tends to produce instabilities and nonlinear behaviour in the output characteristics. On the other hand, the weakness of the mode confinement is advantageous in very high-powered lasers where the optical density is sufficiently high (1 MW cm^{-2}) to produce damage in the facets of index-guided lasers. Thus high-power lasers tend to be gain-guided.

In index-guided lasers, such as the buried-heterostructure laser illustrated in Figure 3.21(b), the laser stripe is bounded by regions of different refractive index. Usually the refractive index of the active layer is higher than that of the surrounding material to produce a waveguide for the optical mode; this improves the efficiency of the device considerably, reducing threshold currents and producing better mode behaviour. Index-guided lasers are most commonly used in low-power applications such as compact disc players, laser printers and optical communications.

Buried-heterostructure lasers are formed by etching a narrow (1–2 μm) mesa in the top layer of a double-heterostructure laser to form the stripe region and then depositing the material with a lower refractive index around it. Current is prevented from leaving the stripe region either by depositing an insulating layer on the top (as in gain-guided lasers) or by suitable doping to produce a pn junction between the mesa and the surrounding material that is reverse-biased in normal operation. This combination of optical and current confinement makes buried-heterostructure devices very efficient.

In addition to buried-heterostructure lasers, there are many other types of index-guided devices, including ridge waveguide, channelled-substrate and buried-crescent lasers, which are illustrated in Figure 3.22.

Ridge waveguide lasers (Figure 3.22a) have a ridge above the active stripe whilst the rest of the material is etched close to the active layer (typically within 0.2 μm). The edges of the ridge reflect light guided within the active layer, thereby forming a waveguide, and oxide coatings on the surrounding areas help to confine the current flow.

Channelled-substrate lasers (Figure 3.22b) have an active layer which is grown on a substrate with a channel etched into it. The substrate has a high refractive index but the material which forms the cladding between active layer and substrate has a low index. The thick cladding in the channel region, which creates a planar surface for the growth of the active layer, also isolates the optical field from the substrate in the channel area and thus creates a region of low loss. The thinner cladding in the surrounding area means that losses are higher there and the resulting difference in the refractive index between the two regions produces a waveguiding effect.

Buried-crescent lasers (Figure 3.22c) are similar to channelled-substrate lasers in that they have a groove or channel etched into the substrate. However, the channel is not filled completely with cladding to provide a planar surface for active layer growth, but instead the active region is grown within the groove by liquid phase epitaxy to form a stripe in the active layer with a crescent-shaped cross-section. This serves to provide lateral confinement of the optical mode.

3.4.6 Quantum well lasers

Single quantum well (SCH) lasers

The concept of a semiconductor quantum well was made viable by the development of sophisticated growth techniques, such as molecular beam epitaxy (MBE), which facilitate growth control on the scale of atomic monolayers. As we saw in Chapter 2, a typical *single* quantum well (SQW) structure comprises a thin layer of one semiconductor sandwiched between two thicker layers of a higher-bandgap material, similar to the structure of a double-heterostructure (DH) laser except that in this case

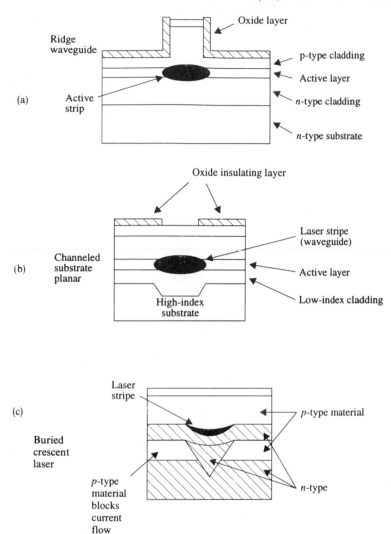

Figure 3.22 Important types of index-guided lasers [3.1].

the low-bandgap (quantum well) material is thin enough (typically < 200 Å) for *quantum size effects* to become important. Thus the behaviour of devices in which the active region is a quantum well is substantially different from that of normal DH lasers, and we have to take into consideration the energy quantization in the conduction and valence bands and the consequent two-dimensional density of states (2-D DOS). The most important consequence of this two-dimensionality is that maximum gain always occurs at an energy equal to the lowest interband transition energy, so that all injected carriers contribute to the gain. This is in contrast with the 3-D case where, as we have seen, maximum gain shifts to higher energy as pumping

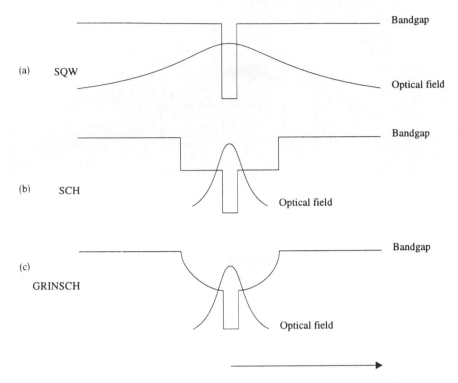

Figure 3.23 Bandgap variation and optical confinement in various quantum well heterostructures: (a) single quantum well, (b) separate confinement heterostructure (SCH) and (c) graded refractive index separate confinement heterostructure (GRINSCH).

increases with carriers below g_{max} not contributing, so that quantum well lasers are more efficient at low injection. However, we know better than to expect something for nothing; the price that has to be paid for the increased efficiency at low injection is that the gain saturates at a finite value, corresponding to the total inversion of both electron and hole states, whereas the gain never saturates in a DH laser because the 3-D DOS increases without limit. However, the transparency current in a QW laser is smaller because the DOS to be inverted is much smaller, and the overall result is to reduce threshold currents to about one-half or one-third of those in comparable DH lasers.

Clearly, with so small a gain volume some sort of structure has to be incorporated into the device to prevent the optical field from spreading out too far. This is done by providing layers adjacent to the quantum well of intermediate bandgap (and hence refractive index), which act as a waveguide for the optical field, whilst the quantum well itself confines the carriers. This type of structure, where the optical field and the carriers are localized in different regions, is known as a *separate confinement heterostructure* (SCH), two of which are illustrated in Figure 3.23. In the more complex of the two, the *graded refractive index separate confinement heterostructure* (or

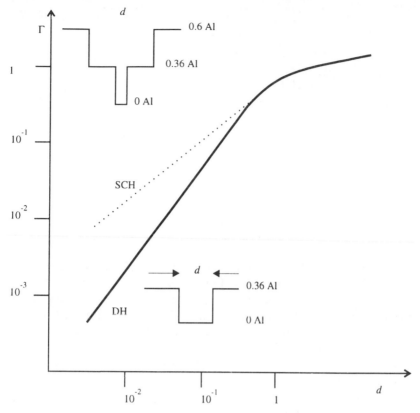

Figure 3.24 A comparison of the optical confinement factors of DH and SCH $Ga_{1-x}Al_xAs$/GaAs lasers for a given active layer thickness. The separate confinement layer is 0.2 μm wide. The Al content of each layer is indicated.

GRINSCH), the intermediate layer is usually a ternary material and is fabricated by gradual variation in the composition. The increase in the confinement factor, Γ, of separate confinement structures over conventional double-heterostructures is typically a factor of 10 for an active region thickness of 50 Å, as illustrated in Figure 3.24.

A comparison of the modal gain in a GRINSCH laser with that of a DH laser is illustrated in Figure 3.25. We see that reducing the active layer thickness has the effect of reducing the transparency current and increasing the differential gain at low injection currents.

It turns out, however, that there is an optimum thickness at which the current threshold is a minimum; this arises because as the wells get thinner the ground state levels are pushed to higher energies in the wells and begin to spill out into the confinement cavity, causing an increase in J_0. Even in lasers at the optimum width, pumping at high currents can shift the quasi-Fermi level in the conduction band close to the barrier band edge, producing some population in the confinement cavity (this

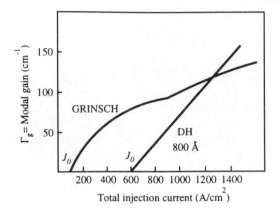

Figure 3.25 The gain–current curves for a 120 Å QW GRINSCH laser and an 800 Å DH laser, showing the difference in the transparency currents and differential gains.

does not occur in the valence band because of the much larger effective mass). The improvement in the performance of GRINSCH lasers over other SCH lasers in terms of threshold current, illustrated in Figure 3.25, can largely be accounted for by improvements in the *carrier* confinement, rather than any improvement in optical confinement. It has been calculated [3.9] that, at threshold, the cavity population in the conduction band in a SCH laser is roughly equal to that in the well, whereas in a GRINSCH laser it is only about 20% because of the smaller DOS in a non-rectangular cavity.

Multiple quantum well (MQW) lasers

Multiple quantum well lasers have an active region that contains several quantum wells separated by high-bandgap material barriers of sufficient width that no quantum-mechanical tunnelling occurs between them. Broadly speaking, the properties of an MQW system can be deduced from those of a similar single quantum well, assuming homogeneous carrier injection. To a first approximation, the confinement factor, Γ, can be considered to be proportional to active layer width so that the gain–current relation of an MQW can be obtained by simple scaling.

The use of MQW lasers overcomes the problem of gain saturation that occurs in SQW lasers and thus they are better suited to situations where the losses in the laser system are high, since operation even at high currents still occurs on the steep part of the gain–current curve. On the other hand, if the system losses are low then SQW lasers offer the clear advantage of substantially low transparency current and lower *intrinsic* losses.

Quantum well lasers have other advantages over DH lasers in addition to their low threshold currents. Because the active layer is very thin compared to the width of the optical cavity, degradation of the facets of the optical cavity at high powers is reduced and thus quantum well lasers have a much lower propensity to catastrophic failure and are intrinsically more reliable. This property, allied to high quantum efficiency and low internal losses, makes these devices very suitable for use in high-power diode arrays which are used to pump other solid state lasers, such as described in Chapter 8.

In principle, both modulation speed and spectral width should be much improved in quantum well devices because both these parameters are a function of the differential gain dg/dJ which, as we have seen, is high in quantum well lasers because of the 2-D DOS.

Strained layer structures

Although the market for semiconductor diode lasers is dominated by lattice-matched material systems like $Ga_{1-x}Al_xAs/GaAs$, more recently devices have become available which are fabricated from materials where the lattice mismatch is rather large. In fact, it has been shown that a large mismatch ($> 1\%$) between epilayer and substrate can be accommodated without the generation of large numbers of strain-relieving dislocations provided that the epilayer does not exceed a certain *critical thickness*. Clearly this opens up a wide variety of opportunities for laser design which were previously unavailable because of lattice-matching considerations.

In most quantum well heterostructures fabricated from lattice mismatched materials the only part that is strained is the quantum well itself. Thus, if we take the $Ga_{1-x}In_xAs/GaAs$ system as an example, the $Ga_{1-x}In_xAs$ quantum well is grown on an unstrained GaAs epilayer (which forms the barrier on one side of the QW) such that the atoms in the quantum well are in one-to-one registry with the atoms in the GaAs layer below it. Since the lattice constant of the ternary material is (on average) greater than that of the binary in this case, the quantum well material is thus *compressively strained*. The other GaAs barrier is then laid down on top of the quantum well and is again unstrained.

In addition to the greater variety of materials that become available when the lattice-mismatch restriction is lifted, the strain present in the quantum well has the significant effect of actually *modifying the band structure*. To see how this affects the laser performance we consider the relative quasi-Fermi levels in conduction and valence bands during laser operation, and it turns out that E_F^c is always very much greater than E_F^v, for two reasons. First, in the valence band the confinement energies are lower and the density of states is greater than in the conduction band because of the larger effective mass of the holes. Furthermore, the two hole states (*light* and *heavy*) that are degenerate at the top of the valence band in bulk material are still very close together in energy in confined systems, so that many hole states are available at a given energy. However, in the $Ga_{1-x}In_xAs/GaAs$ system, for example, the light and heavy holes are split further by the strain so that they are much further apart in energy and thus only the lowest of these is populated by carrier injection at low currents. This means that E_F^v shifts to higher energy more quickly and inversion is achieved more easily.

3.4.7 Edge-emitting arrays

Several factors effectively limit the output power available from individual index-guided diode lasers to around 200 mW, namely the inherent restriction of the active

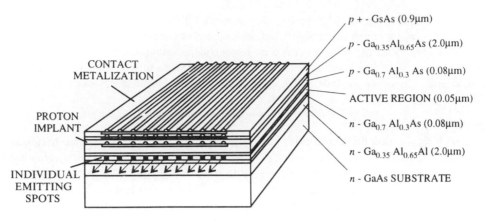

Figure 3.26 An edge-emitting array of diode lasers [3.1].

volume, the need to remove heat and the potential damage to the laser facets at high powers. Gain-guided lasers offer more power for reasons explained earlier, but for powers in the region of watts or more multiple diode arrays are needed. A typical edge-emitting diode array is shown in Figure 3.26 and we see that it is made up of many parallel laser stripes of width a few μm, typically 10 μm apart. In practice, up to 200 stripes can be fabricated on a chip 1 cm wide, giving cw powers of up to 10 W, above which heat dissipation becomes a problem – see Chapter 8.

The extent to which individual stripes are coupled together differs from array to array. Some are optically phase coupled so that the overall beam is partially coherent, whereas others emit incoherently. Clearly the beam shape will be unusual, particularly for the highest power arrays which emit from an area 1 cm wide but only 1 μm high, but this beam shape can be improved by phase-locking to an external single-mode laser.

3.4.8 Edge-emitting resonant cavity structures

Both the output power and frequency of diode lasers can be significantly altered by changing the nature of the optical cavity 'seen' by the optical field. For example, output power can be increased by a factor of two by applying an antireflection coating to the output end, reducing its reflectivity typically from 30% to 5 or 10%.

Simple Fabry–Perot cavities do have some drawbacks, however, the major one being that the laser gain curve is often sufficiently wide to encompass several longitudinal modes, resulting in multimode oscillation. This can be avoided to a certain extent in narrow-stripe lasers operating cw, but even these oscillate multimode when the injection current is modulated at high frequency, as it is in optical communications. In addition, variations in operating temperature and drive current can cause 'mode hopping', giving the laser a much broader effective bandwidth and short coherence length.

The short coherence length and broad bandwidth do not cause problems in certain applications, such as compact disc playback, but this is often not the case. Thus

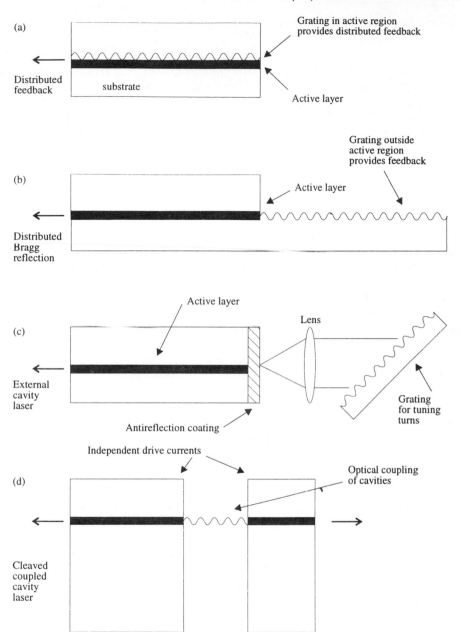

Figure 3.27 Cavities used to produce single longitudinal mode operation in diode lasers: (a) distributed feedback laser, (b) distributed Bragg reflection laser, (c) external cavity laser, and (d) cleaved-coupled-cavity (c³) laser [3.1].

several methods have been developed to ensure single-frequency operation; these are illustrated in Figure 3.27.

Distributed feedback lasers incorporate a grating next to the active layer in the region where current flows; this scatters the optical mode in the active layer, producing optical feedback without the need for reflecting facets. The frequency of operation is then related to the grating spacing.

Distributed Bragg reflection lasers operate along similar lines to distributed feedback lasers but this time the grating is 'external' in the sense that it is placed in a region where no current flows. This is essentially to simplify the fabrication, since no growth is then required on top of the grating.

External cavity lasers include external optics to select and also *tune* the laser frequency. In the example illustrated in Figure 3.27(c), one facet of the laser is antireflection-coated so that light can reach the diffraction grating which can be turned to adjust the wavelength. Very narrow linewidths (< 100 kHz) can be achieved in this way.

Cleaved-coupled-cavity (c^3) *lasers* consist of two lasers that are coupled optically but pumped by independent drive currents. These can be frequency-tuned but are difficult to fabricate.

To date, the most popular method of achieving single-mode operation has been the distributed feedback method, although external cavity lasers are also available commercially.

Thus, amongst these lasers, distributed feedback lasers are currently the most important. We now treat the basic principles at greater length.

3.4.9 Distributed feedback lasers

This form of the heterostructure junction laser has become important owing to the possibility of restricting laser action to a single mode. Single-mode operation is, in turn, of significance in long-range fiber transmission and tunable laser spectroscopy.

The use of a corrugated grating in a laser structure was first suggested as a novel feedback mechanism based on a periodic perturbation of the refractive index or in the gain (or loss) of the guiding medium. This can give rise to reflection and to lasing without end mirrors. It was first suggested and demonstrated by Kogelnik and Shank [3.9] for a dye solution.

The principle is illustrated in Figure 3.28 in which reflectivity of a corrugated filter thin-film waveguide is plotted against deviation from the 'Bragg condition' namely

$$\lambda_0/n \approx 2\Lambda/m \tag{3.82}$$

This applies when there is in the bulk of the guide harmonic modulation of index n or gain g, i.e.

$$n(z) = n + n_1 \cos 2\beta_0 z \tag{3.83}$$

$$g(z) = g + g_1 \cos 2\beta_0 z$$

where $\beta_0 = l\pi/\Lambda$.

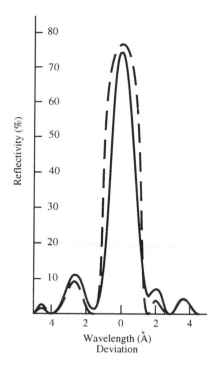

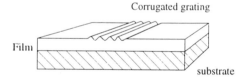

Figure 3.28 Plot of reflectivity of a corrugation filter in a thin-film waveguide as a function of deviation from the Bragg condition [3.10].

Very similar results are obtained for a passive guide with corrugation of one of the surfaces and leads to an identical result to the gain modulated bulk result above.

We define $\Delta\beta = \beta - \beta_0 \cong (\omega - \omega_0)n_{eff}/c = 2\pi n_{eff}/\lambda_0 - 2\pi/\Lambda$ when the expression for oscillating mode frequencies [3.11] is given from

$$(\Delta\beta m)L \cong -(m + \tfrac{1}{2})\pi \tag{3.84}$$

as

$$\omega_m = \omega_0 - (m + \tfrac{1}{2})\frac{\pi c}{n_{\text{eff}}L}.$$

The use of n_{eff} indicates an average value over the corrugated structure and we note that no oscillation takes place exactly at the Bragg frequency ω_0. The frequency spacing of the modes is

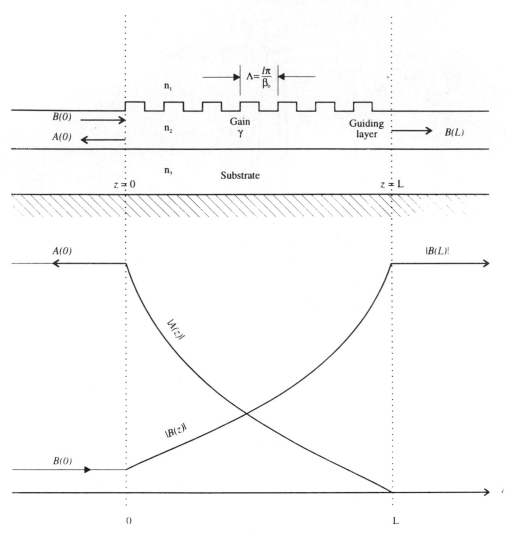

Figure 3.29 Incident and reflected fields inside an amplifying periodic waveguide near the Bragg condition $\beta \approx l\pi/\Lambda$ [3.10].

$$\omega_{m-1} - \omega_m = \tau c/(n_{eff} L) \tag{3.85}$$

i.e. approximately the same as the two-reflector resonator of length L. The threshold gain g_m is obtained as

$$\frac{e^{2g_m L}}{g_m^2 + (\Delta\beta)_m} = \frac{4}{\kappa^2} \tag{3.86}$$

where the coupling coefficient κ depends on the nature of the periodic perturbation. This indicates an increase in threshold with increasing mode number and so facilitates

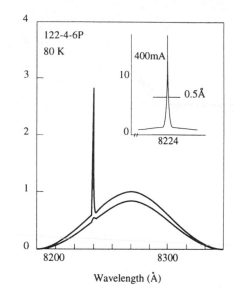

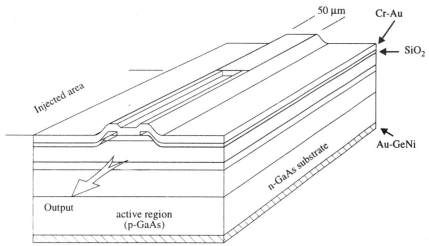

Figure 3.30 A schematic structure of a DFB laser, together with a typical emission spectrum [3.11].

single-mode operation at a well-defined wavelength. The behaviour of the internal waves in a periodic amplifying waveguide is shown in Figure 3.29.

Fabrication of DFB lasers

The oscillation wavelength is determined approximately by the Bragg condition which implies for GaAs with $\lambda_0 \sim 800$ nm a period Δ of order 0.1 μm. This is prepared by etching the crystal through a periodically slotted mask. Such a mask can be prepared

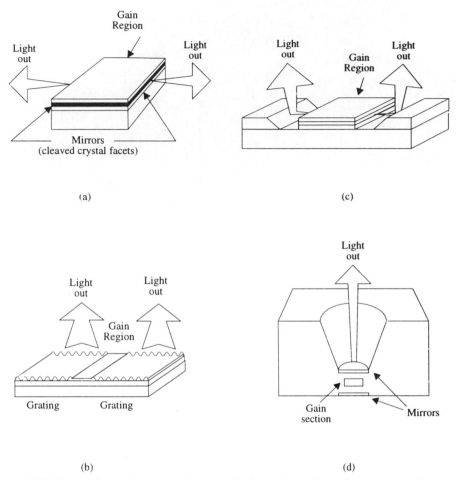

Figure 3.31 Various laser schemes: (a) standard edge emitter, (b) surface emitter with grating couplers, (c) surface emitter with end mirrors and (d) vertical cavity surface emitter [3.9].

by holographic techniques in a thin (100 nm) photoresist film or by electron-beam patterning followed by reactive-ion beam etching. Developed heterostructures are then finalized by growth-back over the corrugations. A typical structure is shown in Figure 3.30 together with a typical spectral output. Such lasers date from 1974/75 in collaborative work between the Hitachi company and California Institute of Technology [3.11] in addition to Bell Telephone Laboratories [3.12].

3.4.10 Surface-emitting lasers

Surface-emitting lasers have long been a goal of researchers because they offer considerable advantages over conventional devices, the main one being the possibility of high packing density in 2-D arrays with high powers and low thresholds. Potential applications of such lasers include optical interconnections, optical computing, 2-D

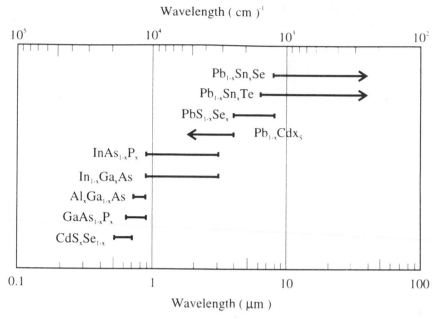

Figure 3.32 Compositional tuning ranges of some semiconductor alloys [3.10].

imaging and fiber-optic communications.

Several schemes have been outlined, illustrated in Figure 3.31, and these fall into two main categories: those in which the laser cavity is planar, and those in which it is vertical with respect to the substrate.

Planar-cavity surface-emitting lasers have structures (either end mirrors or gratings) which divert the laser output into the vertical direction, so that the laser occupies at least as much area on the semiconductor chip as a convention edge-emitting laser.

Vertical-cavity surface-emitting lasers (VCSELs) have a very short gain region, limited by the growth process, so that the mirror loss factor in one round trip of the cavity is relatively large. Thus reflective multilayers or dielectrics are deposited above and below the active layer to increase the 'effective cavity length' so that the losses can be overcome. Each of these layers occupies an area on the chip no more than a few microns across, so that large numbers ($\sim 10^6$) can be accommodated on an individual chip. Vertical-emitting lasers of this type usually have quantum well active regions so that the gain is very high and threshold currents are extremely low (<1 mA) in these devices. In addition, because of the very short cavity length the longitudinal mode spacing is much greater than in conventional diode lasers and narrow-bandwidth single-mode operation is normally obtained.

3.4.11 High-resolution spectroscopy with tunable diode lasers

Semiconductor diode lasers are used for high-resolution spectroscopy of molecules

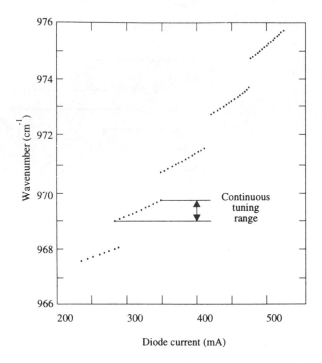

Figure 3.33 Fine tuning of modes from a $Pb_xSn_{1-x}Te$ diode laser as a function of injection current. Tuning is continuous within each mode [3.10].

in the gas phase in the middle and near-infrared spectral regions. With the appropriate compound and/or alloy composition any desired wavelength between 0.6 and 30 μm can be chosen, as illustrated in Figure 3.32. Any individual diode laser can then be fine-tuned by a magnetic field, hydrostatic pressure or by temperature – including the effects of injection current. The method most often used is tuning by current. The output of most diodes consists of several longitudinal and transverse modes, which increase in number and shift in frequency with increased drive current. This fine tuning is illustrated in Figure 3.33. Essentially, the heating of the diode causes a change in refractive index, which alters the effective cavity length. Continuous tuning of a single mode occurs over several cm^{-1}, after which a sudden 'mode jump' occurs to the next mode. Nevertheless, complete coverage of the spectrum can be obtained, and a grating spectrometer in series with the laser is used to select one mode. The resolution is very high – a linewidth of as little as 100 kHz is obtainable – and the scanning range is quite adequate for scanning, say, a pressure broadened molecular absorption line, typically 0.1 cm^{-1} (3 GHz) wide. Figure 3.34 shows a typical diode laser spectrum of a water vapour doublet near 1942 cm^{-1}.

Theoretically, the output frequency of a laser is given by [3.10]:

$$\nu = \frac{\nu_c \Gamma_s + \nu_s \Gamma_c}{\Gamma_c + \Gamma_s} \tag{3.87}$$

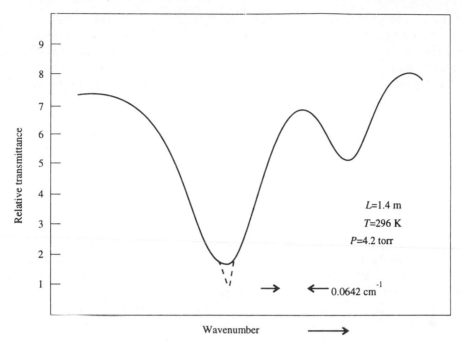

Figure 3.34 Relative transmittance of a water vapour doublet near 1942.5 cm^{-1}.

where v_s is the frequency at the peak of the gain, Γ_s is the full width at half-maximum (FWHM) of the gain, and v_c and Γ_c are the corresponding cavity parameters. The cavity linewidth is given by

$$\Gamma_c = \left[\alpha - \frac{\ln R_1 R_2}{2L} \right] \bigg/ \left[n + v \frac{\partial n}{\partial v} \right] \tag{3.88}$$

where n is the refractive index, α is the absorption coefficient and R_1 and R_2 are the reflectivities.

For diodes, $\Gamma_s \gg \Gamma_c$, whence

$$v_c = cM/2L \left[n + v \frac{\partial n}{\partial v} \right] \tag{3.89}$$

where M is a large integer and $n = n_0 + (dn/dT)\Delta T$.

Output powers of spectroscopic lasers are usually of the order of milliwatts, and they are available commercially in the USA and in Germany.

3.5 Semiconductor modulators using *pn* junctions

3.5.1 Introduction

The importance of modulators as interfaces between optics and electronics is emphasized in several sections of this book. The commonest technology has been that of liquid crystals (Chapter 5), primarily due to their widespread application in displays. For very fast modulators electro-optic effects have been exploited (Chapter 7) and in general these are relatively large (cm) bulk devices requiring high voltages ($\sim$ kV) for operation. Up to the present, semiconductor technology *per se* has made only a modest contribution to modulator devices, although, in principle, there are a number of advantages that could lead to more widespread use.

The first fundamental physical consideration is the ability of the depletion region of a *pn* junction to sustain a high electric field without passing a current so large as to cause excessive heating. In essence, the *pn* junction produces in the semiconductor a region of low conductivity. The electric field, $\sim 10^4$–10^5 V cm^{-1}, is then large enough to cause usable electro-optic effects in the thin (μm thick) depletion region but has the associated advantage of being generated by an absolute voltage in the range 3–5 V that is compatible with logic circuits. Further advantages include the fact that established semiconductor microfabrication techniques, such as photolithography, are applicable to semiconductor devices (making the number of pixel elements almost independent of cost), and large improvements in the speed of modulation are possible compared with liquid crystal technology.

At present, disadvantages include the wavelength restriction and the degree of contrast. In this respect liquid crystal devices give a clear advantage since birefringent methods of modulation allow both a wide wavelength range (the complete visible spectrum) and contrasts of the order of 1000:1 to be obtained.

In this section we illustrate the principles and possibilities by means of a discussion of the *self-electro-optic effect device* (SEED). The earliest references to this concept can be traced to Ryvkin [3.11] although its effective implementation awaited the application of multiple quantum wells in GaAs by Miller *et al.* in 1984 [3.12]. Miller further applied the principle of optical bistability using increasing absorption, which gives a similar effect functionally to that described for all optical refractive bistability in Chapter 4. With these devices it is possible to obtain logic operation and memory as well as the useful property of 'time sequential' gain. In logic switching a very low optical switching energy for a device has been obtained, in the region of 100 fJ. The problem of relatively low contrast was circumvented by Lentine *et al.* [3.13] with the introduction of the 'symmetric SEED', which allows the implementation of dual-rail ratio operation.

3.5.2 Physical principles of the exciton absorption

Figure 3.35 shows the absorption spectrum near the band edge of a $Ga_{1-x}Al_xAs$/GaAs multiple quantum well for 0, 6 and 10 applied volts, in which the main feature is the exciton absorption. This peak is caused by an electron and hole orbiting around each other in a manner very similar to the electron and proton in a hydrogen atom, as described in Chapter 2. A series of absorption lines is expected, but with the short electron scattering times in a semiconductor and the low effective mass, there are broadening effects and a low binding energy. At room temperature there is, in fact,

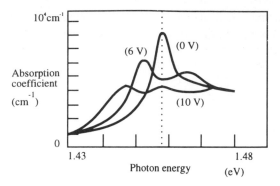

Figure 3.35 Absorption spectrum near the band edge in GaAs showing exciton structure [3.14].

no distinct exciton absorption feature in bulk material and this limits the extent of modulation possible. For wider-gap materials such as ZnSe and CdS, the exciton binding energy is higher so that the feature is stronger and thus the possibility of room-temperature modulator action exists. The solution to the problem of room-temperature operation in GaAs is found in the use of two-dimensional confinement of the electron–hole motion by means of quantum wells.

3.5.3 Quantum well electro-absorption and modulators

The physical mechanism by which SEEDs operate is known as the quantum-confined Stark effect (QCSE), which gives rise to large electro-absorption near the bandgap of quantum well semiconductor structures. The wells consist of alternating layers of two different semiconductors such as GaAs and $Ga_{1-x}Al_xAs$, as shown in Figure 2.24. The GaAs layers are generally very thin (<10 nm) while those of $Ga_{1-x}Al_xAs$ are somewhat thicker (~ 50 nm), and both electrons and holes have minimum energy in the GaAs layer. The $Ga_{1-x}Al_xAs$ layers on either side of the GaAs behave as *barriers* and, because the layers are so thin, the electrons behave like *particles in a box* – this is a well known quantum-mechanical problem (Chapter 2), which imposes boundary conditions on the carrier wavefunctions. The resulting confinement leads to discrete energy levels for the electrons and holes as far as their motion perpendicular to the layers is concerned. The levels have energy separations in the region of ~ 10–100 meV and therefore the optical absorption edge breaks up into a series of steps, corresponding to transitions between different *subbands* associated with the confined carriers. Modified exciton absorption peaks appear at the edges of these steps and these peaks appear in GaAs heterostructures and related systems *even at room temperature*. The physical explanation for this result lies in the effective increase in the binding energy caused by the confinement, and it is this that renders the mechanism usable for modulation purposes.

Application of an electric field perpendicular to the layers pulls the electron toward one barrier and the hole towards the other in a quantum well, and this reduces the energy of the electron–hole pair. As a consequence, the optical absorption associated with creation of the pair is reduced and the absorption *red shifts*; i.e. it moves to longer wavelengths. In addition, the electron and hole are prevented from escaping by the walls of the well, preventing *ionization* of the exciton. This aspect of the mechanism compares favourably with the bulk situation where the effect of an electric field on the exciton is predominantly a *broadening* of the absorption edge. Thus the QCSE mechanism gives more effective absorption modulation by suppressing this *lifetime broadening*.

The structure for a QCSE modulator as used in SEEDs is a *p–i–n* diode, illustrated in Figure 3.36, in which the intrinsic region contains the quantum wells. By reverse-biasing the diode we apply an electric field perpendicular to the layers and the resulting modulator can be used with the incident light either perpendicular to or in the plane of the layers in a waveguide configuration.

The electrical energy required in the electro-optic effect is the energy needed to charge up the volume of the device to give the operating field. The device can be treated as a capacitor and SEED devices with typical dimensions of $10 \times 10 \ \mu m$ have a capacitance of about 10 fF. The fields required are in the range $5 \times 10^4 - 2 \times 10^5$ V/cm corresponding to voltages of 5–20 V for devices 1 μm thick, which in turn correspond to energies of ~ 1 fJ/μm^2 or 100 fJ per device.

The microscopic physics of device operation shows that the devices can be fast – the mechanism reacts in times less than 1 ps if sufficient power can be applied quickly. Measurements on electro-absorption itself have given experimentally limited times of 370 fs.

The devices are made by molecular beam epitaxy (MBE) and have already been developed in a variety of forms, as we set out below.

3.5.4 The self-electro-optic effect device (SEED)

The basic SEED comprises an electrically controlled absorption modulator described above and a photodetector which is provided by the *p–i–n* diode itself. It then operates with an optical input and output and can act as an optical logic gate (similar to the all-optical devices in Chapter 4) or as a temporary memory.

The simplest version is the resistor-based SEED, or R-SEED, as shown in Figure 3.36, which is operated at a precise photon energy. The device is illuminated at wavelengths in the band edge close to the exciton absorption peak: such photons create extra charge carriers, causing current to flow in the circuit of Figure 3.36. The voltage across the *p–i–n* junction is itself reduced because the created photocarriers reduce the 'effective resistance' of the *p–i–n* device and so cause a higher current to flow in the load resistor, R. Thus the voltage across the MQW changes and consequently modulates the absorption. The reduced voltage across the junction means of course a *reduced* electric field which shifts the position of the exciton peak by the QCSE. The directions of the shift (Figure 3.35) indicate that the absorption is thereby *increased*: This results in an *increase* in photocurrent so that the effect of increasing incident intensity is reinforced by *positive feedback* and the device switches into a high absorption state and shows bistability. The situation is very similar to that which exists in all-optical refractive bistability, Chapter 4 (where the effect is perhaps more easily understood in a tutorial manner).

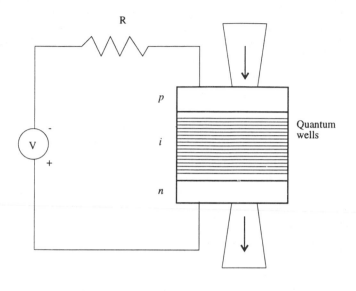

Light in

Light out

Figure 3.36 A quantum well *p-i-n* diode used as an R-SEED.

Bistability by increasing absorption is not unique to SEEDs and, indeed, was observed in purely optical effects at an earlier date. The electronic microcircuit contribution to the SEED is, however, distinct and it opens up a variety of engineering possibilities. It is best appreciated by using a *load line* analysis, which is applied in electronics to the theory of transistors and is illustrated graphically in Figure 3.37. Two simultaneous equations are being solved. The first is the relation between photocurrent, I, generated in the diode for incident light power P, plotted as a function of reverse voltage V (solid line). As V increases, I falls because the optical absorption decreases as the exciton peak moves to lower photon energies. As the diode is driven into forward bias (to the far left in the diagram) the current decreases because of the growth of forward current for a forward-biased diode and, more importantly, due to the fall in quantum efficiency with decreasing reverse bias. This gives rise to the curve peaked near zero-bias.

To a reasonable approximation in the operating region, the current at any given voltage is proportional to the light power, P. We define a responsivity $S(V)$, via

$$I = I(V, P) = PS(V) \tag{3.90}$$

The second equation represented in Figure 3.37 is that of the current through the load resistor, R, which is simply the straight line

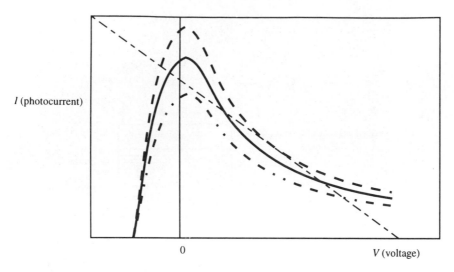

Figure 3.37 Load-line analysis for R-SEED operation (after [3.14]).

$$I = (V_0 - V)/R \qquad\qquad (3.91)$$

where V_0 is the applied voltage. This linear plot in the figure is known as the 'load line'. Intersections with the curves give the allowed values of steady state voltages. Curves are plotted for different light powers, P, and these have different numbers of intersections. At low power there is just one intersection at high voltage and at high power, and there is also one at low voltage. At intermediate powers there are three intersections, the middle one corresponding to an unstable point and the remaining two to stable points. For the latter, any departure from the points results in resetting towards the points and hence we have bistability. For an R-SEED the voltage across the *pn* junction swings from zero to V_0.

A typical bistable input–output curve for a SEED is shown in Figure 3.38 and we see that the contrast is of the order 2 or 3:1. The device can also be viewed as a capacitor, with the photocurrent acting to discharge it and the resistor and power supply tending to recharge it. Hence the time constant is effectively RC and the device can be switched by providing charge $Q = CV_0$. This requires an absorbed optical energy

$$E_a = (\hbar\omega/e)CV \qquad\qquad (3.92)$$

where e is the electronic charge. Writing $C = \varepsilon A/d$, where A is the device area and d is its effective thickness, and substituting appropriate parameters for GaAs/Ga$_{1-x}$Al$_x$As, gives $E_a \sim 1.7$ fJ μm^{-2} at a field of 10^5 V cm^{-1}. Actual devices have measured optical energies of the order of 7–10 fJ/μm^2. Switching speed is subject to the phenomenon of *critical slowing down* – a general property of systems undergoing phase transitions – if the switching power slightly exceeds the critical value near the transition. Increasing the switching power by a factor of two results however in the device switching at close to its intrinsic limit, in this case $t = RC$. Thus, the variation

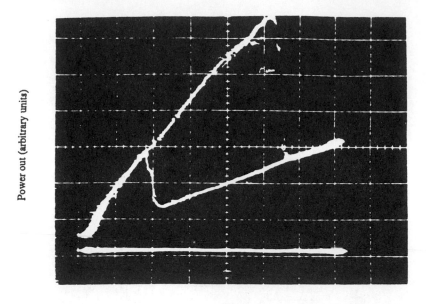

Power in (arbitrary units)

Figure 3.38 Optical output characteristics of a SEED device, showing optical bistability [3.14].

of R allows either slow, low-power switching or fast, high-power operation. For actual practical devices of dimension $\sim 10 \times 10\ \mu$m, microwatts are required for microsecond operation and milliwatts for nanosecond operation.

3.5.5 Symmetric SEEDs

An interesting development proposed by A.L. Lentine *et al* [3.13] is the symmetric SEED (or S-SEED), which is illustrated in Figure 3.39. Two *p–i–n* quantum well diodes are grown next to each other and connected back-to-back, each acting in place of the load resistor for the other. The load-line diagram is shown in Figure 3.40 for this case, and this shows that the arrangement is bistable in the ratio of the two optical powers. This enables an optical logic plane operating on *dual-rail logic* to be constructed, i.e. logic states correspond to these ratios. Advantages accrue in that this overcomes the intrinsic lack of contrast of the SEED, reduces the need to use a sharp *threshold* for logic operation, and thus reduces tolerances on optical parameters. Finally, if the power is reduced on *both* diodes, memory of the state is retained. Thus one can operate in a cyclic mode, establish a state and then read out at higher power. This provides time-sequential gain, which is useful for logic circuits.

Further development of S-SEEDs has yielded usable arrays of up to 64×64 and other variants of components have been tried out in place of the load resistor. The most recent uses a field effect transmitter (FET) to provide gain, thereby allowing lower optical switching energies to be reached. The integration of electronics to give

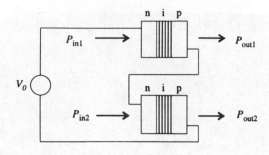

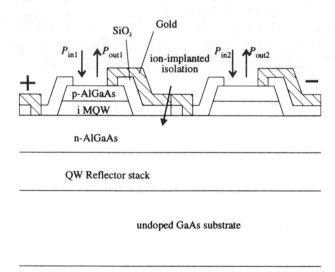

Figure 3.39 Schematic diagram of an S-SEED [3.13].

artificially enhanced optical nonlinearity also allows the possibility of flexibility – this is sometimes known as a 'smart pixel'.

As a general device for interfacing electronics to optics, the SEED concept suffers from the disadvantage of having to operate exactly at the exciton wavelength to ± 1 or 2 nm. This wavelength is about 850 nm (i.e. in the near-infrared) for $GaAs/Ga_{1-x}Al_xAs$ devices. Recently, the emergence of *pn* junction technology in MBE ZnSe has allowed the realization of ZnSe *p–i–n* quantum well modulators and SEEDs operating at 500 nm in the visible region. However, contrast problems currently preclude application in displays where liquid crystal devices using birefringence have a large advantage.

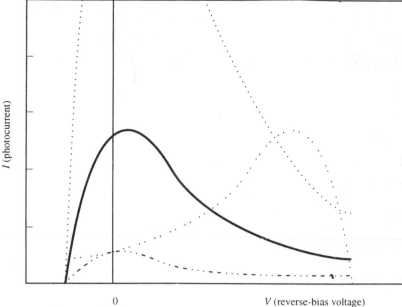

Figure 3.40 Load-line analysis for S-SEED operation.

References

[3.1] J. Hecht, *The Laser Guidebook* (Tab Books, Blue Ridge Summit PA, 1992), p. 335.

[3.2] See, for example, D.A. Fraser, *The Physics of Semiconductor Devices* (Oxford University Press, 1979).

[3.3] A. Einstein, *Phys. Z.* **18**, 121 (1917).

[3.4] H.C. Casey and M.B. Panish, *Heterostructure Lasers, Part A: Fundamental Principles* (Academic Press, New York, 1978), p. 122.

[3.5] K. Gillessen and W. Schairer, *Light Emitting Diodes: An Introduction* (Prentice Hall International, London, 1987).

[3.6] J.E. Goell, *Bell Syst. Tech. J.* **48**, 3445 (1969).

[3.7] G.H.B. Thompson, *Physics of Semiconductor Laser Devices* (Wiley, Chichester, 1980), p. 95.

[3.8] H.C. Casey, Jr. and M.B. Panish, *Heterostructure Lasers, Part A: Fundamental Principles* (Academic Press, New York, 1978), p. 165.

[3.9] C. Weisbuch and B. Vinter, *Quantum Semiconductor Structures: Fundamentals and Applications* (Academic Press, London, 1991), p. 174.

[3.10] A. Mooradian, in *Very High Resolution Spectroscopy*, ed. R.A. Smith (Academic Press, London, 1976), p. 77.

[3.11] B.S. Ryvkin, *Sov. Tech. Phys. Lett.* **10**, 239 (1979).

[3.12] D.A.B. Miller, D.S. Chemla, T.C. Damen, A.C. Gossard, W. Wiegmann, T.H. Wood and C.A. Burrus, *Appl. Phys. Lett.* **45**, 13 (1984).

[3.13] A.L. Lentine, H.S. Hinton, D.A.B. Miller, J.E. Henry, J.E. Cunningham and L.M.F. Chirovsky, *IEEE J. Quantum Electron.* **25**, 1928 (1988).

[3.14] D.A.B. Miller, *Opt. Quantum Electron.* **22**, S61 (1990).

4

Optically controlled devices

4.1 Introduction

There are a range of devices that use 'light to control light'. Usually these devices rely upon the optical generation of carriers for their operation. The number of carriers generated per cm^3, N, is described (see Section 2.2.2) by

$$N = \tau_R \alpha \frac{I}{\hbar\omega} \qquad (4.1)$$

where

N is the carrier density (in cm^{-3});
I is the intensity of the incident light (in $W\ cm^{-2}$);
α is the absorption coefficient for those photons responsible for carrier generation (in cm^{-1}) – see. Eq. (2.22);
τ_R is the carrier *recombination time* (or lifetime), which is a measure of how long the optically excited carriers remain, before returning to their initial states (either radiatively or nonradiatively) or becoming trapped in other states; and
$\hbar\omega$ is the incident photon energy.

In order to give a general estimate of the carrier density we can substitute some values into Eq. (4.1). Typical intensities are of order $MW\ cm^{-2}$. Therefore, with $I/\hbar\omega = 10^{26}$ photons/s, and a recombination time τ_R which may vary between 10^{-4} and 10^{-12} s, and given complete absorption ($\alpha \sim 10^4\ cm^{-1}$), the optically generated carrier densities are $N = 10^{15}–10^{19}\ cm^{-3}$. Carrier densities of this magnitude are sufficiently high to modify the optical properties of semiconductor materials signifi-

132

cantly. The simplest model is the plasma effect in refraction (Eq. (2.32)) and in absorption (Eq. (2.35)).

Equation (4.1) describes an *equilibrium* situation where the production of carriers is balanced by their decay. However, one of the most challenging and potentially most useful applications of such a carrier generation effect is in picosecond optical and electronic technology; for example, with sampling gates and photodetectors that have response times in the picosecond range. When picosecond time scales are considered the equilibrium case described above does not necessarily apply. For excitations above the bandgap where the energy of the photons is greater than the bandgap energy, $\hbar\omega > E_g$, the generation of free carriers is limited only by the rate of the incoming photon flux (i.e. on how rapidly the system is pumped) and this therefore has the possibility of being very fast. Thus reflectivity, absorption, refraction and photoconductivity which depend on the density of the free carriers, can exhibit changes that have rise times of this order. The latter can provide, for example, with amorphous silicon, a carrier generation mechanism (known as an Auston switch [4.1]) useful as a pulse generator input to an electro-optic modulator, which can be used to generate fast electrical pulses.

To terminate the response equally quickly (i.e. in a few picoseconds) is more difficult since the decay time depends upon the value of the recombination time, τ_R. This can be very much greater than the response time. A fast termination time can be accomplished by rapid recombination, trapping or sweeping out of the excited free carriers by an electric field. There exist a variety of solutions, such as the use of amorphous material (silicon film), doping or proton bombardment to vary this time. The recombination time τ_R can thus be varied from hundreds of milliseconds to a few picoseconds. For a discussion of some of the basic properties of α and τ_R see Chapter 2.

4.2 Absorption processes in semiconductors

The fundamental interband absorption processes have already been described in Chapter 2. We now briefly review these, and a number of further processes which can be used for carrier generation in devices. The absorption of photons will occur when an incident photon possesses enough energy to excite an electron from a lower to a higher energy state. The absorption can be quantified by the coefficient of absorption, α, which is a measure of the change in light intensity in a material per unit length. This coefficient is proportional to the square of the probability of a transition occurring between the initial and final states of the electron that is affected by the absorbed photon, and is also related to the electron densities in the two states (see Chapter 2, Eq. (2.116)), Figure 4.1.

There are several different ways in which an absorbed photon can be involved in electron excitation; however these processes can be divided into two main groups, *linear* and *nonlinear* processes.

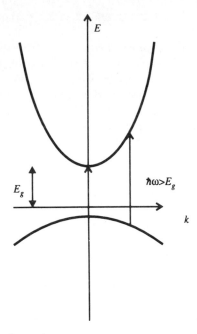

Figure 4.1 Direct band gap absorption.

4.2.1 Linear absorption

Direct interband absorption
The most common form of absorption is known as *direct interband absorption* and is described in Chapter 2. Direct absorption is strong compared with the other possible processes. The absorption coefficient for semiconductors has a value $\alpha \sim 10^4$ cm^{-1} for photon energies above the energy gap E_{g}. Thus all light is absorbed within a few microns of the surface.

The absorption α rises very sharply as the energy of the incident photons exceeds the energy gap of the material, i.e. when $\hbar\omega_{\mathrm{L}} > E_{\mathrm{g}}$. The coefficient of absorption is given by Eq. (2.129) in Chapter 2 and can be substituted into the equation for transmission (Eq. (4.2)), which describes the variation in the transmitted power T through distance d of the material, where I is the final intensity and I_0 is the initial intensity. The absorption depth $d \sim 1$ μm and so the carriers are generated very close to the surface:

$$T = \frac{I}{I_0} \exp(-\alpha d) \tag{4.2}$$

The carriers reach greater depths (and hence volumes) only by diffusion and this is governed by the diffusion length, L, which is given by

$$L = D\tau_R \tag{4.3}$$

where

$$D = \frac{\mu kT}{e} \tag{4.4}$$

and

$$\mu = \frac{e\tau_S}{m^*} \tag{4.5}$$

The parameter D is the diffusion coefficient, μ the mobility of the carriers, τ_S is the intraband scattering time (of order picoseconds) and m^* their effective mass. Clearly the diffusion length depends upon the type of material being used and so it is important to 'engineer' materials to give the required properties.

Indirect bandgap absorption
This occurs (see Section 2.4) in cases where a single photon cannot provide a sufficient change in momentum to allow the electron transition to occur. To conserve momentum a phonon (normally acoustic) interaction is also required and the phonon can be either emitted or absorbed. The coefficient of absorption is proportional to the square of the probability of transition between the initial and final electron states as well as the probability of a phonon interaction occurring: it is therefore relatively weak.

Indirect bandgap absorption is an important absorption mechanism in silicon and germanium: however, it is much weaker compared with direct bandgap absorption and $\alpha \sim 10\text{--}100 \text{ cm}^{-1}$, so that the depth of absorption is much greater.

Free carrier absorption
This type of absorption occurs when an electron is excited to a higher energy state in the same band. A phonon (acoustic or optical) or some other interaction is required to provide conservation of momentum. The absorption increases with wavelength to a power that depends on the type of interaction involved. This mechanism is discussed in Section 2.1.2. The absorption of the incident photon energy is said to produce 'hot carriers' in a material, i.e. the carriers are temporarily in different states within the same band and have different properties such as mobility. It is in this manner that the 'holes' in germanium become 'hot carriers' upon absorption of infrared radiation, and allow fast infrared detection from fast laser pulses in what is known as the 'photon drag' effect (Chapter 9).

Exciton effects
Exciton effects are discussed in Section 2.1.7 for bulk materials and in Section 2.6 for low-dimensional structures.

Optical excitation can 'bleach' the exciton feature and lead to both absorptive and refractive effects. A simple model is discussed in Section 2.2.2 and accounts for the nonlinear refraction by 'exciton-like' features which are 'saturated' by optical excitation. Note that since the scattering time (i.e. the lifetime in a state) is short, the effect must be integrated over many states and is 'band filling' not 'state filling' in the case of real semiconductors – in contrast with the similar effect in atoms in the gas phase.

Band-tail effects

A *tail* on the absorption spectrum of semiconductor materials can be observed and is due to both thermal broadening and impurity effects. Thus the strictly parabolic form of the energy–wavevector relationship is not obeyed in practice. It is possible to excite carriers into conduction states on the band tail with photons of sufficient energy. Within such a tail region there will be absorption saturation and nonlinear refraction. Usually band-tail effects lead to absorption over a wider range of wavelengths than exciton absorption and produce greater dispersion. An empirical expression for such an absorption tail is given by

$$\alpha(\omega) = A_0 \exp\!\left(\frac{Bc}{kT}\,(\hbar\omega - \hbar\omega_0)\right) \tag{4.6}$$

where A_0, B and ω_0 are temperature-independent fixed fitting parameters of the system, k is Boltzmann's constant and T is the sample temperature. This empirical expression models the optical absorption coefficient at the band edge as an exponential function of both photon energy and inverse temperature T^{-1}, and is known as the 'Urbach' tail [4.2–4.5].

Impurity states

Impurities in the crystal lattice can produce energy levels which lie in the bandgap of a perfect lattice. Absorption occurs due to electrons being excited from the impurity states. Both intra-impurity state transitions (line-like) or impurity–band transitions (continuum-like) are possible and both can give rise to optical control mechanisms related to the level of excitation. Unfortunately, impurity states can also trap free electrons or holes and hence terminate their role in modifying optical properties.

4.2.2 Nonlinear absorption

Nonlinear absorption involves the absorption of more than one photon per electron excitation in, for example, semiconductor material. An example of nonlinear absorption is two-photon absorption, where the magnitude of the semiconductor bandgap lies between the energy of one photon and the energy of two photons combined, i.e. $\hbar\omega < E_g < 2\hbar\omega$. Other nonlinear absorption processes are two and three stepwise photon absorption processes in which an electron makes a transition from the valence band to the lowest conduction band via a linear (one photon) indirect absorption process and is then raised to a higher conduction band state by the absorption of one or more additional photons.

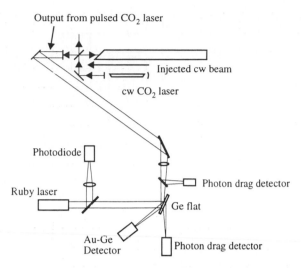

Output from pulsed CO_2 laser

Injected cw beam

cw CO_2 laser

Photodiode

Ruby laser

Photon drag detector

Ge flat

Au-Ge Detector

Photon drag detector

Figure 4.2 Experimental configuration used to investigate the reflection switching of CO_2 laser radiation.

Free carriers generated as a result of nonlinear absorption can modify some of the properties of the material such as the absorption spectrum and hence refractive index. These changes in the semiconductor material can be used in order to provide optical switching and optical bistability. For example, quasi-steady state optical bistability in an InSb etalon at room temperature has been observed as a result of the generation of free carriers through two-photon absorption of 10 μm radiation [4.5; 4.6].

4.3 Semiconductor optical switching and modulation

In this section we consider how the properties of semiconductor materials, such as reflectivity, can be altered on a very short timescale (nanoseconds to picoseconds) by firing an optical pulse (also of short duration) onto the material [4.7; 4.8]. This change in reflectivity can be used to reflect a separate optical beam and hence to switch it on a very short timescale.

We consider the specific case of CO_2 laser radiation meeting a slab of germanium at the Brewster angle. The reflectivity of the semiconductor is effectively zero at this angle provided that the CO_2 laser radiation is linearly polarized in the plane of incidence. By irradiating the germanium with a high-energy ($\hbar\omega \geqslant E_g$) optical pulse (e.g. from a YAG or ruby laser) it is possible to generate a free carrier plasma (Eq. (2.32)) of sufficient density within the germanium to cause complete reflection of the CO_2 laser radiation that is incident. This reflection will occur on a timescale that is similar to the driving pulse (whereas mode-locked pulses from a YAG laser are of picosecond duration, CO_2 pulses are slower).

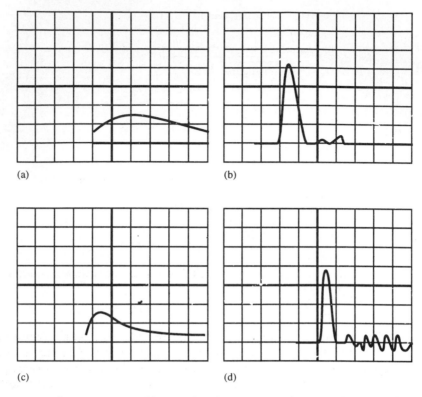

Figure 4.3 Oscilloscope traces of laser pulse shapes. (a) Incident CO_2 laser pulse, 50 ns/div; (b) incident ruby laser pulse, 5 ns/div; (c) transmitted CO_2 laser pulse, 100 ns/div; (d) reflected CO_2 laser pulse, 10 ns/div. (The noise after the main pulse is due to electrical interference from a laser-triggered spark gap in the ruby laser system).

Reflection switching of CO_2 laser radiation on a nanosecond timescale using polycrystalline germanium and ruby laser illumination is illustrated in Figure 4.2. The transversely excited atmospheric pressure (TEA) CO_2 laser operates on a single transverse mode and can provide a relatively long pulse (~ 200 ns) of plane-polarized radiation with a peak power of 2 MW. Modulation of the output of this laser due to axial mode beating is minimized by injecting a small amount ($\sim 1\%$) of radiation from a 1 W cw CO_2 laser into the TEA laser's cavity. A typical pulse is shown in Figure 4.3a.

The polished slab of n-type germanium is aligned close to Brewster's angle so that the quantity of reflected radiation is minimized. The transmitted beam is monitored by a photon drag detector and the reflected radiation with a Ge:Au photoconductive detector (with a response time ~ 1 ns).

A short pulse (~ 2 ns) of ruby laser radiation is fired onto the semiconductor surface and overlaps the region that is illuminated by the 10 μm beam (Figure 4.3b). At a sufficiently high intensity of ruby laser radiation there is complete reflection of the

CO_2 radiation. The rise time of the reflected 10 μm pulse is approximately equal to that of the ruby laser radiation, although its decay time is somewhat slower than that of the ruby laser (Figure 4.3d). There is also a substantial decrease in the transmitted CO_2 laser signal (Figure 4.3c).

4.4 Optical logic and optical memory

4.4.1 Introduction

The starting point from which electronics began its take-off as an information-processing industry can be traced to the publication by Lee de Forest of the patent for the triode valve in 1907. This defines the point at which we could say 'electricity controls electricity'. The device described as 'apparatus for amplifying feeble electrical currents' gave the possibility not only of amplification but also high-speed switching so necessary for signal processing. The era of vacuum tube electronics lasted just 40 years until the invention of the transistor by Shockley, Bardeen and Brattain in 1947.

The transistor reduced both size and power by factors of the order of hundreds and within a few years, greater reliability allowed the construction of much more practical computing devices. The ability to utilize a large number of electronic switches encouraged the transition to binary digital methods, which in comparison with analogue techniques, have the advantage of being indefinitely extensible without error. Digital techniques, however, have the disadvantage of being relatively clumsy in that all numbers, instructions and calculations must be encoded by a series of zeros and ones. This demands an ever-increasing rate of sequential operations.

In extrapolating to the future it is instructive to consider the physics of such computation. A good account is given in Mead and Conway's, *An Introduction to VLSI Systems* [4.9]. Information is essentially stored as energy and switching a device from logic-0 to logic-1 requires a definite switching energy. This energy must be greater than kT (and will usually be several hundred times kT) and increases with the size of the device. There will usually also be a trade-off between speed and power. A further absolute requirement for electronic logic is the necessity to restore a logic level after each switching action so that errors due to imperfect devices and signals do not accumulate. Restoring logic must have power gain and this power is normally drawn from a power supply and not the signal channel. These considerations appear to be applicable to all systems of signal processing and computing. The requirement of power gain is obviously necessary, i.e. one switching device must be capable of driving at least the next in the series so that a circuit containing multiple elements can be constructed.

Comparison of optics with electronics
The response of semiconductor microelectronics to the high data rate requirements of digital signal processing and computing has been to increase switching speeds and further miniaturize components in the form of very large scale integration (VLSI).

Transistors made from GaAs have been reported with effective switching times as fast as 12 ps [4.10]. However, this will not necessarily solve the problem of coping with a high data rate, since the processing time in conventional computers is many times the logic switching time, due to the necessity to transfer information to the next part of the circuit. This involves capacitance time-constant limits, as pointed out by Huang [4.11] and Mead and Conway [4.9]: VLSI does not *solve* the *RC* time constant problem since, as the length of a wire shrinks by a factor α and the cross-sectional area of the wire is reduced by a factor α^2, the capacitance of the wire decreases by this factor α, while the resistance increases by the same amount. Thus the time constant remains the same and the input charging time remains unaltered independent of scaling [4.9].

Turning to computers, the standard method of communication in use today connects the logic unit with the memory through an address device. This reduces the number of interconnections but can only address one storage element at one time. This widely used scheme was first advocated by John von Neumann but, rather than being given credit for this most practical innovation, he is now rather undeservedly blamed for the so-called 'von Neumann bottleneck' (Figure 4.4). The timing problems associated with circulating logic signals around a one-dimensional processor of this type ('clock skew') combine to indicate that future problems in digital computers are likely to be those of communication. This may apply at the architectural, bus and chip levels, and stems from the use of time multiplex to compensate for the inability of electrical methods to communicate many channels of information in parallel.

Photonics, optronics or digital optics
We pose the question 'can optics help to further the cause of high-speed computing?' Present practice has seen the invasion of electronically based communication by optical methods through the use of optical fibers in long-range telephone lines. The higher carrier frequency used gives potentially higher bandwidth, although *electronic* limitations on modulation techniques have restricted our ability to exploit fully this greater information carrying capacity. Currently it is the long-range feature made possible by low signal attenuation that has been utilized. Optics is responsible for the man–machine interface at the input and output of a computer. The use of optics for processing information has so far been handicapped by the absence of optical circuit elements or efficient modulators to interface with electronics. However the qualities of non-interfering propagation at the speed of light, the high available bandwidth and, perhaps even more important, the ease of use of *parallelism* would seem to give usable advantages. A simple lens costing very little can readily transmit millions of resolvable spots equivalent, when combined with optical circuit elements, to millions of electrical wires – a difficult micro-engineering task even for VLSI techniques. The advantage of optical methods has promising implications in those areas that are currently undergoing difficulties with existing technologies: these include image processing and recognition, sorting, radar array signal processing, machine vision and artificial intelligence.

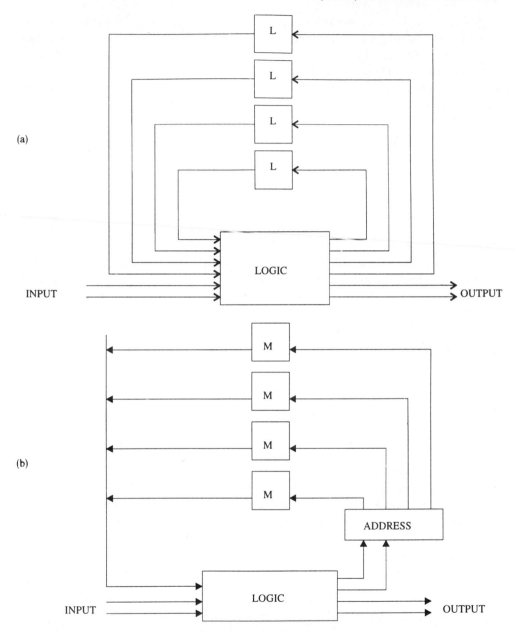

Figure 4.4 (a) A classical finite state machine does not suffer from the 'von Neumann bottleneck' since it can update all of its memory in parallel without the need for addresses: (b) A modified finite state machine suffers from the von Neumann bottleneck since it can only update one memory at a time and consequently needs an address to do so.

Over the period 1970–90 there has also been a body of work often called 'optical computing' practised by optical researchers using linear processing devices, e.g. the use of spatial filters and Fourier transform processes in image processing. A missing link has been the absence of *nonlinear all-optical* or *optoelectronic circuit elements*, preferably capable of fabrication in the form of two-dimensional arrays and of small enough size to have small switching energies and high speeds. One example, discussed in Chapter 3, is the SEED, which operates through a 'hybrid' optoelectronic nonlinearity. Here we shall describe *all-optical* nonlinear logic switches, amplifiers and memories.

In comparing optical circuit elements with electronic components we note that electrons interact with each other very strongly at short range (indeed, they need insulation to separate their motion), whereas photons only interact in the presence of *nonlinear optical material*. Until fairly recently it was thought that nonlinear optical effects implied the use of high-power laser beams, with intensities of the order of MW cm^{-2}, implying oscillating electric fields of the order of 10^7 V cm^{-1} – comparable to interatomic fields. It has now been shown that nonlinear devices can be operated with power densities less than W cm^{-2} and input powers for an individual device of the order of microwatts or less [4.12].

4.4.2 All-optical circuit elements

The recent progress, which has produced prototype all-optical circuit elements, is mostly based on the combination of *optical nonlinearity* and *feedback*. This has led to the concept of 'optical bistability', and hence to a whole family of devices based upon a common set of physical and mathematical principles. The series of devices includes optical logic gates, bistable memories, amplifiers – sometimes known as optical transistors or transphasors – and power limiters. An account is given by Smith as *An Introduction to Optically Bistable Devices and Photonic Logic* [4.13].

A seminal paper was presented by Szöke *et al.* [4.14] who in 1969 proposed that a Fabry–Perot optical resonator containing a saturable absorber as its spacer layer could exhibit two states of transmission for the same input intensity. This simple bistable action was predicted to arise from the existence of a high internal optical field at constructive interference given that sufficient intensity had been incident on the resonator to bleach the absorber. To reach this condition required a greater input intensity than that required to maintain it. By contrast, at low input intensity, the non-bleached absorption held the transmission of the device at a low level. In practice this condition is quite hard to achieve experimentally and the experiments described do not in fact show optical bistability. Observation of optical bistability was not made until 1976 when Gibbs *et al.* [4.15], using an interferometer containing sodium vapour, observed bistable transmission but deduced that the dominant mechanism was refractive – involving a shift in resonator frequency – rather than absorptive. Effective refractive nonlinearity nevertheless resulted from a saturation of the atomic absorption. Such a device, although using only milliwatts of power, was relatively large (centimetres in length) and relatively slow (milliseconds) compared with electronic circuit components.

In the same year, 1976, a surprising discovery of giant nonlinear refraction was reported. This is defined from

$$n = n_0 + n_2 I \tag{4.7}$$

(where the nonlinear refractive index n_2 can be measured in units of $cm^2\,kW^{-1}$). The material used was the narrow-bandgap semiconductor InSb. A simple investigation of beam propagation by Weaire *et al.* [4.16] indicated that an incident Gaussian beam developed a twin peak output and doubled its output width with 30 mW input power. This surprising result led to the deduction of the existence of a very large negative nonlinear refractive index n_2, of the order of 0.1–1 $cm^2\,kW^{-1}$. The immediate implication was that a bistable resonator could be made of micron dimensions and, since the effect was shown to be electronic, it would be fast, on a nanosecond time scale. A second implication was that one beam could modulate the optical properties of a small slice of semiconductor and affect a second beam, thus making an optically controlled modulator or optical transistor. The microscopic explanation was given by Miller *et al.* [4.17]. Both the device possibilities described above had been practically realized in InSb by 1979 (Miller *et al.* [4.18]) in which *continuous wave laser beams* were used, leading to *steady-state* operation and true optical bistability, as well as the observation of *gain* in an optical transistor [4.19]. Simultaneously and independently, optical bistability was reported by Gibbs *et al.* [4.20] using pulsed dye laser radiation in the larger-gap semiconductor GaAs. Larger absorption and a smaller nonlinearity in this material, however, prevented steady-state operation and this observation was therefore quasi-dynamic and did not permit the demonstration of differential amplification.

Origin of giant nonlinearities
The physical explanation of the large nonlinearities in both these semiconductors involves the excitation of electrons to give some degree of saturation or 'blocking'. In the case of InSb, exciton effects are negligible in the conditions of the experiment and a plausible explanation has been given by Miller *et al.* [4.17] in terms of a 'dynamic Burstein–Moss' shift of the band edge. Physically, a number of electrons (ca. 10^{15}–10^{16} cm^{-3}) are excited into lower conduction states by the laser photons, as described by Eq. (4.1) and Section 4.2.1.

Subsequent band filling, following thermalization, modifies the absorption edge and by application of the Kramers–Kronig relationship causes a change in refractive index given by the principal part of the integral for Δn,

$$\Delta n = \frac{\hbar c}{\pi} \int_0^\infty \frac{\Delta\alpha(\hbar\omega')}{(\hbar\omega')^2 - (\hbar\omega)^2}\, d(\hbar\omega') \tag{4.8}$$

where $\alpha(\hbar\omega)$ is the interband absorption coefficient. Hence, using Eqs. (4.7) and (4.8) and standard semiconductor band theory, with the approximation $n_2 \cong \Delta n/I$, we obtain:

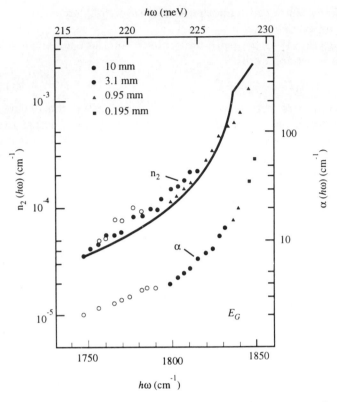

Figure 4.5 The nonlinear refractive index coefficient (n_2) and absorption coefficient (α) plotted against wavenumber. The solid line is a semi-empirical fit similar to Eq. (4.9).

$$n_2 = \frac{-4}{3}\sqrt{\pi}\,\frac{e^2 P^2}{n_0 kT}\,\frac{\alpha \tau_R}{(\hbar\omega)^3}\,F(\hbar\omega/E_g) \tag{4.9}$$

where P^2 is a constant for most semiconductors and F is a function resonating at the bandgap E_g.

Experimental results are shown in Figure 4.5 and indicate a good fit to this semi-empirical theory. The magnitude of n_2 and the coincident band-tail absorption are important. n_2 is ~ 0.1 cm^2 kW^{-1} and $\alpha \sim 10$ cm^{-1} at a typical operating point. This means that (if the nonlinearity is constant) an intensity of 1 W cm^{-2} can change n by 0.001 and can be obtained from a milliwatt power laser focused on a spot of diameter of about a third of a millimeter – a very modest power density. This change in n of only $\sim 10^{-3}$ is sufficient for device action. The absorption level means that, in a sample 100 μm thick, negligible temperature rises occur. In GaAs the mechanism seems to involve the saturation of a discrete exciton peak as well as the band edge effect and leads to an $n_2 \sim (10^{-4}–10^{-5})$ cm^2 kW^{-1}; also usefully large. However, this is accompanied by a high absorption level ($\alpha \sim 10^3–10^4$ cm^{-1}), which means in practice that devices are restricted to a thickness of a few microns. Since this early work many materials have been shown to exhibit giant nonlinearity and they can be compared by noting that n_2 is related to the third-order susceptibility χ_3 through the relationship:

$$\text{Re}\,\chi_3 = \frac{9 \times 10^8}{4\pi}\, 4\varepsilon_0 n_0^2 c n_2 \text{ (in V m}^{-1}) \tag{4.10}$$

This quantity defines the third-order polarization, P, in the expansion:

$$P = \varepsilon_0 \chi_1 E + \varepsilon_0 \chi_2 E{\cdot}E + \varepsilon_0 \chi_3 E{\cdot}E{\cdot}E$$

The value of the χ_3 for the case discussed is $\sim 10^{-7}$ V m^{-1}. Usual values of χ_3 in solids are in the range 10^{-17}–10^{-20} V m^{-1}. A simple explanation of this very large range of nonlinearity has been given by Wherrett [4.21; 4.22].

A quasi-dimensional analysis provides a first assessment of χ_3. Consider N_0 discrete atoms/unit volume with just two energy states with energy difference E and transition dipole moment ex. Introducing the electromagnetic field interaction three times to obtain the third-order polarization, N_0ex, one obtains:

$$\chi_3 \propto N_0 e^4 x^4 E^{-3} F(\hbar\omega/E) \tag{4.11}$$

The factor F, determined by the closeness to resonance, can be set to unity for purposes of comparison, and with N_o representing the density of dipoles at NTP (10^{18} cm^{-3}), we obtain a value of $\chi_3 \sim 10^{-26}$ V m^{-1} – a very small nonlinearity.

Transferring the argument to solids, the density of dipoles rises to $\sim 10^{23}$ cm^{-3} with N_0 replaced by a sum over semiconductor k-states per unit volume and E replaced by the k-dependent energy gap $E(k) = E_g + \hbar^2 k^2/2m_r$. In semiconductor notation it is usual to express the dipole moment through a momentum operator described by the Kane P-parameter proportional to the momentum matrix element for interband transitions, thus expressing the reduced effective mass, m_r, in terms of P through $\mathbf{k}{\cdot}\mathbf{p}$ theory (see Section 2.3.2) we have:

$$\chi_3 = e^4 P E_g^{-4} F(\hbar\omega/E_g) \tag{4.12}$$

If we again assume $F = 1$, then $\chi_3 \sim 10^{-17}$ V m^{-1}. The advantage of this notation is that the quantity P is essentially constant over a large range of semiconductor materials. The dependence of χ_3 as E_g^{-4} shows that small-gap semiconductors are strongly favoured: between visible wavelengths and 5 μm in the infrared, factors of the order of thousands are involved. The quantity F will contain resonance enhancement factors. A four-stage transition scheme describes the three electromagnetic field interactions and the emission of a photon via the generation polarization [4.21]. For the case of refractive nonlinearity (real part of χ_3) all frequencies are the same, leading to 'multiple resonance'. The magnitude of the resonance enhancement will depend on the energy mismatch (ΔE) between the intermediate (virtual) state of the system and the initial state. With degenerate frequencies $\Delta E = 0$ and the contributions diverge to infinity. In practice each intermediate state exists for a time

Δt leading to a state broadening following the uncertainty principle ($\Delta t \sim \hbar/\Delta E$). Up to nine orders of magnitude increase in χ_3 can be derived from F.

Near to resonance the nonlinearity becomes 'active' with real excitations which modify the optical properties of the material over a range of timescales from picoseconds to microseconds. The 'virtual' nonlinearities are enhanced in the ratio of the carrier lifetime, τ_R, to the state-broadening interval, the latter being of the order of picoseconds.

An analytical expression for n_2 has been given by Miller *et al.* [4.23], Eq. (4.9). This expression gives a good description of the resonant refractive nonlinearity in InSb, InAs and CdHgTe; α is determined empirically from experiment. For experiments that integrate over carrier relaxation times (tens of picoseconds) the mechanism then becomes the same as that discussed above as 'band filling'.

Device physics

The simplest configuration that provides optical feedback is a simple Fabry–Perot etalon containing a nonlinear refractive material (Figure 4.6). Its optical thickness is given by:

$$nL = (n_0 + n_2 I_c)L \tag{4.13}$$

and this changes with the internal intensity I_c. Consider now (Figure 4.6) the transmission of such an interferometer as a function of intensity: if we start in an initial condition where the illuminating wavelength is detuned from maximum transmission by a wavelength increment $\delta\lambda$ (Figure 4.6a), we see from Figure 4.6(b) that the relation between output and input would give rise to, in the case of a linear device, a straight line of low slope: if the devices were tuned to resonance and there were no absorption loss, output would be related to input by a line at 45°. If we now increase the intensity from the initial condition the *nonlinear resonator* tends towards resonance as its optical thickness changes with intensity. This would give rise to a nonlinear relation between output and input. However, as we approach resonance the internal field circulating within the resonator itself builds up according to:

$$I_c = I_i T(1 + R)/(1 - R) \tag{4.14}$$

where I_i is the incident intensity, T is a function of frequency (as in Figure 4.6a) and R is the (constant) reflectivity of the resonator mirrors. Thus, at resonance, the internal intensity is at its maximum where $T = 1$ and is amplified by the term $(1 + R)/(1 - R)$. This gives rise to positive feedback. As resonance is approached the internal field builds up and the rate of approach to resonance depends on the change in optical thickness which *itself* depends on the magnitude of the internal field. The rate of approach to resonance thus speeds up. This can be readily expressed through the expression:

$$d\,\frac{(\delta\lambda)}{dI_i} = T(\lambda)\bigg/\left(\frac{T_{max}}{2n_2 L} - I_i\,\frac{dT}{d\lambda}\right) \tag{4.15}$$

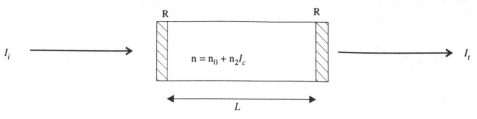

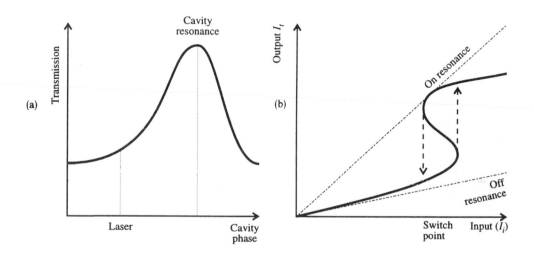

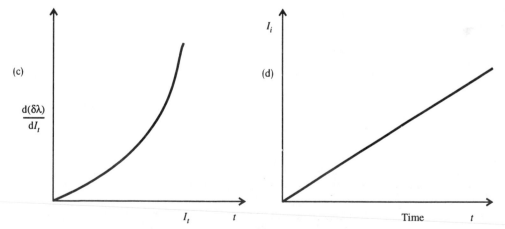

Figure 4.6 The dynamics of switching in a Fabry–Perot etalon.

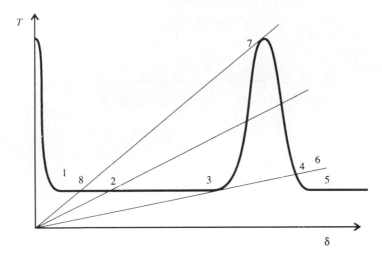

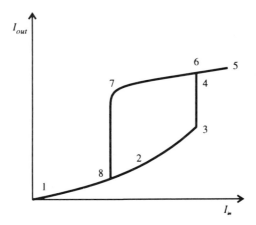

Figure 4.7 The graphical solutions of Eqs. (4.16) and (4.17): the sequence 1 to 8 is equivalent to increasing and decreasing the incident power.

Figure 4.6(c) shows a plot of the rate of approach to resonance as a function of either incident intensity I_i (or as a function of time if we assume a linear ramp of I_i against time, Figure 4.6(d)). As I_i increases, $dT/d\lambda$ also varies, the denominator of Eq. (4.15) tends to zero and the rate of approach to resonance becomes infinite, leading to rapid switching. This gives a physical feel for the dynamics of the switching process.

A simple description of bistable switching can be made by considering a plane wave model of the linear Fabry–Perot resonator. In this model the intensity is assumed constant throughout the resonator, both transversely and longitudinally. The Airy sum of multiple beams for such a resonator, modified for intensity dependent optical thickness is given in Eq. (4.16).

$$T(\lambda, I_i) = \frac{I_t}{I_i} \propto \frac{1}{1 + F \; \sin^2 \; \delta/2} \tag{4.16}$$

where I_t is the transmitted intensity and the intensity-dependent phase change $\delta = 2\pi n L/(\lambda/2)$ so that $\delta(I_c) = 4\pi n_0 L/\lambda + 4\pi n_2 L I_c/\lambda$.

A second representation for T can also be obtained by relating the internal and external intensities I_c and I_i:

$$T(\lambda, I_i) = \frac{I_c (1 - R)}{I_i (1 + R)} \tag{4.17}$$

Only certain values of I_c for certain I_i have satisfactory self-consistency of Eqs. (4.16) and (4.17). The solution is best found graphically (Figure 4.7). Equation (4.16) gives a straight-line relationship of vertical slope at zero intensity and lessening slope as the intensity is increased. Intersections with Airy's sum, Eq. (4.16), give the solutions. As the slope is reduced, intersections eventually show two simultaneous solutions. This represents optical bistability, i.e. two values of the transmission for one of the intensity. Two features are noteworthy:

(1) A large value of finesse (ratio of spacing to half-width of the transmission peaks) makes it easier to obtain bistability in 'first order'; otherwise, sufficient intensity and nonlinearity to reach higher orders is necessary.
(2) The positions of intersection and the form of the curve implied by intersections change with the initial conditions, i.e. depend upon $\delta\lambda$.

We illustrate such a set of characteristics in Figure 4.8, which is in fact an early experimental result using a CO laser and a plane parallel interferometer – 100 μm thick – constructed from InSb. The family of curves shows the characteristics obtained for different values of the initial detuning parameter $\delta\lambda$ if the initial condition is close to resonance, power limiting action is obtained (curve a) since the transmission can only fall as intensity rises. As $\delta\lambda$ is increased, the curve b begins to kink, and c shows a greater change in output than for an incremental change in input, thus exhibiting differential gain. This is responsible for optical transistor or transphasor action. A further increase in $\delta\lambda$ leads to bistable loops of varying width. Thus by a simple change of initial conditions a whole family of optical devices can be produced.

Figure of merit for optical circuit elements
An analysis of the factors involved in the design of such optical circuit elements has been given by Miller [4.24]. He shows that the quantity that gives a figure of merit is $n_2/\lambda\alpha$, where:

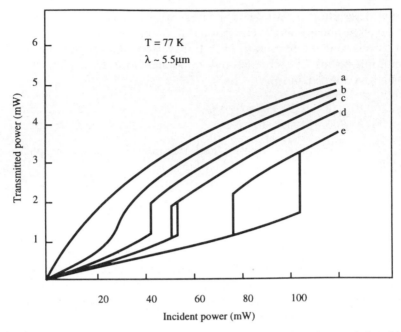

Figure 4.8 Experimental observations of a family of characteristics of an InSb bistable device obtained by changing the initial detuning at resonance of the etalon from 0 to π.

$$I_s = \frac{\lambda\alpha}{n_2} f(R, \alpha L) \tag{4.18}$$

This equation determines the lowest critical value of input irradiance I_s for a device of given size to produce bistable switching or a nonlinear characteristic, where $f(R, \alpha L)$ is a function describing the cavity properties.

The result is physically sensible since the switching power will be lower for a larger nonlinearity n_2; the shorter the wavelength, the smaller a refractive change $\delta(nL)$ will be required to effect a change from constructive to destructive interference (i.e. $\delta(nL) = \lambda/2$), and for a smaller absorption (assumed linear in this analysis), the longer the device may be for a given loss. Results show that a value of $n_2 \sim 0.1–1$ cm^2 kW^{-1} and $\alpha \sim 100$ cm^{-1} gives useful devices. The nonlinearity described is for the case of electronic origin and does not depend upon the volume of the medium. For nonlinearities of thermal origin I_s may take the form [4.25]:

$$I_s = \left[\lambda\alpha \bigg/ \left(\frac{\delta n}{\delta T} L'\right)\right] \frac{x_s f(R, \alpha L)}{\alpha L} \tag{4.19}$$

where x_s is the thermal conductivity of the substrate, L is the cavity (or active layer) thickness and L' is the substrate thickness. Thus in this case

$$n_2 = \frac{\delta n}{\delta T} \frac{\alpha L L'}{x_s} \tag{4.20}$$

and device dimensions are included.

To obtain favourable values for $n_2/\lambda\alpha$, effects resonant with the energy gap E_g in semiconductor materials are used. It is found that $n_2 \sim 1/E_g^3$, so that $I_s \sim \lambda^2$.

In designing such devices with electronic origin n_2 we must then combine these trends with the figure of merit $n_2/\lambda\alpha$. Diffraction limits suggest that the area limit will be $\sim (\lambda/n^2)$. Thus, although the nonlinearity is clearly larger at longer wavelengths for small-gap materials, the interference conditions and device size favour shorter wavelengths. This indicates the type of compromises necessary in device design.

Historically, two materials have been extensively investigated: InSb (Figure 4.5) with a typical working wavelength of 5.5 μm yields devices with $n_2 = 0.1$ cm^2 kW^{-1} and $\alpha = 10$ cm^{-1} at 77 K. This material is one of the few in which there are enough laser frequencies to allow a detailed examination of the frequency and hence resonant behaviour of n_2 and α near the bandgap. GaAs has also been investigated [4.20; 4.26; 4.27] but differs from InSb in that it has a strong exciton feature near the absorption edge. The nonlinearity $n_2 \sim 10^{-3}$ cm^2 kW^{-1} is practicable for devices, but the absorption coefficient in epitaxially grown material is of the order of 10^4 cm^{-1}. Thus thicknesses are restricted to a few microns and it was only recently that thermal stability problems were solved. More recently, GaAs waveguide structures have shown optical bistability.

Both of the above materials show negative values of n_2 caused by electronic effects. The carrier lifetime is typically 10–100 ns. The nonlinearities can be 'switched on' more quickly than this time interval by rapidly introducing carriers with relatively intense pulses. The relative figures of merit between GaAs and InSb favour InSb by a considerable factor in switching energy, which has given InSb the advantage of steady-state operation, although the device still needs to be operated at a temperature of 77 K. GaAs has been used at room temperature and at shorter wavelengths.

There is a second useful form of nonlinearity involving thermal shift of the band edge due to a temperature rise in the bulk of the nonlinear material. This effect also resonates with the band edge and is associated with moderate values of the absorption coefficient. Examples in bulk and thin-film material include ZnSe. Such thermal effects when associated with thin-film devices are particularly useful: the devices take the form of interference filters where the film thickness (~ 2 μm) implies a very small heat capacity so that in small devices microsecond time constants will occur. The important physics lies in the control of heat-sinking by means of the conductivity of a relatively massive substrate.

Requirements for optical logic suitable for processing and computing
It is possible to perform both all-optical and 'hybrid optoelectronic' logic operations using the optical circuit elements described here and in Chapter 3. Some of the requirements necessary for constructing optical logic circuits or a prototype digital optical processor can be defined as follows:

(1) *High contrast.* A logic device must show a large change in output between logic levels zero and one. Some devices such as the symmetric SEED however use *ratios of signals* in a low contrast dual-rail scheme (see Section 3.5).

(2) *Steady-state bias.* In order to make a variety of different logic gates it must be possible to alter optical bias levels in a controllable manner. In terms of optical bistability this means that it must be possible to 'hold' the device indefinitely at any point on the characteristic with a cw laser beam – the 'holding beam' – and implies a degree of stability. Devices based on InSb at 77 K and on ZnSe at 300 K were the first to show this behaviour. Dual-rail devices such as the S-SEED minimize this requirement for a steady state bias.

(3) *External address.* For logic functions it is clearly important that separate external signal beams can be combined with the holding beam in order to switch the device. The switching energy is derived from the holding beam, and this switched beam propagates in transmission or reflection as the output beam to the devices next in line in an optical circuit.

(4) *The elements must be cascadable.* This means that the output of one device must be sufficient to switch at least one succeeding device. The ability to set a cw holding beam near to switch point in fact fulfils this condition since the extra increment is then small compared with the change in output even in the presence of loss. Since each device has its own 'power supply', i.e. holding beam, logic levels are restored. The bistable memory function of the S-SEED allows this feature to be obtained by 'time sequential gain' – an alternative method.

(5) *Fan-out and fan-in.* The advantage of parallel processing in optical devices allows one device to drive a large number of succeeding devices, using free-space propagation for address. The summed effect of several elements can readily be focused on one device to achieve fan-in; however, it is difficult to differentiate between many signals.

Both ZnSe optothermal logic elements and S-SEEDs were used in early prototype free-space digital optical processors. Cascading, restoring logic and cyclic operation were demonstrated in a primitive optical computer in 1987 [4.28]. More recently, optical architectures have investigated 'smart pixels' combining Si processing with III–V input and output devices to address the 'pin-Hertz' limitations of electronics.

4.5 Optical disk data storage

Optical data storage is an example of light controlling light which, in contrast to the concepts of Section 4.4, has already achieved mass commercial exploitation. In the case of the compact audio disk (CD), or the video disk, the information is *written optically* and *read optically* with, of course, a long delay between these operations. As we will describe, optical storage has a number of advantages over magnetic storage, the mainstay of electronic information storage worldwide since the 1940s.

Currently, both magnetic tapes and optical disks are used in audio and video recording as well as back up archival computer files, as opposed to the interactive processor-type memory referred to in Section 4.4.

It is noteworthy that in both cases *mechanical drive* is used to reach different memory locations with consequent long *access time*, τ_a. This is defined as the average time spent going from one randomly selected spot on the disk to another. It is the sum of the *seek time* τ_S, the average time to acquire a target track, and a latency τ_1, the time spent waiting for the required sector. The recent advent of spatial light modulators (SLMs) actintg as dynamic diffraction gratings (Chapter 5) now allows optical beam steering without moving parts controlled by silicon backplanes. The optically interconnected logic planes of Section 4.4 could also read and write to a 2D optical memory. Optics may therefore be able to break this bottleneck in the future. Large volumetric capacity and low cost characterize both magnetic and optical storage technology which is still successful despite its sequential access drawback. Magnetic hard disks and removable flopy disks are of course ubiquitous, with the latter providing 2 Mbytes of storage on 3.5 in diameter disks. Optical disks can also provide recording densities in the range of 10^7 bits/cm^2 and beyond; they combine the functions of the hard disk (high capacity, high data transfer rate, rapid access) with those of the floppy (back-up storage, removable) in a single optical disk drive.

Optical recording can support read-only, write-once and erasable/rewritable modes all in one unit, unlike magnetic technology. Information can be recorded on a master disk, in the form of a series of 'holes' ablated from the 'track' by the write laser. This master is thus used as a stamper for inexpensive reproduction. The write-once, read many times (WORM) and magneto-optical erasable media can be handled by one unit. A thermo-magnetic process is used for both recordings and erasure. At ambient temperature, the magnetic film has a strong resistance to magnetization reversal but, when heated by a laser beam, a hot spot can be 'switched' in direction by an external magnet. Removing the laser beam brings the temperature down and 'freezes in' the reverse magnetical domain. The changed direction of magnetization alters the optical polarization properties of the surface which can be read optically.

4.5.1 Disks and tracks

The circular disk is located on a hub and spins (typically) at 1200–3600 rpm. The distance between tracks (which can be concentric or spiral) varies between 100 μm (magnetic floppy), 10 μm (magnetic hard) and 1 μm (optical). Figure 4.9 shows micrographs of various optical recording media. In the magnetic case the head weighs about 5 g and seek times are ~ 10 ms. For optical disks, the laser and optical components give the head a weight of 50–100 g and a τ_S of 20–100 ms. Integration and miniaturization are likely to reduce this to a few milliseconds.

The typical recording dimension for optics is ~ 1 μm in both spot size and depth of focus. It is impossible to make mechanical systems with such tolerance (better than 100 μm) and thus the objective lens is mounted in an actuator (typically a voice coil) provided with optical feedback signals to maintain tracking. Disks are in general not

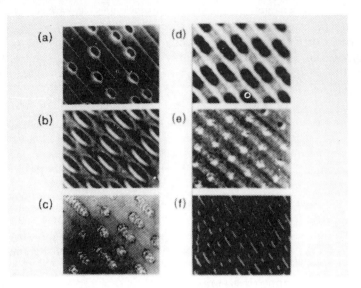

Figure 4.9 Micrographs of several types of optical storage media. The tracks are straight and narrow with a 1.6 μm pitch, and are diagonally oriented in each frame. (a) Ablative, write-once tellurium alloy, (b) Ablative, write-once organic dye. (c) Amorphous-to-crystalline, write-once phase-change alloy GaSb. (d) Erasable, amorphous magneto-optic alloy GdTbFe. (e) Erasable, crystalline-to-amorphous phase-change tellurium alloy. (f) Read-only CD-Audio, injection-moulded from poly-carbonate with a nickel stamp. [4.29].

flat and moments away from focus of up to 50 μm may occur. Thus automatic focusing is also required. Figure 4.10 shows an example of a focus error detection optical system.

Performance of the optics is limited by diffraction, e.g. beam diameter at focus, d, is

$$d = \lambda/\text{NA}$$

where NA is the numerical aperture (typically ~ 0.6) and λ is the wavelength. The depth of focus, δ, is likewise given by

$$\delta = \lambda/(\text{NA})^2$$

Clearly the development of practical blue lasers will change both tolerances and storage capacity from those currently limited by GaAs-based lasers with $\lambda \sim 700$ nm.

4.5.2 Magneto-optical readout

Readout is accomplished by polarization-sensitive optics with the aid of the magneto-optic Kerr effect (rather similar to the Faraday effect described in Chapter 2) which rotates the plane of polarization $\sim 10°$ between the directions of magnetization of the domains. This is a small effect and noise limitations become important. A variety of materials and magnetic effects have been considered and will provide directions for future research.

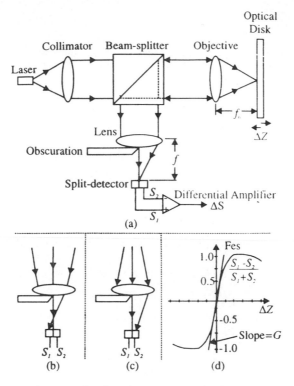

Figure 4.10 Focus error detection by the obscuration method. In (a) the disk is in focus, and the two halves of the split detector receive equal amounts of light. When the disk is too far from the objective (b) or too close to it (c), the balance of detector signals shifts to one side or the other. A plot of the focus error signal versus defocus is shown in (d), and its slope near the origin is identified as the FES gain, G.

4.5.3 Future prospects

The brief review of the advantages already displayed by optical storage suggests that the future will be fruitful. The possibility of multiple-track read/write with diode laser arrays brings advantages in proportion to the number of parallel channels and requires only proliferation of the lasers with the optics remaining common.

The extension of this concept to massively parallel optical read/write would have major impact on computer memory technology. En route to this objective, the use of optically steered beams with no moving parts appears to be a possibility: a ferroelectric liquid crystal SLM with a silicon backplane can in principle produce a dynamic diffraction grating which could reconfigure in ~ 1 μs.

The use of holographic/diffractive optics in the more modest context of the optical head could reduce the weight of the present assembly of discrete components. This evolving technology is likely to lead to many innovations and improvements [4.29].

References

[4.1] D.H. Auston, 'Picosecond optoelectronic switching and gating in silicon', *Appl. Phys. Lett.* **26** (1975).

[4.2] F. Urbach, *Phys. Rev.* **92**, 1324 (1953).

[4.3] S.W. Kurmick and J.M. Powell, *Phys. Rev.* **116**, 597 (1959).

[4.4] E.J. Johnson and H.Y. Fan, *Phys. Rev.* **139**, A1991 (1965).

[4.5] W. Ji, A.K. Kar, J.G.H. Mathew and A.C. Walker, 'Quasi-cw optical bistability in InSb at room temperature', *IEEE J. Quantum Electron.* **QE-22** (1986).

[4.6] A.C. Walker, 'A comparison of optically nonlinear phenomena in the context of optical information processing', *Opt. Computing and Processing* **1**, 91–106 (1991).

[4.7] A.J. Alcock, P.B. Corkum and D.J. Jones, 'A fast scalable switching technique for high power CO_2 laser radiation', *Appl. Phys. Lett.* **27** (1975).

[4.8] P.B. Corkum, 'High-power, sub-picosecond 10 μm phase generation', *Optics Lett.* **8** (1983).

[4.9] C. Mead and L. Conway, *An Introduction to VLSI Systems* (Addison-Wesley, London, 1980).

[4.10] Z.C.P. Lee *et al.*, *Dig. Tech. Papers IEEE* GaAs IC Symposium (1983) p. 162.

[4.11] A. Huang, *Proc. IEEE* **72**, 780 (1984).

[4.12] S.D. Smith and A.C. Walker, Paper B5-10, *Conference Digest* ICO-13, Sapporo (1984).

[4.13] S.D. Smith, B.S. Wherrett and A. Miller (eds.), *Optical Bistability, Dynamical Nonlinearity and Photonic Logic* (1984); *Phil. Trans. R. Soc.* (London), March 1984.

[4.14] A. Szöke *et al.*, *Appl. Phys. Lett.*, **15**, 376 (1969).

[4.15] H.M. Gibbs, S.L. McCall and T.N.C. Venkatesan, *Phys. Rev. Lett.* **36**, 1135 (1976).

[4.16] D. Weaire *et al.*, *Opt. Lett.*, **4**, 331 (1979).

[4.17] D.A.B. Miller, C.T. Seaton, M.E. Prise and S.D. Smith, *Phys. Rev. Lett.* **47**, 197 (1981).

[4.18] D.A.B. Miller, S.D. Smith and A. Johnston, *Appl. Phys. Lett.* **35**, 658 (1979).

[4.19] D.A.B. Miller and S.D. Smith, *Opt. Commun.* **31**, 101 (1979).

[4.20] H.M. Gibbs *et al.*, *Appl. Phys. Lett.* **36**, 6 (1979).

[4.21] B.S. Wherrett, *Proc. R. Soc.* (London) **A390**, 373 (1983).

[4.22] B.S. Wherrett, *Phil. Trans. R. Soc.* (London), Proc. Mtg. for Discussion, Optical Bistability, Dynamical Nonlinearity and Photonic Logic, London (1984).

[4.23] D.A.B. Miller, S.D. Smith and B.S. Wherrett, *Opt. Commun* **2**, 35 (1980); D.A.B. Miller *et al.*, *Phys. Rev. Lett.* **47**, 197 (1981).

[4.24] D.A.B. Miller, *IEEE J. Quantum Electron.* **QE-17**, 3 (1981).

[4.25] B.S. Wherrett, D. Hutchings and D. Russell, private communication, 1985.

[4.26] H.M. Gibbs *et al.*, see Ref. [21].

[4.27] D.A.B. Miller, A.C. Gossard and W. Wiegmann, *Opt. Lett.* **9**, 169 (1984).

[4.28] S.D. Smith, A.C. Walker, B.S. Wherrett and F.A.P. Tooley, 'The demonstration of restoring digital optical logic', *Nature* **325**, 27–31 (1987).

[4.29] M. Mansipur, 'Optical information and image processing', in *Handbook of Optics 1*, ed. M. Bass, ch. 31 (McGraw-Hill, New York, 1995).

Further reading

A. Miller, D.A.B. Miller and S.D. Smith, 'Dynamical nonlinear optical processes in semiconductors', *Adv. Phys.* **30**, 697–800 (1981).

5

Liquid crystals in optical technology

Fliessende Krystalle! Ist dies nicht ein Widerspruch in sich selbst – wird der Leser der Überschrift fragen –,

Otto Lehmann
Z. Phys. Chem. IV (1889)

In this chapter we consider a variety of liquid crystal materials. We start by looking at how these materials were discovered and then consider some of the compounds that are used today. The chemical and physical properties of liquid crystals are considered briefly. We then present an outline of a theory developed in order to model the dynamics of nematic liquid crystals and consider the uses of a variety of different liquid crystal properties in devices. Finally, we describe how the basic liquid crystal cell is manufactured and indicate its use in more complex devices such as spatial light modulators.

5.1 Historical background

Liquid crystals were discovered more than 100 years ago but have only become widely used relatively recently. This is a good example of a long gestation time between laboratory and application. Several important steps of experimental discovery and technical progression can be identified en route. Traditionally, matter is thought to exist in three well defined states, *viz.*, solids, liquids and gases, each of which is associated with a certain degree of atomic or molecular order. Solids are rigid, highly ordered arrays of molecules or atoms, whereas liquids and gases do not exhibit any such ordered structure. However, it is known that some organic compounds form an intermediate state between the liquid and solid phase where a well defined long-range structure exists in one or two dimensions whilst still appearing to be in the liquid state.

In 1888 Friedrich Reinitzer [5.1] reported from Prague that cholesteryl benzoate melted at 145°C into a cloudy brown liquid which then suddenly became clear at 178.5°C. He stated that the sample appeared to be liquid but when studied under high magnification under a microscope some crystalline character could be distinguished. Studies carried out between 1899 and 1922 established that this and other compounds actually represented a new state of matter. Lehmann [5.2, 5.3], from Aachen, between 1889 and 1900, coined the name *liquid crystal*. In 1922, Friedel [5.4] referred to it as a *mesomorphic state of matter*, mesomorphic meaning of intermediate form. Both these terms have subsequently been used more or less synonymously. A great variety of materials have since been found to exhibit mesomorphic states.

The foundations of an understanding of the structure and properties of liquid crystals were laid in the 1920s, and progress continued thereafter. In 1958 F.C. Frank [5.5] presented 'a continuum theory of curvature elasticity' that was based on the work of Oseen. A comprehensive review of the theory of liquid crystals is given in P.G. de Gennes' book *The Physics of Liquid Crystals* [5.6], which provides a detailed treatment, including theory, on a molecular scale. However, the macroscopic theoretical approach of the continuum theory is sufficient to understand device applications. The interest in liquid crystals for displays dates from the 1960s. There exist several thousand compounds that have been found to exist in liquid crystal states, many with useful properties. Important progress was made by George Gray [5.7] and his co-workers at Hull University and the Royal Signals and Radar Establishment (RSRE), Malvern, in the late 1960s by the synthesis and evaluation of materials that were stable at ambient temperatures under exposure to light, air and electricity. A useful review of the physics and display applications of liquid crystals is given by Shanks [5.8, 5.9].

5.2 Liquid crystal physics

Some organic compounds, and a smaller number of inorganic ones, possess this intermediate state between the solid and liquid. It is characterized by a high degree of long-range orientational and/or translational ordering of molecules. Most liquid crystal molecules are uniaxial and dipolar (for example, ferroelectric liquid crystals). They can be like rods, or like disks, or they can organize themselves into assemblies (micelles) having these shapes, which are then ordered to give the liquid crystal. They can be formed by melting a solid (thermotropic liquid crystal) or by adding a liquid

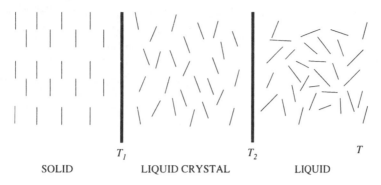

T_1 T_2 T

SOLID   LIQUID CRYSTAL   LIQUID

Figure 5.1 Molecular ordering in solid, liquid crystal and liquid phases.

to the solid (lyotropic liquid crystal). For most practical purposes only thermotropic liquid crystals having rod-like molecules need be considered, and only these will be considered in the remainder of this section. Figure 5.1 indicates the nature of the ordering through various temperature regimes of such a thermotropic liquid crystal.

There are three main types of liquid crystal, which can be described according to their structure:

(i)   smectic (soap-like): stratified liquid crystals with a variety of possible configurations of the molecules within strata;
(ii) nematic (thread-like): molecules with only orientational ordering along locally preferred directions;
(iii) cholesteric or chiral nematic (like compounds of cholesterol): these molecules have an orientational ordering, but with the locally preferred direction rotating about an axis to give a helical structure with a defined pitch.

These are illustrated in Figure 5.2. We note that the smectic type has some eleven variations, of which three are illustrated. It is this class which can show ferroelectric properties if the molecules are tilted with respect to the normal to the strata and they are either (a) chiral or (b) doped with a chiral additive.

5.2.1 Smectic liquid crystals

Smectic liquid crystals may show varying degrees of positional order in each layer. The classification of the smectics is based on the nature of the ordering within and between these layers. The smectics are labelled from A through to K, smectic A being the least ordered. The main area of interest within the smectic liquid crystals are those in the chiral C group, S_c^*, which are also known as ferroelectric liquid crystals; these are discussed in Section 5.6.8. Other chiral, tilted smectics (e.g. S_I^*) can also be ferroelectric.

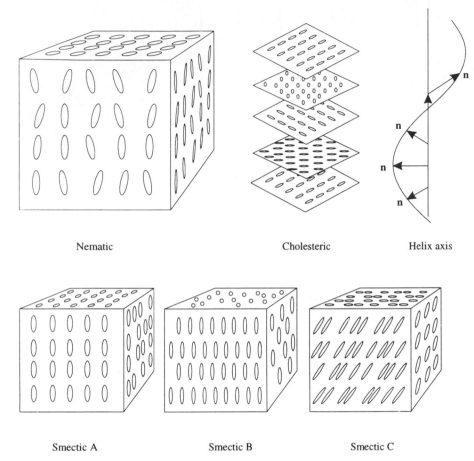

Nematic   Cholesteric   Helix axis

Smectic A   Smectic B   Smectic C

Figure 5.2 Illustrations of liquid crystal groups.

5.2.2 Nematic liquid crystals

These are mobile liquids comprised of achiral (i.e. superposable onto their mirror image) anisotropic molecules and are the least highly ordered of the liquid crystals. The molecules are more or less parallel in orientation and the liquid crystal is optically uniaxial (see Section 6.1.3) with a strong birefringence, i.e. the refracture index for polarized light varies with direction. The orientation is along one common direction that is described by a unit vector **n**, called the *director* (Figure 5.3), and represents the average orientation of the molecules. Thermal motion causes the molecular direction to fluctuate statistically, with centres of gravity and orientations distributed about their mean values. The direction of **n** can be influenced by external factors such as the boundary plates or electric fields. Of the three types of ordering, the nematics

Figure 5.3 The director.

have the fewest constraints on molecular orientation and the lowest viscosity; they therefore respond most easily to applied fields.

Nematic liquid crystals generally consist of long organic molecules. The example shown in Figure 5.4 has two aromatic rings connected by some bridging element. Permanent dipoles associated with the group (R') of the chain, or induction of a dipole by neighbouring molecules can give each molecule a dipole moment. The molecules are long and fairly rigid, exhibiting a larger polarizability along the molecular axis than perpendicular to it.

The degree of ordering about the preferred direction, **n**, is defined by Tsvetkov [5.10] as

$$S = \tfrac{1}{2}\langle 3\cos^2\phi - 1 \rangle \tag{5.1}$$

where ϕ is the angle that the long molecular axis of the individual molecule makes instantaneously with the preferred axis, defined by the director. The brackets $\langle \; \rangle$ denote the statistical average. This gives a measure of the long-range order of the molecules. For perfect parallelism, $S = 1$; a totally random distribution would have the value $S = 0$.

For the nematic case the value of S is intermediate (e.g. 0.5–0.7) and depends strongly on temperature, dropping abruptly to zero at the temperature above which the liquid crystal becomes an isotropic liquid. S can be related to the optical anisotropy, i.e. to the birefringence. The mean field theory in which each molecule is considered to move in the mean field of all the others is used to discuss the dynamics of liquid crystals. This was proposed by Born [5.11] in 1916, and used by Maier and Saupe [5.12] in 1958.

Quartz, which is often used for its birefringent properties, has a Δn of ~ 0.01. However, liquid crystals, owing to their structure, show a much larger birefringence

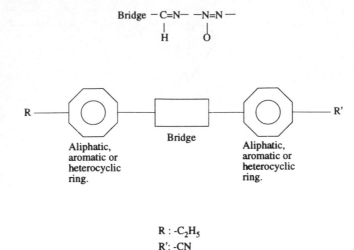

Figure 5.4 Typical liquid crystal structure.

with Δn of the order of 0.01 (for cyclohexanes) to 0.4 (for benzenoids). Through Eq. (2.19), the refractive effects are related to the dielectric constants of the material. In terms of the parallel ($\varepsilon_\parallel$) and perpendicular ($\varepsilon_\perp$) components of the dielectric constant the dielectric anisotropy, ε_a, is defined by Eq. (5.2) and can be either positive or negative:

$$\varepsilon_a = \varepsilon_\parallel - \varepsilon_\perp \tag{5.2}$$

Figure 5.5 shows the field alignment of the director, **n**, for two different cases (a) $\varepsilon_a > 0$ and (b) $\varepsilon_a < 0$. For the case of an applied alternating electric field **E**, the liquid crystal responds to the rms value of the field at frequencies above about 30 Hz.

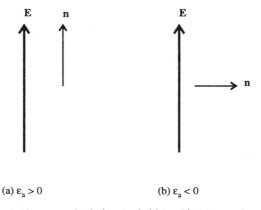

Figure 5.5 The director **n** in an applied electric field **E** with (a) $\varepsilon_a > 0$ and (b) $\varepsilon_a < 0$.

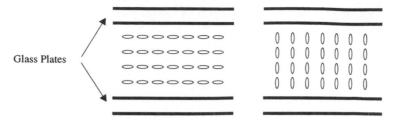

Figure 5.6 Two types of nematic crystals obtained between parallel glass plates.

The conductivity is also anisotropic but almost always has a positive value since $\sigma_\parallel > \sigma_\perp$. For use as an electro-optic device the liquid crystal is sandwiched as a thin layer between two glass plates with separations which vary between 1 and 20 μm. In a bulk sample of liquid crystal, the director **n** varies from region to region and from time to time. As a consequence of such spatial variations of the optic axis, light is scattered, and so the sample is *usually not transparent*. This can be changed by imposing a definite order, called a texture, on the sample. The properties are especially marked in 'monocrystalline' samples obtained, for example, when a thin nematic layer is trapped between two glass plates ∼20 μm apart. The layer can be viewed using polarized light in order to exploit the birefringence effects it introduces. There are two cases of molecular orientation that are of particular interest from a device point of view (Figure 5.6).

The effect of surfaces on alignment of the molecules was discovered quite early in the history of liquid crystal effects. The molecules next to the glass plates can have an orientation imposed on them by using surface 'scratches' (e.g due to rubbing) or coating effects [5.13]. This is illustrated in Figure 5.7.

More energy is required for **n** to lie across the grooves than along them, so the molecules line up along grooves. The anchoring of the molecules adjacent to the glass plates is quite firm. These then determine the orientation of the neighbouring molecules through long-range interaction and so on into the liquid crystal. Both plates must be clean, since dust particles can be of the same size as the desired plate separation. An alternative to the use of 'scratches' is to use thin-film evaporation at non-normal incidence where the structure of the film can play a similar role to the 'scratches'.

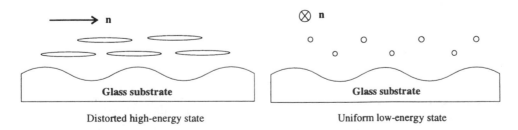

Figure 5.7 The effect of microgrooves ('scratches') on the orientation of the molecules.

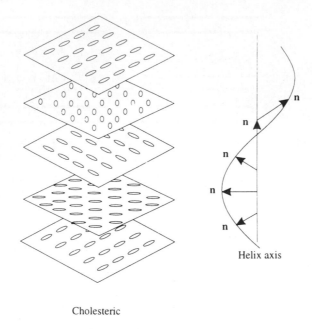

Cholesteric

Figure 5.8 The structures of cholesteric planes.

The initial alignment may be changed or disturbed by the application of a voltage (usually alternating at 30 Hz–100 kHz) to the transparent electrode layers (e.g of indium–tin oxide) on the inside surfaces of the glass plates. This is the basis of device action with typical voltages of the order of 5 V. The details of such processes depend on the type of nematic liquid crystal material, and there are several possible effects that can be exploited.

5.2.3 Cholesteric liquid crystals

Cholesteric liquid crystals are a special case of nematic ordering where a helical twist exists as well as the nematic ordering. They are also known as *chiral nematic liquid crystals*. The direction of the optical axis is perpendicular to the twist axis. The distance for a 360° turn is called the pitch. In essence, the name means 'screw-like'. Cholesterics are formed naturally by materials that would normally form a nematic liquid crystal phase, but whose molecules have a handedness (chirality) which allows them to be distinguished from their mirror image. They can also be formed by dissolving optically active chiral (left- or right-handed) compounds in nematic liquids. The pitch can thence be controlled by the amount of this additive. These phases were first observed in pure cholesterol esters – hence the name cholesterics. Figure 5.8 illustrates the different planes in the continuously twisting structure.

The pitch length P is usually in the region 0.1–50 μm and it can consist of around 1000 molecular thicknesses. The pitch can thus be of the order of an optical wavelength depending on the particular choice of molecule.

5.3 The chemistry of liquid crystals

Until the 1970s no practical, stable liquid crystals that existed at room temperature were available to any of the would-be liquid crystal device manufacturers. There were some high-temperature liquid crystals, such as *para-azoxyanisole* (named PAA by its manufacturers [5.14]), which exists in the nematic phase between 118 and 135.5°C. From about 1970 there were also products known as MBBA (para-methoxy benzylidene butylaniline) and EBBA (p-ethoxy benzylidene butylaniline) which, although they were used in some early devices, neither of these was very chemically stable, and they had negative values of the dielectric anisotropy, ε_a.

The credit for the first reliable and stable liquid crystal materials goes to Gray *et al.* [5.15], working at Hull University, who showed that the main cause of chemical instability in such molecules was the weak central bridge. They found a group of organic molecules classed as 'cyanobiphenyls' that did not have this bridge group, their aromatic rings being directly linked. These formed reliable nematic cells. The discovery was backed by the RSRE (for Ministry of Defence purposes), where a team was involved in the evaluation of the chemical and physical properties of 'cyanobiphenyls' [5.16]. Many families of stable liquid crystals are now known to exist and the manufacture of high-performance room-temperature devices is now possible. From 1973 onwards biphenyls started to be produced on a larger scale for research purposes and a year later commercial production began for use in displays. So called 'twisted nematic' cells would be produced by using surface effects and rotating one of the glass plates. These are used in displays.

5.3.1 Nematic and smectic biphenyls

Figure 5.9 illustrates some nematic liquid crystal molecules. Structures (a) and (b) are cyanobiphenyls with n-alkyl or alkoxy chains from 1–12 carbon atoms, and (c) is a

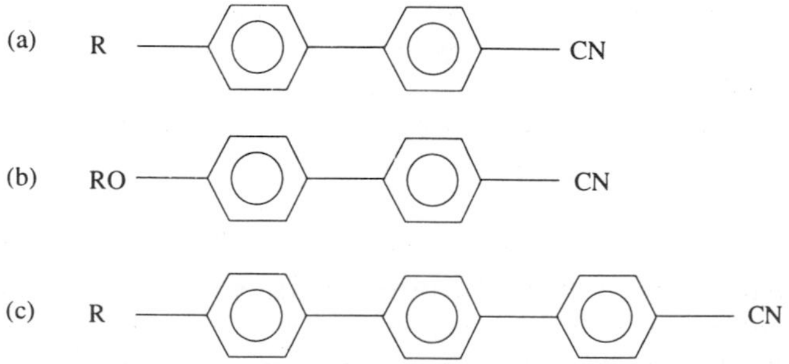

Figure 5.9 Some liquid crystal chemical structures (R and RO are different terminal substituents).

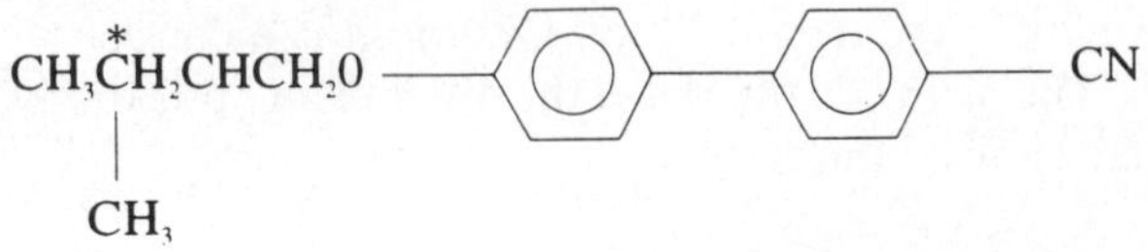

$$CH_3CH_2\overset{*}{C}HCH_2O \quad\text{——}\bigcirc\text{—}\bigcirc\text{——}\quad CN \qquad C15$$

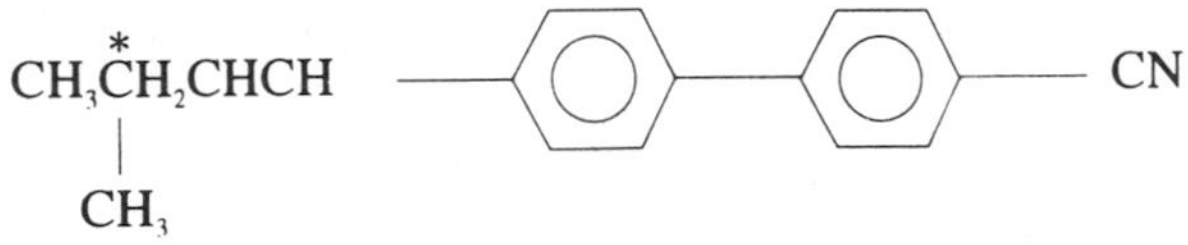

$$|$$
$$CH_3$$

$$CH_3\overset{*}{C}H_2CHCH \quad\text{——}\bigcirc\text{—}\bigcirc\text{——}\quad CN \qquad CB15$$

$$|$$
$$CH_3$$

Figure 5.10 Examples of cholesteric molecules (* indicates the chiral centre in the molecule).

cyanoterphenyl with an n-alkyl chain. All of these molecules are nematic liquid crystals. If, however, a branched alkyl or alkoxy chain is used, then cholesteric liquid crystals are produced.

Smectic and nematic liquid crystals are used in simple watch displays and are biplexed, which allows the liquid crystal segments to be switched 'on' and 'off' using two distinct voltage levels. It has been possible to produce mixtures in which the liquid crystal segments or electrodes can be sequentially addressed; these are used for calculators and computer terminal displays.

5.3.2 Cholesteric (chiral nematic) biphenyls

Figure 5.10 shows the structure of two cholesteric molecules. These are often used in twisted nematic cells to prevent the formation of areas of reverse twist. An area of reverse twist is one where the liquid crystal molecules rotate in a direction opposite to that required, but still meet the orientation conditions imposed on them at the cell walls. For example, in a 90° twisted nematic cell some areas may twist −90° in the opposite sense to the 90° twisted molecules. These −90° twisted areas still meet the orientation requirements of the cell boundaries and have an energy equal to that in the 90° twisted areas. In order to avoid this it is possible to use molecules with a definite handedness to force the molecules to twist along the required direction and not form areas of reverse twist. The molecule CB15 twists in the opposite direction to C15, and they are both short-pitch chiral nematics.

5.4 The physical properties of liquid crystals

The structure of the liquid crystal determines its physical properties, although the manufacturers have a certain amount of control. The transition temperature is the temperature at which a material enters and leaves a particular liquid crystal phase. The crystal to nematic transition temperature is written as T_{K-N}, and the nematic to isotropic liquid transition temperature can be written as T_{N-I} or T_c (critical temperature). There are various structural factors which affect the critical temperature:

- the terminal substituents R and R',
- the central component or bridge, X,
- the general length of the molecule,
- the nature of the molecular ring.

The structure of the molecular ring affects the transition temperatures due both to Van der Waals interactions between the molecules and to steric hindrance effects. Cyclohexane ring materials tend to have the highest transition temperatures, then bicyclooctane and benzene rings, in that order.

A sharp rise in the transition temperature is observed with increases in molecular length. This could occur, for example, through the insertion of benzene (or other) rings into the chain. However, the introduction of the lateral substituents into the benzene rings, or into the central bridge, generally breaks up the nematic structure and can lead to altered thermal stability.

The colour or absorption properties of liquid crystals are determined by the electronic properties of the particular molecule and depends on the distribution of double and single bonding. Where there is a conjugated chain of atoms (the bonding alternates between single and double bonding) delocalization of the π electrons occurs along the length. This induces an absorption of radiation at relatively long wavelengths. When there is little or no conjugation the optical absorption is caused mainly by the σ bonds and this is in the UV region so the liquid crystal appears colourless in the visible.

Any chain with a conjugated benzene ring has its absorption edge moved towards the visible region. Further lengthening of the conjugated chain increases the absorbed light in the visible region. By selecting suitable chromophores (a chromophore is a central conjugating group in a molecule, e.g. $-N=N-$) and auxochromes R and R' (the auxochromes are the push/pull substituents, e.g. NO_2, which enhance the strength of the absorption), coloured liquid crystals can be obtained.

Two types of non-thermal physical effects have been used for displays: 'dynamic scattering', caused by a weak applied current; and, more importantly, effects caused by an electric field. Examples of the electric field effects include twisted nematic, electrically controlled birefringence and cholesteric-to-nematic phase change displays.

5.5 Dynamics of nematic liquid crystals

We shall consider the use of nematic liquid crystals in a variety of devices and so at this stage it is useful to consider nematic liquid crystal dynamics and derive expressions for some of their key parameters. The continuum theory, proposed by Frank [5.5] and explained by de Gennes [5.6], shows that the details of the molecular structure

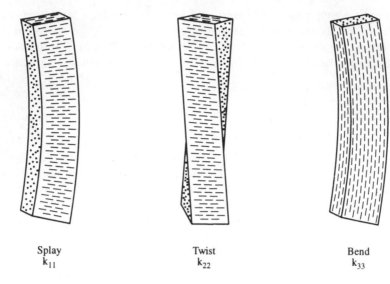

Splay
k_{11}

Twist
k_{22}

Bend
k_{33}

Figure 5.11 Splay, twist and bend in a nematic crystal.

can be disregarded and the system considered macroscopically. The system is described using the director $n(r)$ but allows for distortions imposed by the limiting surfaces of the sample (e.g. the container walls) and by external fields (magnetic and electric). For a weakly distorted system the local optical properties remain the same as those of a uniaxial crystal (see Section 6.1.3) and the magnitude of the molecular anisotropy is unchanged. However, the optical axis is rotated.

Two approaches have been developed, both of which use the coupling between the orientation and flow of matter within the liquid crystal:

1. a macroscopic approach based on classical mechanics (Ericksen [5.19], Leslie [5.20], and Parodi [5.21]), and
2. a microscopic approach developed by a Harvard group [5.22] using a study of correlation functions.

The two approaches give in essence the same results. The macroscopic approach assumes that in a nematic liquid crystal the dynamic situation of any point r, can be specified by a velocity field $v(r)$, which gives the flow of matter about the point r, and a unit vector (the director) $n(r)$, which describes the local state of molecular alignment. The dynamics of the system depends on the orientational effects introduced by the free energy, viscosity η, and elastic constants k_{11}, k_{22}, k_{33}. These constants describe the splay, twist and bend of the molecules, respectively, as illustrated in Figure 5.11.

We now obtain expressions for the response and decay times of a liquid crystal device, and the critical electric field and voltage needed to alter its state. The torque Γ on a molecule, due to a molecular field $\mathbf{h}$, is given by

$$\Gamma = \mathbf{n} \times \mathbf{h} \tag{5.3}$$

$\mathbf{n(r)}$ is the director, whilst $\mathbf{h}$ is related to the viscous stress in the liquid and arises from all other adjacent molecules.

In the work done by Leslie for incompressible nematics, a typical component of $\mathbf{h}$ can be expressed in the following form:

$$h_\mu = \gamma_1 N_\mu + \gamma_2 n_\alpha A_{\alpha\mu} \tag{5.4}$$

where γ_1 and γ_2 are components of the viscosity, n_α is a component of the director perpendicular to the viscosity components, N_μ is a component of the vector that represents the rate of change of the director with respect to the background fluid (i.e. $\partial \mathbf{n}/\partial t$), and $A_{\alpha\mu}$ is a component of the symmetric part of the velocity gradient tensor.

Note that the components of the viscosity around our molecule, γ_1 and γ_2, are in fact related to coefficients known as the Leslie coefficients of viscosity (of which there are six in total) [5.20]:

$$\gamma_1 = \alpha_3 - \alpha_2 \tag{5.5}$$

$$\gamma_2 = \alpha_2 + \alpha_3 = \alpha_6 - \alpha_5 \tag{5.6}$$

The tilt angle $\phi(z)$ is the angle between the long axis of each of the molecules and the director. The length z is the distance between a molecule and one of the glass plates. When an electric field is applied to the cell $\phi(z)$ varies across the cell and can be expressed as

$$\phi(z) = a \sin\left(\frac{\pi z}{d}\right) \quad \text{where } 0 \leqslant z \leqslant d \tag{5.7}$$

d is the fixed distance between the plates and it has been assumed that $\phi(0) = \phi(d) = 0$, i.e. at the cell walls the molecules are in a well-anchored state.

To calculate the rise or decay time of molecular motion we need to form an equation with ϕ expressed as a function of position, z, and time, t. We can consider a simplified case in which we assume that $k_{11} = k_{22} = k_{33}$ and that back-flow effects (i.e. the flow of liquid that accompanies the reorientation of the director) are disregarded. The torque on a molecule is given by Eq. (5.3) and by substituting Eq. (5.4) into (5.3) we can write the torque (for one component) as

$$\Gamma = \gamma \cdot \frac{\partial \phi}{\partial t} \tag{5.8}$$

It can be shown that the contribution to the torque on a molecule due to an applied electric field $\mathbf{E}$ can be expressed by the term

$$\varepsilon_a E^2 \cdot \sin \phi \tag{5.9}$$

while the contribution due to the viscous forces can be expressed by the term

$$k \frac{\partial^2 \phi}{\partial z^2} \tag{5.10}$$

The torque on a molecule can then be written as

$$\Gamma = \gamma \frac{\partial \phi}{\partial t} = k \frac{\partial^2 \phi}{\partial z^2} + \varepsilon_a E^2 \sin \phi \tag{5.11}$$

First consider the case where the electric field is suddenly switched off. This leads to an abrupt removal of the electric torque on the molecule and at $t = 0$, $E \to 0$. In this case Eq. (5.11) reduces to

$$\gamma \frac{\partial \phi}{\partial t} = k \frac{\partial^2 \phi}{dz^2} \tag{5.12}$$

The equation can be solved by using the technique of separation of variables, and we can write

$$\phi(z, t) = \phi(z) \cdot T(t) \tag{5.13}$$

The solution to Eq. (5.12) thus yields an exponential decay with time of the reorientation angle, ϕ, and hence can write

$$\phi(z, t) = \phi_0(z) \cdot e^{-t/\tau} \tag{5.14}$$

where τ is the decay time constant. Substituting this into Eq. (5.12) and using Eq. (5.7) gives

$$-\gamma \phi(z) \frac{1}{\tau} e^{-t/\tau} = -k \frac{\pi^2}{d^2} \phi(z) e^{-t/\tau} \tag{5.15}$$

Rearranging Eq. (5.15), an expression for the decay time constant is obtained:

$$\tau_{\text{decay}} = \frac{\gamma d^2}{k \pi^2} \tag{5.16}$$

It should be noted that the decay time is proportional to d^2. Substituting some values, $d = 10 \ \mu m$, $\gamma = 0.01$ N s m^{-2} and $k = 10^{-12}$ N, gives $\tau_{\text{decay}} = 0.1$ s. It is worth mentioning here the units that are sometimes used. Viscosity can be stated in poises (P), where 0.1 N s m^{-2} = 1 P and k is sometimes given in dynes, where 1 N = 10^{-5} dynes.

Now consider the case where the electric field has just been switched on. In this case, assuming that ϕ is small, so that $\sin \phi \sim \phi$, Eq. (5.11) reduces to

$$\gamma \frac{\partial \phi}{\partial t} = k \frac{\partial^2 \phi}{\partial z^2} + \varepsilon_a E^2 \phi \tag{5.17}$$

By using the technique of separation of variables again we find the rise time τ_{rise}, which gives the time taken for the director axis reorientation angle ϕ, to change by $1/e$ its original value:

$$\tau_{\text{rise}} = \gamma \left(k \left(\frac{\pi^2}{d^2} \right) - \varepsilon_a E^2 \right)^{-1} \tag{5.18}$$

The critical electric field, E_{crit}, is the field strength required for the electric torque to overcome the viscous force on a molecule. The critical electric field strength is given by Eq. (5.18) as

$$E_{\text{crit}} = \frac{\pi}{d}\left(\frac{k_{ii}}{\varepsilon_a}\right)^{1/2} \quad \text{where } i = 1, 2 \text{ or } 3 \tag{5.19}$$

Since

$$V = E.d \tag{5.20}$$

by definition, this threshold can also be defined by a voltage,

$$V_{\text{crit}} = \pi\left(\frac{k_{ii}}{\varepsilon_a}\right)^{1/2} \quad \text{where } i = 1, 2 \text{ or } 3 \tag{5.21}$$

The rise time τ_{rise} given in Eq. (5.18) can thus be written in terms of the threshold voltage as

$$\tau_{\text{rise}} = \frac{\gamma d^2}{\pi^2 k}\left[\left(\frac{V}{V_{\text{crit}}}\right)^2 - 1\right] \tag{5.22}$$

The movement of nematic liquid crystals in the presence of an electric field was first noted in nematic liquid crystals by Freedericksz *et al.* [5.23]. The free energy of the molecular arrangement for nematic liquid crystals has a minimum in the presence of an external field for a well-defined orientation of the director relative to the field. This is due to the anisotropy of the dielectric susceptibility, ε_a. As already shown in Section 5.2, when ε_a is positive the director will tend to align itself along the field. For negative values the alignment is perpendicular to the field direction. An applied field that is strong enough to counter the elastic forces of the liquid will reorientate the molecules to the state of lowest free energy.

5.6 Physical effects and their application to devices

In this section we consider how a range of physical effects exhibited by various liquid crystal materials can be used in different ways in order to produce devices.

5.6.1 Dynamic scattering devices

Dynamic scattering is an effect that causes a nematic liquid crystal, having negative ε_a, to scatter light [5.17]. There is an anisotropic ionic conductivity σ associated with the shape of the molecules and the presence of ions. This produces flow loops and turbulence in the liquid crystal when a current is passed, which causes the refractive index to change, and the material scatters the incident light instead of being transparent, as it was in its surface-aligned off-state texture. The effect is illustrated in Figure 5.12.

The liquid crystal is doped with impurities to give a resistivity of between 10^9–10^{10} Ω cm. The effect is produced by applying a current of around 10 $\mu A/cm^2$ with a voltage of approximately 20 V. A low-frequency alternating current is used to avoid causing electrolysis to occur. The effect is quickly saturated and has moderate response times ($\sim$100 ms turn-on time and $\sim$500 ms turn-off time).

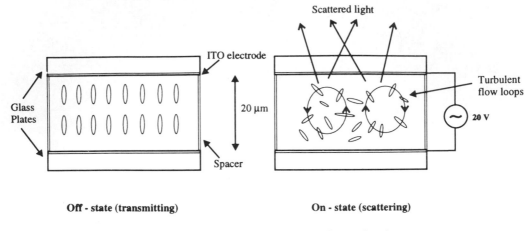

Figure 5.12 Device using dynamic scattering (nematic liquid crystal with $\varepsilon_a < 0$).

5.6.2 Devices using electrically controlled birefringence (ECB)

Application of an electric field which re-orientates the molecules and changes the birefringence, i.e. changes the refractive index with respect to direction of polarization. We first consider the case for which the dielectric anisotropy of the nematic liquid crystal material is negative ($\varepsilon_a < 0$) and then the case for which it is positive ($\varepsilon_a > 0$).

Case 1: Nematic liquid crystal cell with $\varepsilon_a < 0$
This is an effect which uses the applied voltage to tilt the liquid crystal molecules, producing a combination of splay and bend effects (see Figure 5.11). For this effect to be produced there is a high voltage threshold (called the Freedericksz threshold) which must be exceeded. At this threshold the applied electric force is equal to the elastic resistance of the liquid to reorientation of Eq. (5.22). Below this threshold no tilt or splay of the molecules can occur across the cell. Near this threshold the response times of the liquid crystal are slow (around 2 s) but can be rapid when the voltage is changed between two values well above threshold. Any voltage above threshold increases the tilt and splay effects in the liquid crystal cell, which in turn increase the refractive index (strictly, the effective birefringence or the extraordinary refractive index) and so the *optical path length* is effectively changed. This electrical control of the optical path length of the device allows the *phase* of a polarized light beam to be controlled. As a result these devices can be used as spatial light modulators; their use as pure phase modulators is illustrated in Figure 5.13.

The liquid crystal cell is often positioned between two crossed polarizers. When an electric field is applied across the device interference colours can be observed. The colour seen is dependent on the voltage applied across the cell. The nature of the polarization of light that can emerge through the system is dependent on the

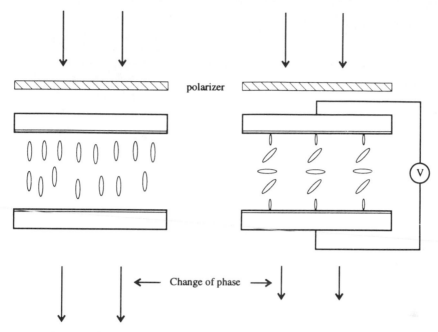

polarizer

← Change of phase →

Figure 5.13 Phase modulation using a nematic liquid crystal cell. Note that to define the direction of tilt uniformly it is necessary in a practical device to use *tilted* homeotropic alignment. This can be obtained by rubbing certain surfactant coatings unidirectionally or by coating a surfactant onto a rubbed polymer alignment coating.

retardation ($2\pi/\lambda \Delta nd$), which changes with wavelength. Only one or a few selected wavelengths will be rotated exactly through $90°$ at any time (producing a retardation of $\pi/2$, $3\pi/2$, $5\pi/2 \dots$). Usually the device is set up so that it can be switched between two voltages corresponding to two colours. The viewing angle is restricted since the observed colour depends upon the angle of incidence. It is possible to use this setup for projection displays but it is not suited to general flat displays (see Figure 5.14).

We can use the theory of the dynamics of nematic liquid crystals, introduced in Section 5.5, in order to write down expressions for the threshold voltage V_{crit} and rise and decay times τ_{rise} and τ_{decay} of a nematic liquid crystal with $\varepsilon_a < 0$. Using Eq. (5.21), the threshold voltage is

$$V_{\text{crit}} = \pi \sqrt{\frac{k}{(-\varepsilon_a)}} \tag{5.23}$$

k is a combination of k_{11} (the elastic constant for splay deformation) and k_{33} (the elastic constant for bend deformation). V_{crit} is around 3–6 V. When $V_{\text{app}} > V_{\text{crit}}$ the sample is deformed.

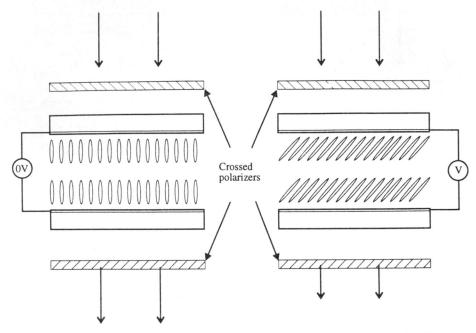

Figure 5.14 Using electrically controlled birefringence to switch a liquid crystal cell between two colours.

The rise and decay times for this case are given by Eqs. (5.22) and (5.16):

$$\tau_{\text{rise}} = \frac{d^2}{\pi^2} \frac{\eta_2}{k_{33}} \left(\left(\frac{V}{V_{\text{crit}}} \right)^2 - 1 \right) \tag{5.24}$$

$$\tau_{\text{decay}} = \frac{d^2}{\pi^2} \frac{\eta_2}{k_{33}} \tag{5.25}$$

where η_2 is another viscosity component.

Case 2: Nematic liquid crystal cell with $\varepsilon_a > 0$

This is an example of splay deformation. The anisotropy ε_a is positive. For this to be the case ε is greater than $\varepsilon_\perp$. Figure 5.15 illustrates this effect, where an undisturbed nematic texture is oriented such that the optical axis is parallel to the surface of the electrodes (5.15a). This orientation is changed by an applied electric field of sufficient strength (5.15b). To obtain reorientation a 'pre-tilt' of the surface is necessary (greater than 1°). This can be achieved by unidirectional rubbing of a polymer, such as polyimide, which gives the tilted homogeneous alignment required.

Under an applied electric field, as for the case where $\varepsilon_a < 0$, there are two competitive actions:

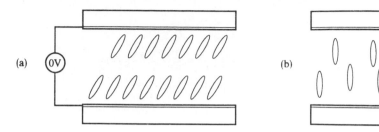

Figure 5.15 Splay deformation.

(i) the torque on each volume element tries to align the optic axis along the field perpendicular to the electrodes; and

(ii) interaction with the boundaries causes elastic torques that try to restore the molecules to the undisturbed state.

Below the threshold voltage V_{crit} (given by Eq. (5.21)) the restoring elastic torque is greater than the applied torque and no deformation occurs:

$$V_{\text{crit}} = \pi \sqrt{k_{11}/\varepsilon_a} \qquad (5.26)$$

where k_{11} is the elastic constant for splay deformation. The threshold voltage is around 1–5 V. When the applied voltage is greater than the threshold voltage, distortion occurs. The angle between the local optical axis and the applied field is a function of V and z, the coordinate perpendicular to the electrodes.

The deformation and relaxation of the molecular orientation is controlled by the viscosity. The rise and decay times are derived from Eqs. (5.22) and (5.16), in this case

$$\tau_{\text{rise}} = \frac{d^2}{\pi^2} \frac{\eta_{11}}{k_{11}} \left(\left(\frac{V}{V_{\text{crit}}}\right)^2 - 1 \right) \qquad (5.27)$$

$$\tau_{\text{decay}} = \frac{d^2}{\pi^2} \frac{\eta_{11}}{k_{11}} \qquad (5.28)$$

where η is the effective viscosity. For $d \sim 10$–20 μm; $\tau_{\text{rise}} \sim 10$–$30$ ms and $\tau_{\text{decay}} \sim 30$–$100$ ms at room temperature. This type of liquid crystal cell can be used in a manner similar to the cells used in case 1 (with $\varepsilon_a > 0$) in order to produce pure phase modulation, spatial light modulators or simple switching devices.

5.6.3 Twisted nematic devices: a second example of the 'field effect'

Twisted nematic devices are manufactured using two plates that are treated by rubbing, or with films, to give a planar or homogeneous texture so that the molecules adjacent to the glass plates will be well anchored along a particular direction determined by the direction of the microgrooves on the plate surface. The two plates are then

No field

Field applied

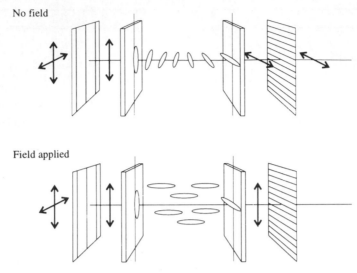

Figure 5.16 A twisted nematic device.

positioned so that their surface alignment directions are *orthogonal*. The cell is then filled with liquid crystal and sealed. The anchoring effect stops the reorientation of the outer molecules near the plate surfaces; the other molecules align themselves so that there is a gradual rotation of the molecules through 90° between the two plates. This produces a similar molecular orientation to the cholesteric ordering.

With no voltage applied to the device the plane of polarization of any incident radiation of wavelength much less than the thickness of the sample and linearly polarized parallel or orthogonal to the surface alignment direction, undergoes a rotation through 90°. Hence if the device is positioned between parallel polarizers there is no transmission. If the anisotropy ε_a of the molecules is positive, then an applied electric field will rotate the molecules in the central region of the cell until they are parallel to the field; then there will no longer be a rotation of the incident polarization and the assembly of cell and polarizers will appear transparent (see Figure 5.16).

When the field is switched off, the molecules reorientate themselves – the molecules in contact with the walls are not moved by the electric field and the intermolecular forces restore them to the 'off' state when the field is removed. This is a high-contrast field effect that requires about 5 V. Just as in the earlier case of electrically controlled birefringence there is a threshold voltage (independent of layer thickness) which must be exceeded if the effect is to be seen. The discrimination between the 'on' and 'off' states is good but there are moderate response times ($\sim$20 ms at 20°C). Pre-tilt in the centre of the layer is also used here to avoid reverse-tilt areas. Using the results of Eqs. (5.21), (5.22) and (5.16) we can write down expressions for the threshold voltage, rise times and decay times for such a device. In this case the dielectric anisotropy

$\varepsilon_a > 0$. The dynamics for this situation involves both the tilt and the twist angles of the director and so all three elastic constants appear in the expression for the threshold voltage:

$$V_{crit} = \pi \sqrt{\frac{k_{11} + (k_{33} - 2k_{22})/4}{\varepsilon_a}} \qquad (5.29)$$

The voltage threshold is normally between 3 and 5 V, but can be as low as 1 V. The rise and decay times for this case are

$$\tau_{rise} = \frac{\gamma d^2}{\pi^2(k_{11} + (k_{33} - 2k_{22})/4)} \left[\left(\frac{V}{V_{crit}} \right)^2 - 1 \right] \qquad (5.30)$$

$$\tau_{decay} = \frac{\gamma d^2}{\pi^2(k_{11} + (k_{33} - 2k_{22})/4)} \qquad (5.31)$$

Twisted nematic cell devices are widely used for displays generally and particularly in colour TV liquid crystal displays (LCDs, in which colour-selective polarizers are used with the twisted nematic cell). A large array of thin-film transistors is usually necessary to control the switching at each picture point (pixel).

5.6.4 Supertwisted nematic devices (STNs)

Conventional LCD devices have usually comprised an array of twisted nematic (TN) cells, each cell forming an individual picture element or pixel. However, there are some disadvantages of using a TN cell, particularly that of the TN cell having a poor selection nonlinearity. The pixels on a display are usually accessed by a matrix arrangement of row and column electrodes, so that all the pixels are interconnected (see Chapter 1). Hence a pixel is ideally required to transmit above a defined background voltage and absorb below it. As can be seen from Figure 5.17, for a TN device the light output versus applied voltage transfer curve shows a very gradual response, and hence a pixel may be half-selected (by an intermediate voltage level) between the 'on' and 'off' values (as it can be when matrix-addressed), and will also transmit some light instead of remaining off. There is a basic problem that the ratio between V_{select} and $V_{nonselect}$ tends rapidly to unity as the number of multiplexed electrodes increases.

TN displays are now being replaced by STN devices in which the inactivated liquid crystal rotates the plane of polarization of the incident light through 270° instead of the 90° of the twisted nematic. As a result the transmission as a function of applied voltage shows a much sharper rise (Figure 5.17) and so, even when matrix-addressed, a pixel will appear to be either transmitting (full on) or non-transmitting (full off). STNs give better contrast, greater viewing angles and the possibility of more rows and columns.

The disadvantages of STN devices are that they are more expensive, have slower response times, have a poor grey scale, lose a great deal of light within the display

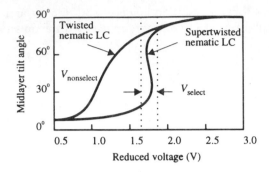

Figure 5.17 Transfer curve of the output light versus applied voltage is much sharper for supertwisted nematic LCs than for twisted nematic LCs.

(rectified by rear illumination) and have yellow and blue casts to the display (which can be avoided by adding one or more additional layers that compensate the light polarization in the opposite direction).

5.6.5 Hybrid field effect: 45° twisted nematic structure

The 'off' state is when the nematic structure is relaxed (Figure 5.18). The twist guides the input polarization plane through 45°, then a reflection occurs, and the light propagates back in the direction it arrived, and is rotated back to its original polarization state. The device is switched 'on' by using an AC signal which tends to concentrate the twist in the middle of the cell. This causes the polarization guiding to break down to give an elliptically polarized output which is then analyzed by the front polarizer to give optical contrast. This method is used in certain light valve systems in order to project the LCD image onto a screen.

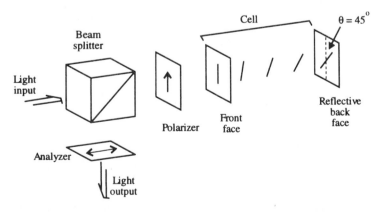

Figure 5.18 The 45° twisted nematic cell.

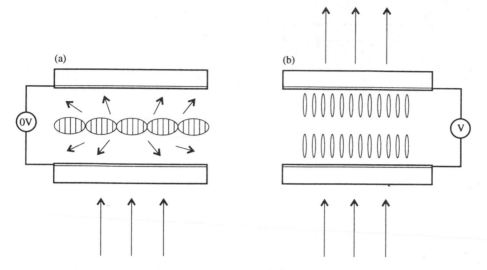

Figure 5.19 Cholesteric to nematic phase change device.

5.6.6 Cholesteric to nematic phase change devices

These devices are sometimes referred to as 'focal conic devices' (see de Gennes [5.6] and Blinow [5.18]) since they exhibit a scattering texture akin to a focal conic texture in their 'off' state. They use a cholesteric or chiral nematic material with $\varepsilon_a > 0$ which possesses a natural helical twist with a pitch of 1–20 μm. The molecules lie in planes perpendicular to the glass surface with the axis of the helical twist parallel to the plane of the electrodes (see Figure 5.19a). When the chiral nematic is placed in an electric field it is possible to unwind the helical structure so that a nematic structure is formed which is transparent, as illustrated in Figure 5.19(b).

This effect requires a positive anisotropy of the dielectric constant, ε_a, so that the molecules line up parallel to the applied alternating electric field. It is a true field effect with an intrinsic field threshold (i.e. the observed threshold voltage scales with the layer thickness). Above the threshold the speed of response is high when a voltage is applied. The effect is suitable for crossbar (matrix) drives but, with typical values being greater than 20 V for threshold, the use of chiral nematics is questionable. The short relaxation time also means that there is no integration effect over several pulses, but the turn-off time can become extremely long in the presence of a field close to the threshold voltage. They can, however, be matrix-addressed using other drive methods and are used in LCDs for overhead projectors.

5.6.7 Bistable devices

It is possible for certain devices to exhibit *memory* rather than simple 'on–off' characteristics. The devices contain a mixture of cholesteric and nematic liquid crystals

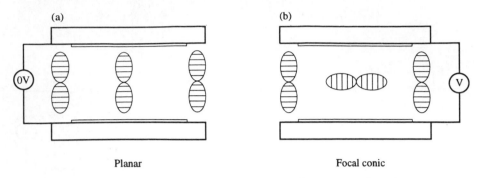

Planar   Focal conic

Figure 5.20 Cross-section showing the two stable states of a memoy effect device.

to give a medium with a long (~ 10 μm) helical pitch. The dielectric anisotropy ε_a is negative and a transition from planar to focal conic is used (Figure 5.20).

   The application of DC or low-frequency (e.g. 50 Hz) AC of ~ 20 V causes a small current to flow, which induces dynamic scattering. When switched off, the device remains in the strongly scattering focal conic state which is metastable and such a display thus has a memory function that can be stored for months. The stored memory is erased by applying a high-frequency (several kHz) AC voltage (100 V) to return the liquid crystal to a planar, clear texture. This effect is rather slow for MBBA/ EBBA (see Section 5.3). The effect has a threshold voltage and the discrimination between on and off is reasonable. The writing time is comparable to the time for dynamic scattering devices and the erase time is of the order of 1 s.

5.6.8 Ferroelectric liquid crystals

Ferroelectricity in liquid crystals was first predicted and demonstrated by R.B. Meyer

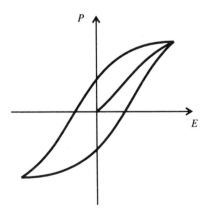

Figure 5.21 Polarization, *P*, of a ferroelectric material as a function of applied electric field, *E*.

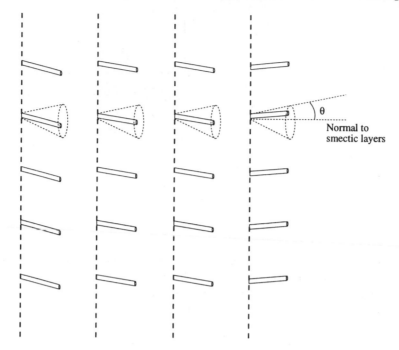

Normal to
smectic layers

Figure 5.22 Chiral-tilted smectic structure.

[5.24] in 1975. Ferroelectric liquid crystals exhibit a macroscopic dipole moment even in the absence of an external field. This is due to the molecules being tilted chiral and dipolar with the centres of the positive and negative charge being noncoincident.

Solid ferroelectric materials exhibit a hysteresis curve above a critical temperature T_c. Typical solid materials include titanates ($BaTiO_3$), TGS (triglycine sulphate), various niobates and KDP (potassium dihydrogen sulphate). All have domains of aligned dipoles like ferromagnets. Figure 5.21 shows a graph of the polarization of a ferroelectric material as a function of an applied electric field.

Liquid ferroelectrics were only discovered fairly recently. Since molecules in liquids are able to move they tend to align as antiparallel dimers; hence the dipole moments tend to cancel out and lead to antiferroelectricity. This prevents ferroelectric effects in high-symmetry cases such as nematic, cholesteric and smectic A liquid crystals. However, in some cases of biaxial crystals the symmetry is lower (this is true for smectic C* cases), giving rise to 'pseudo'-ferroelectric properties due to a degree of ordering of the dipole moment transverse to the molecules.

This type of ordering occurs for chiral molecules which are 'handed', with no mirror symmetry. Due to the strong coupling between the applied electric field and the spontaneous electric dipole the liquid crystals show a fast electro-optic effect,

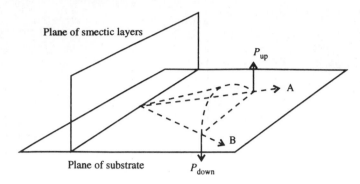

Figure 5.23 Surface-stabilized orientations of the smectic liquid crystal.

which is good for devices. They can also give bistable behaviour that allows a memory effect to be obtained even when the drive voltages are removed.

The tilt away from the smectic layer normal biases rotation along the long molecular axis, Figure 5.22. This gives a macroscopic polarization within the smectic layers if the molecules have a transverse dipole moment. The polarization P is perpendicular to the director $\mathbf{n}$ and the layer normal $\mathbf{v}$ so that

$$P_\perp = P_0 \mathbf{v} \times \mathbf{n} \tag{5.32}$$

Surface-stabilized ferroelectric liquid crystals (SSFELC's)
Most ferroelectric liquid crystal displays made so far have been of the surface-stabilized type. This is an arrangement that selects just two of many possible orientations in which the molecules can exist.

The liquid crystal is enclosed between two narrowly spaced plates about 1–2 μm apart (see Clark and Lagerwall [5.25]). The smectic layers are arranged to be normal to the plates. The plate surfaces stabilize the molecules to usually just two of an infinite number of orientations that are possible around the normal to the smectic layers, thus removing the twist of the smectic C*; this is shown in Figure 5.23. The plates used to form the cell walls are treated so that the smectic molecules will align themselves in the same plane as the plates. To achieve this, and at the same time follow the ordering of the smectic layer, they have to be in either position A or B, as shown. There then exists two optically different stable molecular configurations – 'bistability'. Using an electric field of the appropriate polarity the net polarization of a molecule can be switched 'up' or 'down' between the two orientations.

Switching can be achieved at microsecond rates, which is orders of magnitude faster than the response of traditional nematic liquid crystals. Since both orientations are stable the molecules remain in the switched state until a field of the opposite polarity is applied. Each pixel does not need periodic refreshing; this can lead to brighter displays with lower power consumption.

Table 5.1 Liquid crystal devices

	Dynamic scatterer	ECB	Twisted nematics	Cholesteric to nematic	Cholesteric memory	FELCs
Effect	Current	Field	Field	Field	Current/field	Field
ε_a	−ve	±ve	+ve	+ve	−ve	−ve
Volts (V)	20	5	5	100	20	
Current ($\mu A/cm^2$)	10	1	1	1	1	
Typical response times at 20°C	'on' 10–20 ms 'off' 100–500 ms	10 ms 200 ms	1 ms 200 ms	30 μs 100 ms	10–20 μs 1 s	~10 μs ~10 μs

Prior to the development of SSFELCs, displays were using the change in birefringence, Δn, in the presence of strong elastic restoring torques to alter the optical characteristics of the nematic or smectic liquid crystal material. With the SSFELCs a change in the direction of the optic axis occurs as the director switches between two equal energy states. This mechanism gives faster switching times since it is driven by a field in both directions. This effect has only recently been exploited, but it may be of great importance. Switching times are improved by three to four orders of magnitude and the material is optically bistable, which facilitates the production of larger displays with many switchable elements, all of which consume less power.

Table 5.1 provides a comparison of the liquid crystal devices discussed in this section.

5.7 Manufacture of liquid crystal devices

Twisted nematic devices are used in many everyday applications. Watch and calculator displays are a good example, where clear monochrome displays that are reliable and easy to manufacture are needed. The 90° twist is established by aligning the liquid crystal director correctly at the cell walls. The manufacture of these devices [5.26] can be summarized into five main stages and can be carried out in the laboratory as a 'do-it-yourself' exercise, as follows.

1. Coating one side of glass with indium–tin oxide (ITO). This is done by sputtering the ITO onto the glass using a dedicated evaporator. Alternatively, the glass can be bought ready prepared; it is usually specified in terms of its sheet resistance. For liquid crystals this is of the order of several kΩ per square. The ITO layer is 50–100 nm thick and has a transmission above 75–80%. It can be patterned by etching through a stencil of photoresist, or by etching through resist ink using hydrochloric acid.
2. Cleaning the glass. The glass is first cleaned using layers of tissue soaked in isopropanol or methanol, then rinsed in deionized water and blow-dried. Thereafter the glass is soaked in concentrated sulphuric acid for approximately one minute (not longer since it is important not to remove too much ITO), rinsed in deionized water and checked for cleanliness by observing how the water film behaves. Any tendency for the surface to repel water indicates that the glass is not clean. The glass must then be dried, which is usually performed using air currents.
3. Coating the glass and ITO with an alignment layer and preparing it if necessary. Two methods of coating are suitable. Silicon monoxide can be evaporated at 85° to the plate normal, hence providing alignment of the liquid crystal molecules in the plane of evaporation and a pretilt of 20°–25°. The silicon monoxide layer needs to be about 50 nm thick. Assembling the plates with the pretilts at 90° parallel avoids reverse twist problems.

   An alternative method for preparing large sheets of glass at lower cost uses rubbed polymer alignment layers. The cleaned glass is coated with polyvinyl alcohol (PVA) freshly dissolved in deionized water. The ITO glass sheets are dipped in the PVA solution and left to dry tilted so that any excess can run off. They can be left at 80°C in filtered moving air for a few hours and must be totally dry before the next stage. The alignment direction is defined by rubbing the plates three or four times unidirectionally with lens tissue. The direction of rubbing must be noted since it is important when the cell is constructed; for example, in order to give a uniform pretilt of $\sim 1°$ throughout the cell the rubbing directions of the two plates must be at 90° in the correct sense to match the twist of the chirally doped nematic liquid crystal (see Figure. 5.16).

Alternatively, the surface can be rubbed with a cloth using a suitable paste, but this is a less controlled method. The grooves formed on the surface have a width of about 1 μm, are around 50 nm deep and are spaced less than 10 μm apart.

4. Assembling the cell with the appropriate spacers. For a simple cell the plate separation should be about 10 μm thick. Mylar film, glass rods and glass beads can all be used as spacers. Evaporated layers and photoresists have also been used. The glass plates are arranged so that the ITO layers face inwards, and the rubbing directions are at 90° to each other. Once the cell has been arranged its edges are carefully and completely sealed using Torrseal except that one opening is left in order to allow the cell to be filled with a liquid crystal.

5. Filling the cell with the liquid crystal. The cell is placed in a good vacuum chamber and evacuated so that no air remains inside it. The opening is dipped into the liquid crystal and the chamber restored to atmospheric pressure so that the liquid crystal is forced into the cell. Once the cell has been filled it is removed and sealed. Electrical contacts with the ITO can then be made using a conductive paint.

5.8 Spatial light modulators

Various liquid crystal devices have been discussed in this chapter. One important device application of ferroelectric liquid crystals is in spatial light modulators (SLMs). These devices provide a method of impressing information onto a two-dimensional optical wave front. SLMs modify the phase, polarization, amplitude, and/or intensity of a spatial light distribution as a function of electrical drive information or the intensity of another light distribution. This can include photographic transparencies but usually involves a reusable medium that can alter the amplitude or phase of an output read beam proportional to the information with which it is addressed.

Figure 5.24 summarizes the function of a spatial light modulator. The optical properties of the SLM are controlled by some parameter which can be either optically or electrically varied using an electrode matrix or electron (e-) beams. Electrically addressed SLMs have the capability of converting serial electrical information such as video signals into parallel optical data. Electrical control can be easily interfaced with conventional computers and software. Optical addressing generally inputs a two-dimensional, parallel optical image, projected onto a photosensitive layer that controls the voltage applied locally. Some optically addressed spatial light modulators (OASLMs) are capable of wavelength conversion. With laser beam read they can

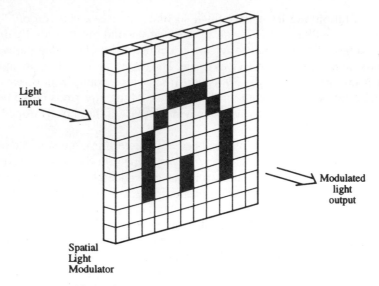

Figure 5.24 A matrix addressed spatial light modulator (SLM).

change an incoherent signal into a coherent readout. The electrical or optical control contains the object information which is transferred to the SLM. This information is then imparted onto the read beam .

The first major liquid crystal SLM was developed by Grinberg *et al.* [5.27] at the Hughes Research Laboratories, California, and is known as the Hughes liquid crystal light valve (LCLV); this is illustrated in Figure 5.25. It is commercially available and has been continually improved over a long period. It is an optically addressed system and when combined with a CRT becomes electrically addressed.

The 'writing' image falls onto a photoconducting photosensor, which in this case is made of cadmium sulphide, CdS. A positive bias on the CdS side of the device depletes the number of local electrons and decreases the fraction of the voltage dropped across it. Optical addressing produces charged pairs; these vary the voltage drop that occurs across the liquid crystal, thereby acting as a 'hybrid field effect' modulator. The magnitude of the voltage drop produced is regulated by the intensity of the writing light to give a grey scale capability. The optical properties of the liquid crystal are altered, and hence so is the intensity of the reading beam, with the possibility of amplification.

The LC light valves use nematic crystals with a negative anisotropy in a tilted homeotropic alignment. The dielectric mirror and absorber are used to isolate optically the writing and reading beams and protect the photoconductor from the intense read beam. The photoconverter can be chosen either to convert a visible, weak and incoherent image into a coherent amplified one, or to render an ultraviolet or an infrared image visible. Liquid crystals with thickness 4–12 μm can be used for colour

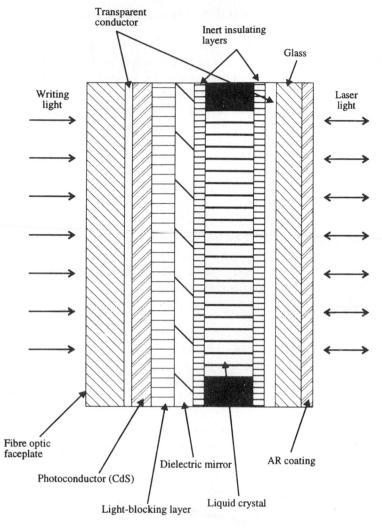

Figure 5.25 A Hughes liquid crystal light valve (layer thickess not to scale).

displays, those of 2 μm thickness are used in black and white projection TV displays.

These devices were very expensive to develop and are difficult to manufacture. The response time is slow for changes in the input intensity, about 10 ms, but there is good contrast and the display is equivalent to 10^5 effective pixels. The speed is marginal for TV projection displays but has been used in superb colour graphics projection units with three large lamps for such applications as military control rooms. Electron beam addressed liquid crystal light modulators [5.28] have been made, which have the potential to have very fast write times.

5.9 Active matrix liquid crystal displays (AMLCDs)

The crossbar address described in Section 1.4.3 is the basis of these devices. An AMLCD device is essentially an array of light valves each controlled by an active switching device such as a thin-film transistor (TFT). Each light valve can represent a pixel and it either transmits or absorbs any incident light from a source placed behind it depending upon the molecular configuration of the liquid crystal. The configuration of the liquid crystal molecules is controlled by the electric field applied across the crystal material. Thus the electric field controls the level of transmitted light.

Passive matrix LCD (PMLCD) devices (see 1.4.3) usually consist of a liquid crystal sandwiched between two glass substrates. Transparent electrode strips are positioned at the front and back of the liquid crystal in order to modulate the electric field across the liquid crystal. The front strips are perpendicular to the back strips so that each intersection point of the front and back electrodes forms a pixel. The field at any pixel can then be modulated by addressing a particular row of the front electrodes and a particular column of the back electrodes.

A disadvantage of using PMLCDs arises from the high degree of interconnection between pixels. As a result, addressing a particular pixel to be 'on' or 'off' can also alter a neighbouring pixel if they are too closely spaced. This limits the resolution and contrast ratio.

AMLCD devices employ electronic switches at each pixel site; these ensure that each pixel is addressed more directly and reduces the likelihood of a pixel being wrongly addressed. This provides better resolution and contrast ratio. Also, since the charge stored in the pixel does not leak away through the transistor to the same extent, a longer time interval between refreshing scans can be tolerated and hence larger arrays addressed.

The switches for each pixel can be combinations of transistors, varistors or diodes. The preferred devices employ thin-film transistors at each pixel site and are known

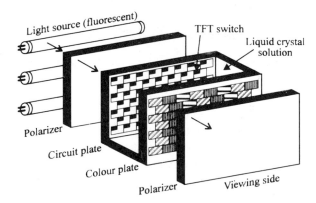

Figure 5.26 An AMLCD structure used to produce a colour display.

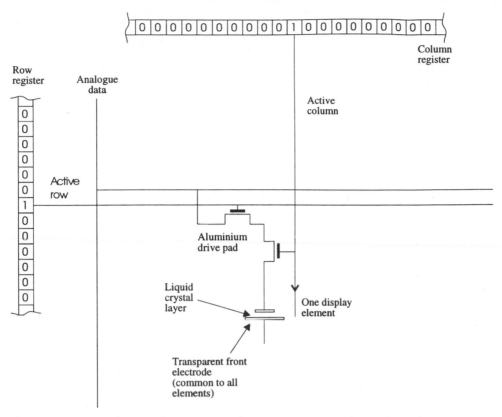

Figure 5.27 Circuit showing the connection between two FETs and a single pixel.

as TFT LCDs. Figure 5.26 shows a 3-D picture of a typical AMLCD (or TFT LCD) that could be used to produce a colour TV display.

Another form of AMLCD consists of a liquid crystal sandwiched between a front glass panel and a large silicon integrated circuit. The glass is coated with a transparent conductor such as ITO, which forms the front electrode of the cell. The rear electrodes are essentially an array of metal pads deposited onto the integrated circuit, each pad having its own switching transistors. Each pixel is provided with two field-effect transistors (FETs) in series. It is only when both FETs are 'on' that a sufficiently high voltage difference is produced across the pixel to induce switching. For both the FETs on a pixel to be switched on they must lie at the intersection of a selected row and column, as shown in Figure 5.27.

A nematic liquid crystal layer can be used to switch between a state which rotates the plane of polarization and one which does not. This, in combination with two polarizers, can produce transmitting and absorbing states. A ferroelectric liquid crystal layer may also be used. Its bistable property allows a particular pixel state (i.e. 'on' or 'off') to be sustained without the continuous supply of an electric field across the pixel. As a result, the display will have lower power requirements and will produce less flicker.

An AMLCD, such as the one described above, can also be used (without a colour filter matrix) as an electrically addressed SLM. In order to do so, an antireflection coating is evaporated onto the front glass panel and it is desirable to minimize all artificial and natural irregularities on the surface of the silicon integrated circuit (IC) – a process known as 'planavization'. We now describe some recent developments in this application of liquid crystals.

5.10 Silicon backplane liquid crystal SLMs

This type of device was first built by STL [5.29]. A liquid crystal about 6 μm thick is enclosed between a front glass panel and a large silicon integrated circuit. The front electrode of the liquid crystal is a transparent conductor that coats the glass panel. The back electrodes are deposited on the integrated circuit in the form of a 192 × 300 array of aluminium pads. Each is connected to a separate switching transistor. The array elements are connected with two field-effect transistors (FETs) as described in Section 5.9. These form the connection between the video signal and the aluminium pads. When both pairs of FETs are switched on an analog voltage is applied by the aluminium pad to activate the liquid crystal.

As an example of such devices, at the University of Edinburgh, UK, a liquid crystal over silicon SLM was developed using a 6 μm nMOS process to fabricate the silicon backplane. The device consists of a 16 × 16 array of pixels spaced 200 μm apart, with each pixel having a flat aluminium pad 100 × 100 μm that acts as a mirror/electrode. The microcircuitry around the mirror has nine transistors which form a gated static memory element with drive circuits for the overlying layer of a nematic liquid crystal. In this way a stable optical output, isolated from the changes at the electrical inputs, is obtained. The SLM has been used as the active switching element in a 16 × 16 cross-bar switch (Figure 5.28), on a project to realize optically connected parallel microprocessors.

The 16 × 16 pixel device, like its 128 × 128 pixel successor [5.30], is addressed as a static random access memory for binary signals, whereas the STL device described above has dynamic addressing for analog video signals. Further work has produced a dynamically addressed binary SLM. This ferroelectric liquid crystal device has a 5.3 × 5.3 mm active array of 176 × 176 pixels with a single transistor at each, and the array is addressed through four shift registers. Positive and negative versions of each image are displayed sequentially to ensure voltage balance, thereby preventing electrochemical degradation of the liquid crystal. The ferroelectric liquid crystal is optically bistable and the image update time, including the voltage balancing frame, is 1 ms [5.31].

Various liquid crystal configurations have been constructed over VLSI silicon backplanes. By exploiting the anisotropic absorption in dichroic dyes, amplitude

Figure 5.28 A 16 × 16 cross-bar switch.

modulation has been achieved using 'guest–host' liquid crystal mixtures. In this case, the light propagating through the mixture with the polarization vector parallel to the long axes of the dye molecules is absorbed preferentially. The strong interactions between molecules of the dye (guest) and the liquid crystal (host) cause the dye molecules to align in an ordered manner throughout the cell so that amplitude modulation can be controlled by driving the liquid crystal orientation. Precise phase-only modulation has also been demonstrated using the electrically controlled birefringence effect in parallel nematic cells and polarization modulation using the hybrid field effect in twisted nematic cells as discussed in earlier sections. Recently, high-speed ferroelectric materials have been used to demonstrate a display with intensity modulation exhibiting 63 grey levels and frame at rates greater than 150 Hz. This is achieved by decomposing the primary image into six binary weighted basis images, which are then exposed at constant intensity for appropriate times or illuminated for fixed durations by light flashes of appropriate intensity.

Of significance for optical systems is the ability to create photodiodes and phototransistors on silicon backplanes. SLMs with photodetectors at each pixel have been developed, and are being used to demonstrate optical image processing at the

University of Colorado, Boulder, USA. Several major international programmes are currently working to develop smart electrically and optically addressed SLMs. These devices can have photodetectors, microcircuitry and modulators at each pixel, and may also have interpixel communication between neighbouring pixels in the micro-circuitry on the backplane. Recent developments are reviewed in Ref. [5.32].

Most of the research is now concentrating on the exploitation of the functionality that can be built into the silicon backplane to produce smart SLMs. However, several issues need to be addressed to improve the optical performance of the liquid crystal light modulating layer. To date, most silicon foundries have optimized the processing to guarantee high electrical performance, together with a high yield of working devices. Little or no consideration has been given to obtaining the surface quality required if the device is to be used in an optical system, especially one using coherent light. Under the auspices of the Scottish Collaborative Initiative on Optoelectronic Sciences (SCIOS), the Edinburgh University Group has investigated the optical quality of fully processed wafers obtained from several silicon foundries. It has been observed that the metal mirrors have hillocks and depressions, or dimples, whose formation can be attributed to the differential thermal expansion between the metal and the substrate during the sintering stage of the fabrication process.

The rough structures on the surface of the metal mirrors absorb and scatter light and they also disrupt the liquid crystal alignment causing a reduction in the contrast ratio between the optical 'on' and 'off' states. The hillocks may also be sufficiently large ($\geqslant 2$ μm) to prevent the fabrication of liquid crystal cells with precisely defined thicknesses (e.g. cells tuned to multiple half-wavelengths of the light) and may also cause nonuniformities to occur over the array. By depositing a constraining layer of silicon oxide (> 250 nm thick) over the metal surfaces before sintering, the initial high optical quality of the metal surfaces can be maintained. Also by depositing layers of silicon oxide about 3 μm thick and polishing the surface of the oxide thereafter, planarized surfaces within mirror finishes may be obtained. This enables the mirror electrodes to be placed above the circuitry, thereby improving the optical efficiency of the device [5.33]. Planarization of the silicon backplanes is essential if the full potential of well aligned liquid crystal structures is to be realized.

By combining the advanced technologies of silicon fabrication and liquid crystal displays powerful processing devices are now being fabricated. Smart SLMs are expected to find applications in telecommunication switching networks, optical implementations of neural networks, image processing and research on optics in computing systems.

5.11 Polymer dispersed liquid crystals (PDLCs)

These materials consist of small spheres of liquid crystals (microspheroids) embedded in a polymer sheet. Normally the microspheroids scatter the light incident upon them, but when an electric field is applied across the sheet it becomes transparent. Since such a device does not require polarizers it is potentially more efficient in using light than conventional liquid crystal device structures. PDLCs can also be made with a dichroic dye dissolved in the liquid crystal microspheroids; they can use the guest–host interaction effect to give a coloured 'off' state which becomes colourless on applying a suitable voltage. PDLCs can also be manufactured in large sheets and then cut and formed into the required shape. However, complex PDLCs do need to be operated using an active matrix addressing system as each pixel needs to be addressed individually. Operating voltages are in the range 10–100 V, significantly higher than those required for other liquid crystal devices. PDLC materials are currently used in large signs and as colour switching overlays on top of membrane switches. In both of these applications their appearance is attractive and they are inexpensive. PDLCs have recently been used at Heriot-Watt University to demonstrate electrically controlled holograms. Application of the electric field alters the effective refractive index of the spheroids, and hence the period, thus redirecting the light beams.

5.12 Conclusions

In the future we expect LCDs to play an increasing role in both commercial and military applications. In September 1993 the *Japan Times* forecast sales of £5 billion in 1995, rising to £12 billion by the year 2000. As flat display panel devices improve, it is feasible that they may be mounted on walls, like paintings, and may incorporate additional features such as built-in touch panels to enable the user to interact directly with the data on screen. These panels will save a great deal of space and will be flicker-free.

The most important attributes of a display device are contrast, grey scale, colour, speed, resolution, size and addressability. At the present time it is difficult to produce a display with all of the above-mentioned attributes. In many respects the cathode ray tube is still superior to many flat-panel displays and so serves as a benchmark against which the performance of new displays can be tested. Some of the shortfalls of LCD devices are: insufficient switching times (typically in the millisecond range), the large dependence of contrast on viewing angle and difficulties in addressing the panel, particularly for large-scale displays.

Despite these problems, progress is still being made. The most widely used LCs to date have been TN devices, phase change devices and devices that incorporate 'guest' dyes in liquid crystal 'hosts'. There have also been a large number of variants of the TN device such as the STN cell. In the future ferroelectric smectic liquid crystals seem promising due to their high switching speeds ($\sim \mu$s) and bistability. Ferroelectric liquid crystals also give high contrast reproduction with only a slight dependence on viewing angle. However, it has been difficult until recently to achieve a uniform alignment of the molecules over the display area.

The switching mechanisms employed in a device should enable each pixel in the system to be addressed efficiently and accurately. Some devices rely on the nonlinear switching characteristics of the liquid crystal, although an increasing number use active matrix addressing (this is effective but the production procedure is complicated and expensive) and the yields of good displays are rather low, particularly for sizes over about 30 cm in the diagonal dimension.

Liquid crystals need not just be used in display systems; they have a variety of other uses such as in SLMs, electrically switchable colour filters, optical computing and signal processing, and optical memory. Clearly, liquid crystals, perhaps in conjunction with other technologies, will rank high in importance in optoelectronic technology in the years to come.

References

[5.1]   F. Reinitzer, 'Beiträge zur Kenntnisse des Cholesterins', *Monatsh. Chem.* **9**, 421–441 (1888).

[5.2]   O. Lehmann, 'Über fliessende Krystalle', *Z. Phys. Chem. IV*, **4**, 462 (1889).

[5.3]   O. Lehmann, *Ann. Phys.* (Leipzig) **2**, 649–705 (1900).

[5.4]   M.G. Friedel, *Ann. Phys.* **18**, 273 (1922).

[5.5]   F.C. Frank, *Disc. Faraday Society* **25**, 19 (1958).

[5.6]   P.G. de Gennes, *The Physics of Liquid Crystals* (Clarendon Press, Oxford, 1974).

[5.7]   G.W. Gray, 'The liquid crystal properties of some new mesogens', *J. Physique (Paris) Coll.* **36**, Cl-337 (1975); G.W. Gray, K.J. Harrison and J.A. Nash, *Electron. Lett.* **9**, 130 (1973).

[5.8]   I.A. Shanks, *The Physics and Display Applications of Liquid Crystals, Contemp. Phys.* **23**, 65–91 (1982).

[5.9]   I.A. Shanks, 'Liquid crystal displays: An established example of molecular electronics', *IEE Proc.* **130** 198–208 (1983).

[5.10] V. Tsvetkov, *Acta Physicochim. URSS* **16**, 132 (1942).

[5.11] M. Born, *Sitzungsber. K. Preuss. Akad. Wiss.* 614 (1916).

[5.12] W. Maier and A. Saupe, *Z. Naturf.* **A13**, 564 (1958).

[5.13] P. Chatelain, *Bull. Soc. Fr. Miner. Crist.* **66**, 105 (1943); D. Berreman, *Phys. Rev. Lett.* **28**, 1683 (1972).

[5.14] *Liquid Crystal materials: Product Information*, BDH Chemicals Ltd, Poole, UK.

[5.15] G.W. Gray, K.J. Harrison and J.A. Nash, *Electron. Lett.* **9**, 130 (1973).

[5.16] C. Hilsum, R.J. Holden and E.P. Raynes, 'A novel method of temperature compensation for multiplexed liquid crystal displays', *Electron. Lett.* **14**, 430 (1978).

[5.17] G. Heilmeier, L.A. Zanoni and L.A. Barton. *Proc. IEEE* **56**, 1162 (1968).

[5.18] L.M. Blinov, *Electro-optical and Magneto-optical Properties of Liquid Crystals* (Wiley–Interscience, New York, 1983).

[5.19] J.L. Erickson, *Arch. Ration. Mech. Analysis* **4**, 231 (1960).

[5.20] F.M. Leslie. *Q. J. Mech. Appl. Math.* **19**, 357 (1966).

[5.21] O. Parodi, *J. Physique* (Paris) **31**, 581 (1970).

[5.22] D. Forster, T. Lubensky and P. Martin,. *Trans. Faraday Soc.* **29**, 919 (1933).

[5.23] V.K. Freedericksz and V. Zolina, *Trans. Faraday Soc.* **29**, 919 (1933).

[5.24] R.B. Meyer, L. Liebert, L. Strzelecki and P. Keller, 'Ferroelectric liquid crystals', *J. Physique (Paris) Lett.* **36**, L69 (1975).

[5.25] N.A. Clark and S.T. Largerwell, 'Submicrosecond bistable electro-optic switching in liquid crystals', *Appl. Phys. Lett.* **36**, 899 (1980).

[5.26] D.J. McKnight, *Practical Notes on Liquid Crystal Cell Fabrication*, Heriot-Watt University, January 1992.

[5.27] J. Grinberg, A.D. Jacobson, W.P. Bleha, L. Miller, L. Francis, D. Boswell and G. Meyer, *Opt. Eng.* **14**, 217 (1975).

[5.28] A.C. Walker, S.D. Smith, R.J. Campbell and J.G.H. Mathew, *Opt. Lett.* **13,** 345 (1988).

[5.29] W.A. Crossland and S. Canter, *A Novel Approach to Large Flat Panel Displays: An Electronically Addressed Smectic Storage Device,* Soc. Info. Display, Int. Symp. Orlando, May 1985.

[5.30] D.J. McKnight, D.G. Vass and R.M. Sillitto, *Appl. Opt.* **28**, 4757–4762 (1989).

[5.31] I. Underwood, D.G. Vass, R.M. Sillitto, G. Bradford, N.E. Fancey, A.O. Al-Chalabi, M.J. Birch, W.A. Crossland, A.P. Sparks and S.G. Latham, 'A high performance spatial light modulator', *Proc. SPIE, Devices for Optical Processing* **1562**, 107–155 (1991).

[5.32] K.M. Johnson, D.J. McKnight and I. Underwood, *IEEE J. Quantum Electron.* **29**, 699–714 (1993).

[5.33] A. O'Hara, J.R. Hannah, I. Underwood, D.G. Vass and R.J. Holwill, *Appl. Opt.* to be published.

6

Crystal and nonlinear optics

In this chapter we start by presenting a basic introduction to linear and nonlinear optical effects in crystals. The electric displacement vector **D** is related to the electric field **E**, first in isotropic media, and then in anisotropic media, leading to a tensor relationship between **D** and **E**. We then consider electromagnetic wave propagation in a crystal and develop an equation for the optical indicatrix (OI). The OI relates the different refractive indices of the crystal with the directions along which they occur. The tensor relationship between **D** and **E** is examined and simplified by the use of *principal* dielectric axes. This relationship allows us to define a crystal as optically isotropic, uniaxial or biaxial. Liquid crystals (Chapter 5) are examples of anisotropic optical media.

The effect of applying an external electric field to some crystals is to change the dielectric permittivity (often referred to as the dielectric constant) and hence the refractive index along one or more of the crystal axes. We consider this effect carefully, and see how a new OI can be obtained for a specific case, quantifying the change in refractive index with applied electric field. This discussion is basic to the devices discussed in Chapter 7.

Finally, we consider the 'mixing' of optical frequencies in a crystal, which arises due to the nonlinear response of the medium to applied fields. The theory of second-order nonlinear optical effects can be developed from the electro-optic effect, described in Section 6.2, but we present a different approach. We start by deriving and then solving the wave equation for nonlinear media. This enables us to obtain solutions that describe how two waves of different frequencies interact within a crystal. We end by considering a variety of different cases of wave mixing. Some results are relevant to the subject of Chapter 8.

6.1 Birefringence and crystal optics

6.1.1 The dielectric tensor

In a vacuum, Gauss's theorem gives:

$$\text{div } \mathbf{E} = \frac{\rho}{\varepsilon_0} \tag{6.1}$$

where $\mathbf{E}$ is the electric field strength, ρ is the density of free charge and ε_0 is the permittivity of free space. In a dielectric medium one must take into account 'polarization charges'. The polarization of a medium is defined as

$$P = \text{dipole moment/unit volume}$$

that is,

$$P = Nqx$$

where N is the number of dipoles per unit volume, q is the charge on a dipole and x is the length of a dipole. Assuming that x is directly proportional to $\mathbf{E}$, we usually write

$$\mathbf{P} = \varepsilon_0 \chi \mathbf{E} \tag{6.2}$$

where the constant χ is called the susceptibility.

Another field can now be defined, which includes both the vacuum electric field term and the response of the medium; this is the electric displacement vector,

$$\mathbf{D} = \varepsilon_0 \mathbf{E} + \mathbf{P} \tag{6.3}$$

Equation (6.1) is then modified in a dielectric and is written as:

$$\text{div } \mathbf{D} = \rho \tag{6.4}$$

In an isotropic medium, Eqs. (6.2) and (6.3) give:

$$\mathbf{D} = \varepsilon_0 \mathbf{E} + \varepsilon_0 \chi \mathbf{E}$$
$$= \varepsilon_0 \mathbf{E}(1 + \chi) \tag{6.5}$$

or

$$\mathbf{D} = \varepsilon_0 \varepsilon_r \mathbf{E} \tag{6.6}$$

where ε_r is the relative permittivity $(1 + \chi)$, or the dielectric constant.

In an anisotropic medium the induced medium polarization is not necessarily parallel to the field direction, $\mathbf{E}$, i.e. the induced dipoles lie at an angle to $\mathbf{E}$. Now, if $\mathbf{P}$ and $\mathbf{E}$ are not parallel, then it follows from Eq. (6.3) that $\mathbf{D}$ and $\mathbf{E}$ will not be parallel. Thus Eq. (6.6), $\mathbf{D} = \varepsilon_0 \varepsilon_r \mathbf{E}$, will no longer be valid. Instead we must write down the effect of each component of $\mathbf{E}$ upon the three components of $\mathbf{D}$, and allow for different constants of proportionality (dielectric constants). However, this can be simplified by defining a tensor, ε_{kl}, which contains all the ε_r coefficients and we can write

$$\frac{1}{\varepsilon_0} \begin{pmatrix} D_x \\ D_y \\ D_z \end{pmatrix} = \begin{pmatrix} \varepsilon_{xx} & \varepsilon_{xy} & \varepsilon_{xz} \\ \varepsilon_{yx} & \varepsilon_{yy} & \varepsilon_{yz} \\ \varepsilon_{zx} & \varepsilon_{zy} & \varepsilon_{zz} \end{pmatrix} \begin{pmatrix} E_x \\ E_y \\ E_z \end{pmatrix} \tag{6.7}$$

where $\mathbf{D}$ and $\mathbf{E}$ have been represented as column vectors.

It can be shown that a property of ε_{kl} is that it is symmetric, i.e. $\varepsilon_{kl} = \varepsilon_{lk}$. In this case the matrix can be further reduced to just three nonzero components by choosing x, y and z to correspond to particular symmetry axes of the crystal field, known as the *principal dielectric axes*. Therefore

$$\varepsilon_{kl} \equiv \begin{pmatrix} \varepsilon_x & 0 & 0 \\ 0 & \varepsilon_y & 0 \\ 0 & 0 & \varepsilon_z \end{pmatrix}$$

where $\varepsilon_x = \varepsilon_{xx}$, etc.

Note that *anisotropic* media are, in general, crystalline, whereas amorphous materials such as glasses, gases and liquids are by nature isotropic.

6.1.2 Electromagnetic wave propagation through a birefringent medium

Consider an electromagnetic (EM) wave of the form

$$\mathbf{E} = \mathbf{E}_0 \exp\left[i\omega\left(t - \frac{n}{c}\mathbf{r}\cdot\mathbf{s} \right) \right] \tag{6.8}$$

This is the general equation for a travelling wave of amplitude E_0, giving a field at position $\mathbf{r}$ and at time t with $\mathbf{s}$ a unit vector along the phase propagation direction (i.e. parallel to the wavevector, and $\mathbf{k} = (\omega n/c)\mathbf{s}$).

Consider Maxwell's equation, $\nabla \times \mathbf{E} = -(\partial \mathbf{B}/\partial t)$. Given the spatial dependence of $\mathbf{E}$ (from Eq. (6.8)), the operator ∇ is formally equivalent to $-(i\omega n/c)\mathbf{s}$ and the time dependence, $\partial/\partial t$, is identical to $i\omega$. So this equation can be written as

$$-\frac{i\omega n}{c}\mathbf{s} \times \mathbf{E} = -i\omega\mathbf{B}$$

or

$$\mathbf{H} = \frac{n}{c\mu_0}(\mathbf{s} \times \mathbf{E}) \tag{6.9}$$

Similarly, using $\nabla \times \mathbf{H} = \partial \mathbf{D}/\partial t$ (assuming that there are no conductors present),

$$\mathbf{D} = -\frac{n}{c}(\mathbf{s} \times \mathbf{H}) \tag{6.10}$$

Now the energy flow is given by the Poynting vector: $\mathbf{N} = \mathbf{E} \times \mathbf{H}$ and therefore $\mathbf{N}$ is perpendicular to $\mathbf{E}$ and $\mathbf{H}$. From Eqs. (6.9) and (6.10) it can be seen that $\mathbf{s}$ is perpendicular to $\mathbf{D}$ and $\mathbf{H}$. So if $\mathbf{D}$ and $\mathbf{E}$ are not parallel, then neither are $\mathbf{s}$ and $\mathbf{N}$ (Figure 6.1).

Clearly the phase velocity is in a different direction to the energy flow (group velocity). It is this 'walk-off' effect that is responsible for the separation of the extraordinary and ordinary ray in double refraction. Using Eqs. (6.9) and (6.10) and eliminating $\mathbf{H}$ leads eventually to *Fresnel's equation:*

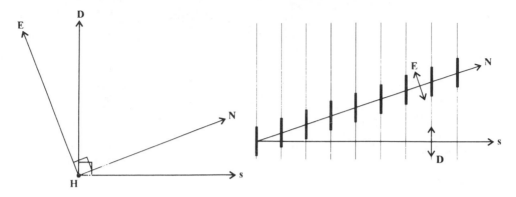

Figure. 6.1 The relative directions of vectors **N**,**E**,**H**,**s** and **D**.

$$\frac{s_x^2}{n^2 - \varepsilon_x} + \frac{s_y^2}{n^2 - \varepsilon_y} + \frac{s_z^2}{n^2 - \varepsilon_z} = \frac{1}{n^2} \tag{6.11}$$

Given a particular propagation direction, $\mathbf{s} = (s_x, s_y, s_z)$, and the dielectric tensor elements, $\varepsilon_x, \varepsilon_y, \varepsilon_z$, then this equation gives the relevant values of the refractive index, n. n^2 is given from a quadratic equation and so *two* positive solutions for n are possible. Further analysis shows that these solutions correspond to two independent (orthogonal) optical polarizations: the so-called *ordinary ray*, and the *extraordinary ray*.

6.1.3 The optical indicatrix (index ellipsoid)

From Fresnel's equation and further analysis it can be shown that the dependence of refractive index upon the propagation and polarization directions can be characterized by

$$\frac{x^2}{\varepsilon_x} + \frac{y^2}{\varepsilon_y} + \frac{z^2}{\varepsilon_z} = 1 \tag{6.12}$$

where x, y, and z refer to the direction of the **D** vector (not the propagation direction). This equation corresponds to a three dimensional surface, an ellipsoid (elliptical in each of the three coordinate planes) which is called the *optical indicatrix* (OI). The precise form of the surface depends on the actual values of ε_x, ε_y and ε_z, which in turn depend upon the crystal symmetry.

If we use the principal dielectric axes we can make the following summary:

$$\varepsilon\text{-tensor} = \begin{pmatrix} \varepsilon_x & 0 & 0 \\ 0 & \varepsilon_y & 0 \\ 0 & 0 & \varepsilon_z \end{pmatrix}$$

(a) $n_e > n_o$

(b) $n_e < n_o$

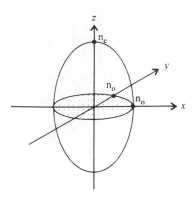

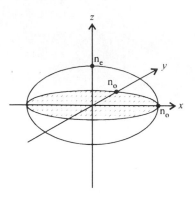

Figure 6.2 The index ellipsoid for: (a) $n_e > n_o$, a positive uniaxial crystal, and (b) $n_e < n_o$, a negative uniaxial crystal.

(i)   For isotropic media and a cubic crystal, $\varepsilon_x = \varepsilon_y = \varepsilon_z$ and there is no birefringence.
(ii) For hexagonal, tetragonal and triagonal crystals, $\varepsilon_x = \varepsilon_y \neq \varepsilon_z$ and so there is *uniaxial birefringence*.
(iii) For orthorhombic, monoclinic and triclinic crystals, $\varepsilon_x \neq \varepsilon_y \neq \varepsilon_z$ and so there is *biaxial birefringence*.

Consider the simplest form of birefringence, uniaxial, where $\varepsilon_x = \varepsilon_y = n_0^2$ and $\varepsilon_z = n_e^2$. In this case the optical indicatrix equation (6.12) can be rewritten as:

$$\frac{x^2}{n_0^2} + \frac{y^2}{n_0^2} + \frac{z^2}{n_e^2} = 1 \tag{6.13}$$

This represents an ellipsoid of revolution, with cylindrical symmetry about the z-axis. (Figure 6.2).

The optical indicatrix (OI) can be used to find:

(i) the two polarization directions, corresponding to propagation as either the ordinary or extraordinary rays; and
(ii) the refractive index values corresponding to these two polarizations, for a specific propagation direction.

This is done in the following manner (see Figure 6.3).

D_a always lies in the x–y plane and a is therefore always independent of θ and corresponds to the *ordinary ray* polarization direction. Hence $n_0 = a$. However, b is a function of θ, and is known as the *extraordinary ray* polarization direction. Hence $n_e(\theta) = b(\theta)$. It can be seen that the ordinary ray is polarized *perpendicular to the z-axis*, whereas the extraordinary ray is polarized *in the k/z plane*. The z-axis, along which propagation is independent of polarization, is known as the *optic axis*.

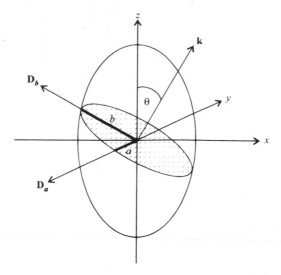

Figure 6.3 Using the OI to determine the ordinary and extraordinary ray functions.

6.1.4 Normal surfaces (index surfaces)

The variation of refractive indices deduced from an analysis of the OI can themselves be represented by 3-D surfaces called *normal surfaces* (Figure 6.4). n_0 is independent of the propagation direction and so the normal surface for the ordinary ray is simply a *sphere*. The angular dependence of $n_e(\theta)$ results in an *ellipsoidal* normal surface for the extraordinary ray. For a positive uniaxial crystal these surfaces are shown in Figure 6.4. In this case, the index values are given by the distance from the origin to

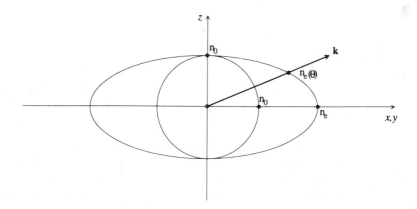

Figure 6.4 The normal surfaces for a positive uniaxial crystal.

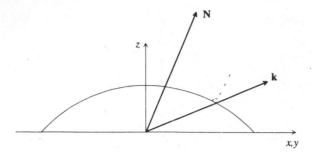

Figure 6.5 Use of the index surface to quantify 'walk-off' effects.

the surface, corresponding to the appropriate component of polarization, in the propagation direction, **k**. Referring back to the OI diagram (Figure 6.3), we can write

$$x^2 + y^2 = (b \cos \theta)^2$$

so that the equation for the e-ray normal surface can be written as:

$$\frac{1}{n_e^2(\theta)} = \frac{\cos^2\theta}{n_0^2} + \frac{\sin^2\theta}{n_e^2} \tag{6.14}$$

The normal surface can be used to quantify the 'walk-off' effect, described earlier, the Poynting vector, **N**, always being perpendicular to it (see Figure 6.5).

Because the n_0 surface is spherical, **k** and **N** are always parallel for the ordinary ray. The extraordinary ray is characterized by **k** and **N** not being parallel, except: (i) when propagating along the optic axis, z, where $n_e(0) = n_0$, and (ii) when propagating at 90° to the optical axis where $n_e(90) = n_e \neq n_0$.

6.2 The electro-optic effect

The application of an electric field to some crystals can cause small changes in the refractive indices. The change in symmetry associated with applying a field along a particular direction can make a non-birefringent medium become birefringent or a uniaxial crystal become biaxial.

When the changes are proportional to the applied electric field, **E**, then the effect is known as a *linear electro-optic effect* or *Pockels' effect.* Microscopically, the effect of electric field on refractive index is due to the 'Stark effect' on the interband absorption transitions which shift the continuum absorption of the solid at short wavelength. Application of the Kramers–Kronig relations then yields the refractive effect where we integrate the absorption over frequency (see Chapter 4, Eq. (4.8)).

6.2.1 The electro-optic tensor ✗

Conventionally, the electro-optic coefficient, r, is defined as relating the change in reciprocal dielectric constant, $1/\varepsilon$, to the applied field $\mathbf{E}$, that is,

$$\left[\frac{1}{\varepsilon}\right]_E - \left[\frac{1}{\varepsilon}\right]_{E=0} = \Delta\left(\frac{1}{\varepsilon}\right) = rE$$

Due to the anisotropy of the medium ε must be replaced by the dielectric tensor ε_{ij}, and consequently r must also be a tensor. So we write:

$$\Delta\left(\frac{1}{\varepsilon_{ij}}\right) = \sum_k r_{ijk} E_k \tag{6.15}$$

where i. j, k can all be x, y or z. r_{ijk} is the *electro-optic tensor*, and because it relates a second-rank tensor to a vector it is itself third rank (three subscripts).

Knowing that $\varepsilon_{ij} = \varepsilon_{ji}$, then it must follow that $r_{ijk} = r_{jik}$. Given that two of the subscripts are interchangeable in this way, there is a standard notation for contracting them into a single subscript. This is known as *matrix notation* and uses numerical subscripts such that:

$$ij : xx = 1, \quad yy = 2, \quad zz = 3, \quad yz = zy = 4, \quad xz = zx = 5, \quad xy = yx = 6.$$

This sequence comes from the matrix arrangement:

$$
\begin{array}{ccc}
xx & xy & xz \\
yx & yy & yz \\
zx & zy & zz
\end{array}
\Rightarrow
\begin{array}{ccc}
1 & 6 & 5 \\
6 & 2 & 4 \\
5 & 4 & 3
\end{array}
$$

The final subscript of r_{ijk} can also be given a number:

$$k : x = 1, \quad y = 2, \quad z = 3.$$

Using this notation r_{ijk} can be represented as a 6×3 matrix, and thus Eq. (6.15) can be written as

$$
\begin{pmatrix}
\Delta(\frac{1}{\varepsilon_1}) \\
\Delta(\frac{1}{\varepsilon_2}) \\
\Delta(\frac{1}{\varepsilon_3}) \\
\Delta(\frac{1}{\varepsilon_4}) \\
\Delta(\frac{1}{\varepsilon_5}) \\
\Delta(\frac{1}{\varepsilon_6})
\end{pmatrix}
=
\begin{pmatrix}
r_{11} & r_{12} & r_{13} \\
r_{21} & r_{22} & r_{23} \\
r_{31} & r_{32} & r_{33} \\
r_{41} & r_{42} & r_{43} \\
r_{51} & r_{52} & r_{53} \\
r_{61} & r_{62} & r_{63}
\end{pmatrix}
\begin{pmatrix}
E_1 \\
E_2 \\
E_3
\end{pmatrix}
\tag{6.16}
$$

The symmetry of r_{ijk}, in the same way as with ε_{ik}, reflects the symmetry of the crystal.

Consider a crystal with inversion symmetry. In such a medium nothing should change if the field is reversed. Thus

$$\Delta(1/\varepsilon_{ij}) = r_{ijk} E_k = r_{ijk}(-E_k)$$

This can only be true for $E \neq 0$ if $r_{ijk} = 0$. Thus *centrosymmetric crystals exhibit no linear electro-optic effect*. Non-centrosymmetric crystals usually have *some* directions that *do* show inversion symmetry and hence fields applied along them again will give no electro-optic (EO) effect. In other directions, which do not have inversion symmetry, there will be an EO effect. Symmetry also requires that certain tensor elements are equal to each other. These considerations permit the form of the r_{ijk} tensor to be predicted purely from the symmetry class into which the crystal falls.

The reason for defining r_{ijk} in terms of $(1/\varepsilon_{ik})$ is that this is the form in which the dielectric tensor comes into the equation for the optical indicatrix:

$$\frac{x^2}{\varepsilon_x} + \frac{y^2}{\varepsilon_y} + \frac{z^2}{\varepsilon_z} = 1 \qquad [6.12]$$

For the moment we must use the more general form of this equation, that is in which the principal dielectric axes are not used as the coordinate system.

$$\left(\frac{1}{\varepsilon_{xx}}\right)x^2 + \left(\frac{1}{\varepsilon_{yy}}\right)y^2 + \left(\frac{1}{\varepsilon_{zz}}\right)z^2 + 2\left(\frac{1}{\varepsilon_{yz}}\right)yz + 2\left(\frac{1}{\varepsilon_{xz}}\right)xz + 2\left(\frac{1}{\varepsilon_{xy}}\right)xy = 1$$

or, more simply,

$$\sum_{\substack{ij \\ =x,y,z}} \left(\frac{1}{\varepsilon_{ij}}\right)ij = 1 \qquad (6.17)$$

With an applied electric field the $(1/\varepsilon_{ik})$ elements will change in value, and hence the OI will be modified:

$$\sum_{\substack{ij \\ =x,y,z}} \left[\left\{\left(\frac{1}{\varepsilon_{ij}}\right) + \Delta\left(\frac{1}{\varepsilon_{ij}}\right)\right\}ij\right] = 1$$

Using Eq. (6.15) then gives

$$\sum_{ij} \left[\left(\frac{1}{\varepsilon_{ij}} + \sum_k r_{ijk}E_k\right)ij\right] = 1 \qquad (6.18)$$

This is the equation for the OI in the presence of an applied field E.

6.2.2 An example of the electro-optic effect

Consider the crystal KH_2PO_4, potassium dihydrogen phosphate, or KDP. This is transparent from about 200 to 1000 nm. The point group symmetry is $\bar{4}2m$ and the crystal is uniaxially birefringent. The r_{ijk} tensor has the form:

$$r_{ijk} = \begin{pmatrix} 0 & 0 & 0 \\ 0 & 0 & 0 \\ 0 & 0 & 0 \\ r_{41} & 0 & 0 \\ 0 & r_{41} & 0 \\ 0 & 0 & r_{63} \end{pmatrix}$$

It follows from Eq. (6.16) that

$$\Delta\left(\frac{1}{\varepsilon_1}\right) = \Delta\left(\frac{1}{\varepsilon_2}\right) = \Delta\left(\frac{1}{\varepsilon_3}\right) = 0$$

$$\Delta\left(\frac{1}{\varepsilon_4}\right) = r_{41}E_1, \qquad \Delta\left(\frac{1}{\varepsilon_5}\right) = r_{41}E_2, \qquad \Delta\left(\frac{1}{\varepsilon_6}\right) = r_{41}E_3,$$

Assuming the conventional axes are used (i.e. the principal dielectric axes for zero field) and noting that KDP is uniaxial, such that $\varepsilon_1 = \varepsilon_2 = n_o^2$, $\varepsilon_3 = n_e^2$, then the new OI, given by Eq. (6.18), is:

$$\frac{x^2}{n_o^2} + \frac{y^2}{n_o^2} + \frac{z^2}{n_e^2} + 2r_{41}E_1 yz + 2r_{41}E_2 xz + 2r_{63}E_3 xy = 1$$

It can be seen that there are, as a result of the applied field, new cross-terms and hence the new OI has axes that no longer correspond to the original crystal axes. To simplify this example, consider what happens if the field is directed along the optical axis, z, that is $E_1 = E_2 = 0$, $E_3 = E$. Then

$$\frac{x^2}{n_o^2} + \frac{y^2}{n_o^2} + \frac{z^2}{n_e^2} + 2r_{63}E_3 xy = 1$$

It turns out that if one transforms to a coordinate system x', y', z' related to x, y and z, as shown in Figure 6.6, then we can write

$$\left(\frac{1}{n_o^2} - r_{63}E\right)x'^2 + \left(\frac{1}{n_o^2} + r_{63}E\right)y'^2 + \frac{z'^2}{n_e^2} = 1$$

the cross-terms are removed and consequently x', y' and z' are the new principal dielectric axes, induced by E. The optical indicatrix is now of the form:

$$\frac{x'^2}{n_{x'}^2} + \frac{y'^2}{n_{y'}^2} + \frac{z'^2}{n_{z'}^2} = 1$$

where

$$\frac{1}{n_{x'}^2} = \frac{1}{n_o^2} - r_{63}E$$

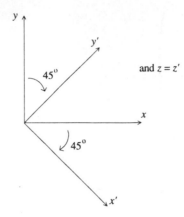

Figure 6.6 Transformation from the *x*, *y* and *z* coordinate system to a new coordinate system *x'*, *y'* and *z'*.

that is,

$$n_{x'} = n_o(1 - r_{63}En_o^2)^{-1/2}$$

thus

$$n_{x'} \approx n_o + \tfrac{1}{2}r_{63}En_o^3$$

Similarly,

$$n_{y'} \approx n_o - \tfrac{1}{2}r_{63}En_o^3 \qquad\qquad (6.19)$$

and

$$n_{z'} = n_e$$

It can be seen that, because $n_{x'} \neq n_{y'} \neq n_{z'}$, the crystal has become biaxially birefringent. If light is propagating down the optic axis, z, then the refractive index will change from being polarization *independent* to being polarization *dependent*. This is shown by the cross-section of the optical indicatrix, normal to the direction of propagation, becoming elliptical (Figure 6.7). This analysis gives the basis of a series of devices in the form of electro-optic modulators described in Sections 7.1.1–7.1.4.

6.3 Nonlinear optics

At low light intensities the optical properties of materials, as described in Section 2.1.2, are independent of intensity. When the electric field of the EM wave increases, such as in powerful laser beams, this is no longer the case and gives rise to a series of nonlinear optical effects.

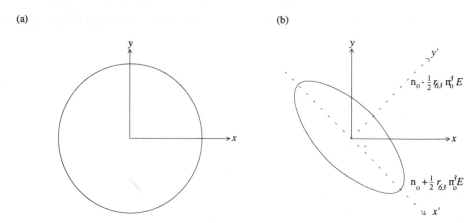

Figure 6.7 Cross-section of the optical indicatrix, normal to the direction of propagation. (a) No birefringence down original optic axis at zero field; (b) birefringence occurs due to applied electric field $E_z = E$.

All nonlinear systems can give frequency mixing. For example, a harmonic oscillator has a restoring force proportional to displacement from equilibrium and oscillates at a well-defined resonant frequency, ω. However, if an additional component is added to the restoring force, proportional to the displacement squared, then when driven at ω, frequency components will also be observed at 2ω (i.e. $\omega + \omega$) and at 0 ($\omega - \omega$).

In principle, the theory of (second-order) nonlinear optics can be developed from the electro-optic effect. Conventionally, however, it is presented differently.

6.3.1 Electromagnetic formulation of nonlinear interactions

Consider the response of a dielectric medium to an applied field. A polarization is induced, $P = Nqx$. In Section 6.1 we assumed that the dipole length $x \propto E$, and consequently that

$$\mathbf{P} = \varepsilon_0 \chi \mathbf{E} \tag{6.2}$$

The relation between $\mathbf{P}$ and $\mathbf{E}$ should not, in general, be restricted in this way. Eventually this simple proportionality will break down. This possibility can be included by using a polynomial expansion in which the higher-order coefficients are small, such that linearity is maintained, but only at low fields. For example, we can write the magnitude of the polarization in terms of powers of the electric field magnitude as

$$P = \varepsilon_0 \chi_1 E + \varepsilon_0 \chi_2 E^2 + \varepsilon_0 \chi_3 E^3 + \dots$$

where χ_1 is the linear susceptibility and χ_2 and χ_3 the second- and third-order nonlinear susceptibilities.

We now examine the effect of this modification upon light propagating through this medium, that is, when the optical field is large enough for the higher-order terms to be significant. Consider Maxwell's four equations, which determine the properties of light waves:

$$\mathbf{V} \cdot \mathbf{D} = \rho \tag{6.20}$$

$$\mathbf{V} \cdot \mathbf{B} = 0 \tag{6.21}$$

$$\mathbf{V} \times \mathbf{H} = \mathbf{J} + \frac{\partial \mathbf{D}}{\partial t} \tag{6.22}$$

$$\mathbf{V} \times \mathbf{E} = -\frac{\partial \mathbf{B}}{\partial t} \tag{6.23}$$

Note that **E**, **H**, **D** and **B** are all *real* variables, and functions of position, time and frequency. The polarization, **P**, comes into these equations within the electric displacement vector **D**:

$$\mathbf{D} = \varepsilon_0 \mathbf{E} + \mathbf{P} \tag{6.3}$$

Using the more general expansion of **P**, this gives

$$\mathbf{D} = \varepsilon_0 \mathbf{E} + \varepsilon_0 \chi_1 \mathbf{E} + \varepsilon_0 \chi_2 \mathbf{E}^2 + \varepsilon_0 \chi_3 \mathbf{E}^3 + \dots$$

$$= \varepsilon_0 (1 + \chi_1) \mathbf{E} + \mathbf{P}_{NL}$$

or

$$\mathbf{D} = \varepsilon_0 \varepsilon_r \mathbf{E} + \mathbf{P}_{NL} \tag{6.24}$$

where the dielectric constant, ε_r, refers to the linear part of the dielectric response. Using this new definition of **D** we now examine the effect that this has on the EM wave equation. The standard approach is to take the curl of Eq. (3.4) and use the relation curl . curl = grad . div − $\mathbf{V}^2$. Thus

$$\mathbf{V} \times \mathbf{V} \times \mathbf{E} = \mathbf{V}(\mathbf{V} \cdot \mathbf{E}) - \mathbf{V}^2 \mathbf{E} = -\frac{\partial}{\partial t} (\mathbf{V} \times \mathbf{B})$$

It can be shown that for most practical situations $\mathbf{V} \cdot \mathbf{E} = 0$, which gives

$$\mathbf{V}^2 \mathbf{E} = \mu_0 \mu_r \frac{\partial}{\partial t} (\mathbf{V} \times \mathbf{H})$$

We can now substitute for $\mathbf{V} \times \mathbf{H}$ using Eq. (6.22), together with Eq. (6.24). Therefore,

$$\mathbf{V}^2 \mathbf{E} = \mu_0 \mu_r \varepsilon_0 \varepsilon_r \frac{\partial^2 \mathbf{E}}{\partial t^2} + \mu_0 \mu_r \frac{\partial^2 \mathbf{P}_{NL}}{\partial t^2} \tag{6.25}$$

This is the wave equation for propagation in a nonlinear medium, where we have assumed that the conductivity **J** is zero. If $\mathbf{P}_{NL} = 0$ then this equation has a familiar plane wave solution for the one-dimensional case:

$$E = \mathrm{Re}(\mathscr{E} \exp \mathrm{i}(\omega t - kz))$$

where $\mathscr{E}$ is the amplitude, which can be complex, in order to include the relative phase term. Suppose a wave of this form enters a medium where $\mathbf{P}_{NL}$ is not zero. To solve this we assume that the effect of the nonlinearity is quantitatively weak and hence a perturbative approach is valid. We substitute this solution for $\mathbf{E}$ into the wave equation (6.25) and then determine the conditions that are imposed. To do this we must write out $\mathbf{P}_{NL}$ fully in terms of $\mathbf{E}$. We simplify the problem by considering second-order effects only, and by assuming that

$$\mathbf{P}_{NL} \propto \mathbf{E}^2.$$

This approximation explains frequency-mixing phenomena such as second harmonic generation and optical parameter oscillators (OPOs), which are discussed in Chapter 8. (The third-order effects, χ_3, proportional to E^3 describe *intensity-dependent* refraction and absorption, as well as stimulated Raman scattering, etc. The effects on refractive index discussed in Chapter 4 can be simply formulated in this way. Here we restrict the discussion to frequency-mixing effects.) Given that $\mathbf{P}_{NL}$ and the electric field are both vectors, and that we will be considering anisotropic crystalline materials, we must write this in tensor form:

$$(\mathscr{P}_{NL})_i = d_{ijk}\mathscr{E}_j\mathscr{E}_k$$

Following convention we have also written the nonlinear susceptibility tensor, d_{ijk}, such as to include ε_0, i.e. $d_{ijk} = \varepsilon_0 \chi_{ijk}^{(2)}$. We have defined d_{ijk} in terms of the field and polarization amplitudes $\mathscr{E}_j$, $\mathscr{E}_k$, $\mathscr{P}$, all of which are complex. Provided that we are working in a spectral region away from any absorption resonances, then we can write this relationship in terms of the instantaneous (time-varying) real fields $\mathbf{E}(t)$ and $\mathbf{P}(t)$. That is,

$$(P_{NL})_i = d_{ijk}E_jE_k \tag{6.26}$$

If this is valid, then in the presence of two oscillatory fields of frequencies ω_1 and ω_2, cross-terms will exist within the overall expression for $\mathbf{P}_{NL}$, since

$$\mathscr{P}_{NL} \propto E^{\omega 1}.E^{\omega 2} \propto \exp(i\omega_1 t).\exp(i\omega_2 t)$$

$$= \exp i(\omega_1 + \omega_2)t$$

there will also be terms like

$$\exp(-i\omega_1 t).\exp(i\omega_2 t) = \exp i(\omega_2 - \omega_1)t$$

New combination frequencies can thus be found in the oscillating medium polarization, and these will be radiated as new electromagnetic waves. In principle, there is no limit to the number of new frequencies because further mixing can occur; for example,

$$\omega_1 + \omega_2 \to \omega_3 \qquad \omega_3 + \omega_1 \to \omega_4$$

To proceed further we must simplify to the minimum number of frequencies compatible with the nonlinear process. In this type of second-order effect this corresponds to

just three interacting fields, with frequencies ω_1, ω_2 and ω_3 such that $\omega_3 = \omega_2 + \omega_1$.]
Instead of a single plane wave, we must thus consider three such waves. Choosing
the propagation direction to be along the z-axis, we can write these as:

$$\omega_1: \qquad E_{1i}(z, t) = \mathrm{Re}(\mathscr{E}_{1i}(z)\mathrm{e}^{\mathrm{i}(\omega_1 t - k_1 z)})$$

$$= \tfrac{1}{2}(\mathscr{E}_{1i}(z)\mathrm{e}^{\mathrm{i}(\omega_1 t - k_1 z)} + \text{c.c.})$$

$$\omega_2: \qquad E_{2k}(z, t) = \tfrac{1}{2}(\mathscr{E}_{2k}(z)\mathrm{e}^{\mathrm{i}(\omega_2 t - k_2 z)} + \text{c.c.}) \tag{6.27}$$

$$\omega_3: \qquad E_{3j}(z, t) = \tfrac{1}{2}(\mathscr{E}_{3j}(z)\mathrm{e}^{\mathrm{i}(\omega_3 t - k_3 z)} + \text{c.c.})$$

Where c.c. denotes the complex conjugates, and i, j and k are the usual general spatial
components. In this case with the vector **k** parallel to z they are restricted to x and
y only.

We want to find out how these three waves are related to each other, using the
nonlinear wave equation (6.25). When we thus expand $\mathbf{P}_{\mathrm{NL}}$ in Eq. (6.25) in terms of
E, we must use the other two frequencies. Using Eq. (6.26) we have

$$(P_{\mathrm{NL}})_i^{\omega_1} = 2d_{ijk}E_{2k}E_{3j} \tag{6.28}$$

The factor of two arises from the assumption that there are two separate fields (ω_2
and ω_3) and therefore we must include $E_2 E_3$ and $E_3 E_2$ terms. Substituting for **E**,
using Eqs. (6.27), and dropping terms in ($\omega_2 + \omega_3$) – a fourth frequency – gives the
component of the polarization at ω_1, due to the presence of ω_2 and ω_3, as

$$(P_{\mathrm{NL}})_{1i} = \frac{d_{ijk}}{2}(\mathscr{E}_{2k}^*\mathscr{E}_{3j}\mathrm{e}^{\mathrm{i}(\omega_1 t + k_2 z - k_3 z)} + \text{c.c.}) \tag{6.29}$$

We are now ready to expand the complete wave Eq. (6.25) in terms of the $\mathscr{E}$ amplitudes.
First consider the $\nabla^2 \mathbf{E}$ term, noting that for plane waves in the z-direction,
$\partial/\partial x = \partial/\partial y = 0$.

$$\nabla^2 E_{1i}(z, t) = \frac{\partial^2}{\partial z^2} E_{1i}(z, t)$$

$$= -\frac{1}{2}\left(k_1^2 \mathscr{E}_{1i}(z) + 2\mathrm{i}k_1 \frac{\mathrm{d}\mathscr{E}_{1i}(z)}{\mathrm{d}z}\right)\mathrm{e}^{\mathrm{i}(\omega_1 t - k_1 z)} + \text{c.c.} \tag{6.30}$$

Where the slowly varying amplitude approximation has been assumed:

$$\frac{\mathrm{d}\mathscr{E}_{1i}}{\mathrm{d}z}k_1 \gg \frac{\mathrm{d}^2\mathscr{E}_{1i}}{\mathrm{d}z^2}$$

After evaluating the two $\partial^2/\partial t^2$ terms on the right, using E_{1i} from Eqs. (6.27) and
$(P_{\mathrm{NL}})_{1i}$ from Eq. (6.29), the wave equation can be written as

$$-\left(k_1^2 \varepsilon_{1i} + 2\mathrm{i}k_1 \frac{\mathrm{d}\varepsilon_{1i}}{\mathrm{d}z}\right)\mathrm{e}^{\mathrm{i}(\omega_1 t - k_1 z)} + \text{c.c.}$$

$$= -\mu_0 \mu_\mathrm{r} \varepsilon_0 \varepsilon_\mathrm{r}(\omega_1^2 \mathscr{E}_{1i}\mathrm{e}^{\mathrm{i}(\omega_1 t - k_1 z)} + \text{c.c.}) - \mu_0 \mu_\mathrm{r} d_{ijk}(\mathscr{E}_{3j}\mathscr{E}_{2k}^* \omega_1^2 \mathrm{e}^{\mathrm{i}(\omega_1 t + k_2 z - k_3 z)} + \text{c.c.})$$

At this point we can drop the complex conjugate (c.c.) terms. They can be regarded as independently satisfied equations written in terms of $e^{-i\omega_1 t}$, instead of $e^{i\omega_1 t}$. It is then possible to cancel the $e^{i\omega_1 t}$ terms throughout, and by dividing by $e^{-ik_1 z}$ and also using $k_1^2 = \omega_1^2 \mu_0 \mu_r \varepsilon_0 \varepsilon_r$, we obtain:

$$\frac{d\mathscr{E}_{1i}}{dz} = -\frac{i\omega_1}{2} \sqrt{\frac{\mu_0 \mu_r}{\varepsilon_0 \varepsilon_r}}\, d_{ijk} \mathscr{E}_{3j} \mathscr{E}_{2k}^* e^{-i(k_3 - k_2 - k_1)z}$$

Similarly, the dependence of ω_2 and ω_3 upon the other two frequencies can be obtained to give:

$$\frac{d\mathscr{E}_{2k}}{dz} = -\frac{i\omega_2}{2} \sqrt{\frac{\mu_0 \mu_r}{\varepsilon_0 \varepsilon_r}}\, d_{kij} \mathscr{E}_{3j} \mathscr{E}_{1i}^* e^{-i(k_3 - k_1 - k_2)z} \tag{6.31}$$

$$\frac{d\mathscr{E}_{3j}}{dz} = -\frac{i\omega_3}{2} \sqrt{\frac{\mu_0 \mu_r}{\varepsilon_0 \varepsilon_r}}\, d_{ijk} \mathscr{E}_{1i} \mathscr{E}_{2k} e^{-i(k_1 + k_2 - k_3)z}$$

These three simultaneous equations describe how the three waves, ω_1, ω_2 and ω_3, interact as they travel through the nonlinear medium. They are the basic equations for many useful optical frequency-mixing processes. Before discussing these we must first consider the properties of the parameter that determines the strength of these effects, d_{ijk}.

6.3.2 The second-order nonlinear susceptibility tensor

The tensor d_{ijk} can be defined in terms of the complex amplitudes:

$$\mathscr{P}_i^{\omega_3} = d_{ijk}^{\omega_3 = \omega_2 + \omega_1} \mathscr{E}_j^{\omega_1} \mathscr{E}_k^{\omega_2} \tag{6.32}$$

As with the linear dielectric constant, d_{ijk} changes in magnitude with ω but only slightly over spectral regions remote from any absorption resonances. A single d_{ijk} tensor can thus describe a region of a particular nonlinear medium.

As with the tensor r_{ijk} to which it is related, d_{ijk} is only nonzero in non-centrosymmetric media, that is, in lower-symmetry crystals. Again this can be seen to be the case purely on symmetry grounds. If full inversion symmetry exists then $P(\mathscr{E}) = -P(-\mathscr{E})$; that is, $P_i = d_{ijk}\mathscr{E}_j\mathscr{E}_k = -(d_{ijk}(-\mathscr{E}_j)(-\mathscr{E}_k))$. Hence it would follow that $d_{ijk} = 0$.

It is also clear that there is nothing special about the order in which we consider the two field components, and so

$$\mathscr{E}_j \mathscr{E}_k = \mathscr{E}_k \mathscr{E}_j$$

and hence it follows that $d_{ijk} = d_{ikj}$. Thus, once again, we can use the matrix notation, for jk: $xx = 1$, $yy = 2$, $zz = 3$, $yz = zy = 4$, $zx = xz = 5$, $xy = yx = 6$. Equation (3.32) can then be written

$$\begin{pmatrix} P_1 \\ P_2 \\ P_3 \end{pmatrix} = \begin{pmatrix} d_{11} & d_{12} & d_{13} & d_{14} & d_{15} & d_{16} \\ d_{21} & d_{22} & d_{23} & d_{24} & d_{25} & d_{26} \\ d_{31} & d_{32} & d_{33} & d_{34} & d_{35} & d_{36} \end{pmatrix} \begin{pmatrix} \mathscr{E}_x^2 \\ \mathscr{E}_y^2 \\ \mathscr{E}_z^2 \\ 2\mathscr{E}_z\mathscr{E}_y \\ 2\mathscr{E}_z\mathscr{E}_x \\ 2\mathscr{E}_x\mathscr{E}_y \end{pmatrix}$$

Under some conditions it turns out that $d_{ijk} = d_{kij} = d_{jik}$ as well; that is, all subscripts can be permutated. This is known as *Kleinman's conjecture*, and was originally deduced by the inspection of actual tensor element values. It reduces the 18 independent values for the elements in the d_{ijk} matrix (6×3) to just 10 (for example, $d_{12} = d_{xyy} = d_{yyx} = d_{26}$, etc.).

Finally, further reduction in the complexity of the tensor results from the actual symmetry of the crystal to which it refers. The form of the d_{ijk} tensors is, as may be expected, identical to that of r_{ijk}, except that the columns and rows are exchanged. For example, for KDP ($\overline{4}2m$), d_{ijk} is given by

$$\begin{pmatrix} 0 & 0 & 0 & d_{14} & 0 & 0 \\ 0 & 0 & 0 & 0 & d_{14} & 0 \\ 0 & 0 & 0 & 0 & 0 & d_{36} \end{pmatrix}$$

and r_{ijk} is

$$\begin{pmatrix} 0 & 0 & 0 \\ 0 & 0 & 0 \\ 0 & 0 & 0 \\ r_{41} & 0 & 0 \\ 0 & r_{41} & 0 \\ 0 & 0 & r_{63} \end{pmatrix}$$

Since $d_{14} = d_{xyz}$ and $d_{36} = d_{zxy}$ we would expect them to be equal on the basis of Kleinman's conjecture, so that in all such $\overline{4}2m$ crystals, there will be only one independent value. The examples in Table 6.1 show that this is the case and that the conjecture is also valid for other examples of optically nonlinear crystals. Note that the units of d can be deduced from $P = Nqx = dE^2$, since dimensionally,

$$\frac{\text{C.m}}{\text{m}^3} = d\,\frac{\text{V}^2}{\text{m}^2}$$

which implies that the units of d are coulombs/volt^2.

In most practical situations the tensor equations containing d_{ijk} can be simplified to non-tensor form in which d_{ijk} is replaced by d_{eff}, the relevant element in the tensor (or combination of elements) corresponding to the specific geometry of the interaction. (This was done when discussing the electro-optic effect in Section 6.2.2, where r_{ijk} was replaced by r_{63}.)

Before leaving this discussion of the properties of d_{ijk}, it should be noted how the requirement of a non-centrosymmetric crystal for d_{ijk} to be nonzero leads directly to

Table 6.1

Crystals	Material	Common name	Second-order nonlinear susceptibility tensor (C/V^2)
$\bar{4}$2m	KH_2PO_4	KDP	$d_{14} = d_{xyz} = 3.9 \times 10^{-24}$ $d_{36} = d_{xyz} = 5.0 \times 10^{-24}$
	KD_2PO_4	KD*P	$d_{14} = d_{xyz} = 4.7 \times 10^{-24}$ $d_{36} = d_{xyz} = 4.7 \times 10^{-24}$
	$NH_4H_2PO_4$	ADP	$d_{14} = d_{xyz} = 5.5 \times 10^{-24}$ $d_{36} = d_{xyz} = 5.0 \times 10^{-24}$
mm2	$BaTiO_3$	Barium titanate	$d_{15} = d_{xxz} = 190 \times 10^{-24}$ $d_{31} = d_{zxx} = 200 \times 10^{-24}$ $d_{33} = d_{zzz} = 71 \times 10^{-24}$
3m	Ag_3AsS_3	Proustite	$d_{22} = d_{yyy} = 250 \times 10^{-24}$ $d_{36} = d_{zxy} = 150 \times 10^{-24}$

the various frequency-mixing processes, such as second harmonic generation.

First consider the potential within which a bound electron lies. Normally, $P = Nqx \propto E$ and hence the potential: $V(x) = \int E.dx \propto \int x.dx = x^2$. This gives the usual symmetric parabolic potential. However, when $P \propto E^2$ the potential corresponds to $V(x) = ax^2 + bx^3$, which is asymmetric. This asymmetry is fundamental to a second-order (χ_2) nonlinearity. Consider the simple model of an electron in an asymmetric potential well (Figure 6.8). Component x_2 consists of a DC shift *plus the second harmonic*. Mathematically, $x \propto E^2 \rightarrow (e^{i\omega t} + e^{-i\omega t})^2 \rightarrow e^0, e^{i2\omega t}$.

6.3.3 Three-wave frequency mixing

The set of three simultaneous Eqs. (6.31) describe all the possible interactions between collinear plane EM waves of frequency, ω_1, ω_2 and ω_3. These include:

[1] $\quad \omega_1 + \omega_2 \rightarrow \omega_3$

[2] $\quad \begin{aligned} \omega_3 - \omega_1 &\rightarrow \omega_2 \\ \omega_3 - \omega_2 &\rightarrow \omega_1 \end{aligned}$

[3] $\quad \omega_3 \rightarrow \omega_1 + \omega_2$

All these processes can occur and it depends on the relative irradiance levels of the three waves as to which actually dominates in any particular instance. (Note that this can also be generalized to noncollinear cases.) We will first consider the simplest example of process [1], sum generation, where $\omega_1 + \omega_2 \rightarrow \omega_3$. If we assume a low

The electron is 'easier to push' to
the left than to the right:

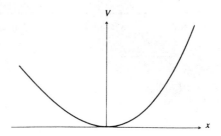

(a)   Thus a sine-wave driving force...

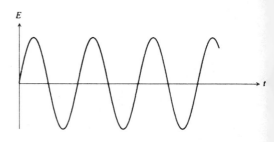

(b)   will cause asymmetric oscillation...

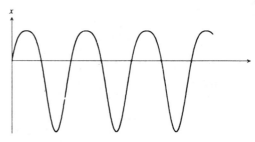

(c)   which can be analyzed into $x = x_1 + x_2$

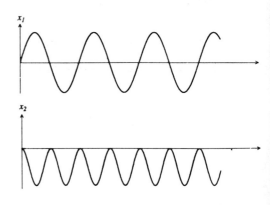

Figure 6.8 The notion of an electron in an asymmetric potential well.

efficiency of generation of ω_3 then very little power will be transferred from the two input waves, ω_1 and ω_2. Hence we can take

$$\frac{d\mathscr{E}_{1i}}{dz} \approx \frac{d\mathscr{E}_{2k}}{dz} \approx 0$$

That is, $\mathscr{E}_{1i}$ and $\mathscr{E}_{2k}$ are constant and consequently we are left with a single equation from (6.31):

$$\frac{d\mathscr{E}_{3j}}{dz} = -\frac{i\omega_3}{2}\left(\frac{\mu_0\mu_r}{\varepsilon_0\varepsilon_r}\right)^{1/2} d_{jik}\mathscr{E}_{1i}\mathscr{E}_{2k}e^{-i\Delta kz}$$

where we have written $k_1 + k_2 - k_3 = \Delta k$. This is easily integrated over a distance, L:

$$\mathscr{E}_{3j}(L) = -\frac{i\omega_3}{2}\left(\frac{\mu_0\mu_r}{\varepsilon_0\varepsilon_r}\right)^{1/2} d_{ijk}\mathscr{E}_{1i}\mathscr{E}_{2k}\int_0^L e^{-i\Delta k.z}dz$$

Thus the amplitude of the generated wave, ω_3, on leaving the crystal of length L, is:

$$\mathscr{E}_{3j}(L) = -\frac{\omega_3}{2}\left(\frac{\mu_0\mu_r}{\varepsilon_0\varepsilon_r}\right)^{1/2} d_{jik}\mathscr{E}_{1i}\mathscr{E}_{2k}\frac{(e^{-i\Delta kL}-1)}{\Delta k}$$

Now the irradiance (W/m^2) of a wave with complex amplitude, $\mathscr{E}$ (V/m), is given by:

$$I = -\frac{1}{2}\left(\frac{\varepsilon_0\varepsilon_r}{\mu_0\mu_r}\right)^{1/2}\mathscr{E}\mathscr{E}* \tag{6.33}$$

Thus the irradiance of ω_3 is given by:

$$I_{3j} = \left(\frac{\mu_0\mu_r}{\varepsilon_0\varepsilon_r}\right)\frac{\omega_3^2}{8}d_{jik}^2\mathscr{E}_{1i}\mathscr{E}_{1i}*\mathscr{E}_{2k}\mathscr{E}_{2k}*\frac{(e^{i\Delta kL}-1)(e^{-i\Delta kL}-1)}{\Delta k^2}$$

Applying (6.33) once more to write $\mathscr{E}_{1i}$ and $\mathscr{E}_{2k}$ in terms of their irradiances, and assuming we have worked out d_{eff} for the orientations of total fields $\mathscr{E}_1$ and $\mathscr{E}_2$, then:

$$I_3 = \frac{\omega_3^2}{2}\left(\frac{\mu_0\mu_r}{\varepsilon_0\varepsilon_r}\right)^{3/2}d_{\text{eff}}^2 I_1 I_2\frac{(2-e^{-i\Delta kL}-e^{i\Delta kL})}{\Delta k^2}$$

or

$$I_3 = \frac{\omega_3^2 L^2}{2}\left(\frac{\mu_0\mu_r}{\varepsilon_0\varepsilon_r}\right)^{3/2}d^2 I_1 I_2\left[\sin\frac{\Delta kL}{2}\bigg/\frac{\Delta kL}{2}\right]^2 \tag{6.34}$$

We can see that (within the limits of the initial assumption that the power in ω_1 and ω_2 is not significantly depleted) the new wave ω_3 has the following dependences:

$$I_3 \propto I_1 I_2$$

$$I_3 \propto \frac{d^2}{\varepsilon_r^{3/2}} = \frac{d^2}{n^3} \quad \text{(used as a figure of merit for the material)}$$

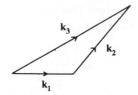

Figure 6.9 The phase-matching vector triangle.

$$I_3 \propto L^2 \left[\sin \frac{\Delta kl}{2} \Big/ \frac{\Delta kl}{2} \right]^2$$

which gives the length dependence with the second term in the product (sometimes called a sinc function) being critical. It could be close (or even equal) to zero and hence give no output at ω_3. If we want to maximize I_3, then the sinc function must have its maximum value of 1. I_3 would then simply increase in proportion to the square of the crystal length.

For sinc($\Delta kL/2$) = 1 we must have

$$\Delta k = k_1 + k_2 - k_3 = 0$$

and therefore

$$k_1 + k_2 = k_3 \tag{6.35}$$

This is known as the *phase-matching* condition, or wavevector requirement – the quantum mechanical equivalent to momentum conservation. It can be generalized to the case of noncollinear interactions and is shown in Figure 6.9:

$$\mathbf{k}_1 + \mathbf{k}_2 = \mathbf{k}_3 \quad \textit{conservation of momentum}$$

Noting that ω is proportional to photon energy, we also have

$$\omega_1 + \omega_2 = \omega_3 \quad \textit{conservation of energy}$$

Note that all three-wave processes governed by Eq. (6.31) are optimized when the phase-matching condition is satisfied.

6.3.4 Phase matching with birefringence

For collinear waves the phase-matching condition (6.35) is $k_1 + k_2 = k_3$ and can be written:

$$\omega_1 n_1 + \omega_2 n_2 = \omega_3 n_3$$

where $|k| = \omega n/c$. Given that $\omega_1 + \omega_2 = \omega_3$, this can only be true if $n_1 = n_2 = n_3$, that is, if there is no dispersion. Normal dispersion makes $n_3 > n_1, n_2$, and hence $\omega_1 n_1 + \omega_2 n_2 < \omega_3 n_3$, as shown in Figure 6.10.

A noncollinear geometry does not help either. To overcome this problem, we can exploit the birefringent properties of the crystal. This gives us two possible values of

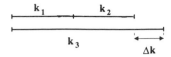

Figure 6.10 The relative sizes of wavevectors k_1, k_2 and k_3.

refractive index for each frequency and their difference can be large enough to compensate for the dispersion.

(a) Phase matching: second harmonic generation

When the 'two input waves' are the same single beam, then the sum frequency corresponds to the second harmonic: $\omega + \omega = 2\omega$. This case leads to a particularly simple condition for phase matching to occur: Instead of $\omega_1 n_1 + \omega_2 n_2 = \omega_3 n_3$, we can write $2\omega n^{(\omega)} = 2\omega n^{(2\omega)}$; that is, $n^{(\omega)} = n^{(2\omega)}$. Thus we only need to arrange that the fundamental and the second harmonic indices are equal. Consider the normal surfaces, which describe the variation of the o- and e-rays, for a negative uniaxial crystal. Dispersion effects will produce a different set of normal surfaces for every frequency. For a particular ω and its second harmonic, 2ω, these might look like those shown in Figure 6.11. Thus propagation of an ordinary ray at ω along this direction will efficiently generate a collinear 2ω beam with the orthogonal (e-ray) polarization. θ_m is known as the *phase-matching angle*.

(b) Phase matching: general three-wave interactions

This solution for second harmonic generation is unique. In general, for three-wave processes in which there are three distinct frequencies involved one must consider wavevector surfaces rather than index surfaces, since it is the k's that must add correctly.

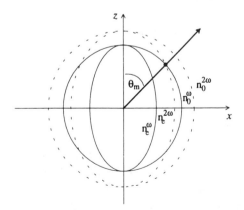

Figure 6.11 The index (normal) surfaces for a negative uniaxial crystal, at frequencies ω and 2ω.

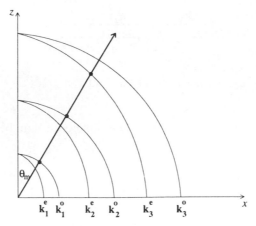

Figure 6.12 The '*k*-surfaces' for three different wavevectors $\mathbf{k}_1$, $\mathbf{k}_2$, $\mathbf{k}_3$ (only a single quadrant has been drawn).

The 'k-surface' can be derived from the index (normal) surface using $|k| = 2\pi n/\lambda$. With such surfaces it can be shown that birefringence can be used to overcome dispersion in these more general interactions as well.

Figure 6.12 shows that the angle, θ_m, can be chosen such that

$$k_1^o(\theta_m) + k_2^o(\theta_m) = k_3^e(\theta_m)$$

This corresponds to the phase matching of ω_1 and ω_2, as o-rays, with ω_3, as an e-ray. It is often possible to find another angle that also satisfies the phase-matching condition, with one of the low-frequency waves also an e-ray, so that

$$k_1^e(\theta_m') + k_2^o(\theta_m') = k_3^e(\theta_m')$$

In general two types of phase-matching are possible for each type of crystal:

- *Type I crystals*, in which the low-frequency waves have the same polarization:
  $k_1^o + k_2^o = k_3^e$ for the negative uniaxial case,
  $k_1^e + k_2^e = k_3^o$ for the positive uniaxial case.
- *Type II crystals*, in which the low-frequency waves have different polarizations:
  $k_1^e + k_2^o = k_3^e$ for the negative uniaxial case,
  $k_1^e + k_2^o = k_3^o$ for the positive uniaxial case.

Having chosen a particular phase-matching geometry, one can then determine what the d_{eff} is (in principle, it could be zero, which means thinking again about phase-matching alternatives). d_{eff} can usually be optimized by adjusting the other angle that fixes the final propagation direction, ϕ, while maintaining $\theta = \theta_m$ (Figure 6.13).

Given θ and ϕ and the specification of o- or e-ray for each beam, the geometry is fully defined. Table 6.2 gives $d_{eff}(\theta, \phi)$ for all the crystal classes. (It is assumed that Kleinman's conjecture does not hold. If it did, these formulae may simplify further in some cases.)

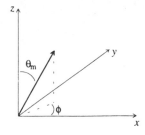

Figure 6.13 The final direction of propagation (θ_m, ϕ).

Table 6.2 Equations to be used for d_{eff} for the 13 uniaxial crystal classes for cases where Kleinman's symmetry does not hold

(a) Two e-rays and one o-ray

Crystal class	Type I (positive uniaxial)	Type II (negative uniaxial)
6 and 4	$-d_{14} \sin 2\theta$	$d_{14} \sin \theta \cos \theta$
622 and 422	$-d_{14} \sin 2\theta$	$d_{14} \sin \theta \cos \theta$
6 mm and 4 mm	0	0
$\bar{6}$m2	$d_{22} \cos^2\theta \cos 3\phi$	$d_{22} \cos^2\theta \cos 3\phi$
3m	$d_{22} \cos^2\theta \cos 3\phi$	$d_{22} \cos^2\theta \cos 3\phi$
$\bar{6}$	$\cos^2\theta(d_{11} \sin 3\phi + d_{22} \cos 3\phi)$	$\cos^2\theta(d_{11} \sin 3\phi + d_{22} \cos 3\phi)$
3	$\cos^2\theta(d_{11} \sin 3\phi + d_{22} \cos 3\phi)$ $-d_{14} \sin 2\theta$	$\cos^2\theta(d_{11} \sin 3\phi + d_{22} \cos 3\phi)$ $+d_{14} \sin \theta \cos \theta$
32	$d_{11} \cos^2\theta \sin 3\phi - d_{14} \sin 2\theta$	$d_{11} \cos^2\theta \sin 3\phi + d_{14} \sin \theta \cos \theta$
$\bar{4}$	$d_{14} \sin 2\theta \cos 2\phi$ $-d_{15} \sin 2\theta \sin 2\phi$	$(d_{14} + d_{36}) \sin \theta \cos \theta \cos 2\phi$ $-(d_{15} + d_{31}) \sin \theta \cos \theta \sin 2\phi$
$\bar{4}$2m	$d_{14} \sin 2\theta \cos 2\phi$	$(d_{14} + d_{36}) \sin \theta \cos 2\phi$

(b) Two o-rays and one e-ray

Crystal class	Type I (negative uniaxial)	Type II (positive uniaxial)
6 and 4	$d_{31} \sin \theta$	$d_{15} \sin \theta$
622 and 422	0	0
6mm and 4mm	$d_{31} \sin \theta$	$d_{15} \sin \theta$
$\bar{6}$m2	$-d_{22} \cos \theta \sin 3\phi$	$-d_{22} \cos \theta \sin 3\phi$
3m	$d_{31} \sin \theta - d_{22} \cos \theta \sin 3\phi$	$d_{15} \sin \theta - d_{22} \cos \theta \sin 3\phi$
$\bar{6}$	$\cos \theta(d_{11} \cos 3\phi - d_{22} \sin 3\phi)$	$\cos \theta(d_{11} \cos 3\phi - d_{22} \sin 3\phi)$
3	$\cos \theta(d_{11} \cos 3\phi - d_{22} \sin 3\phi)$ $+d_{31} \sin \theta$	$\cos \theta(d_{11} \cos 3\phi - d_{22} \sin 3\phi)$ $+d_{15} \sin \theta$
32	$d_{11} \cos \theta \cos 3\phi$	$d_{11} \cos \theta \cos 3\phi$
$\bar{4}$	$-\sin \theta(d_{31} \cos 2\phi - d_{36} \sin 2\phi)$	$-\sin \theta(d_{15} \cos 2\phi + d_{14} \sin 2\phi)$
$\bar{4}$2m	$-d_{36} \sin \theta \sin 2\phi$	$-d_{14} \sin \theta \sin 2\phi$

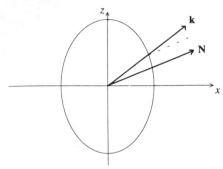

Figure 6.14 The index (normal) surface showing direction of energy flow **N** for phase propagation along the direction of **V**$_p$.

(c) 90° phase matching

As noted earlier, the normal surface gives the direction of energy flow (**N**) for any given phase propagation direction as shown in Figure 6.14. Because **N** and **k** are not in general parallel, 'walk-off' effects can limit the interaction lengths in three-wave mixing processes involving both o- and e-rays. For example, for $\omega_1 + \omega_2 \to \omega_3$ (type II phase-matched) (Figure 6.15). By selecting a crystal that permits phase matching with $\theta_m = 90°$, the walk-off angle (α) can be reduced to zero (whilst still maintaining a difference between the o- and e-rays) and hence long interaction lengths can be achieved, which give a greater efficiency of conversion.

6.3.5 Sum processes $\omega_1 + \omega_2 \to \omega_3$

(a) Simple sum generation

This case was analyzed at the beginning of Section 6.3.3 and a summary is given here (see Figure 6.16)

It was assumed that $I_3(0) = 0$ and that

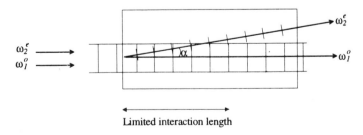

Limited interaction length

Figure 6.15 The limited interaction length between waves of frequency ω_1 and ω_2 due to the 'walk-off' effect.

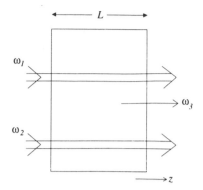

Figure 6.16 Schematic representation of the simple sum generation process.

$$\frac{d\mathscr{E}_1}{dz} \approx \frac{d\mathscr{E}_2}{dz} \approx 0$$

If we had also assumed that this interaction was phase-matched, i.e. $\Delta \mathbf{k} = 0$, then we would have obtained (from Eq. (6.34))

$$I_3 = \frac{\omega_3^2 L^2}{2} \left(\frac{\mu_0 \mu_r}{\varepsilon_0 \varepsilon_r}\right)^{3/2} d_{\mathrm{eff}} I_1 I_2. \qquad [6.34']$$

 (b) Parametric upconversion

We now consider the case where there is a significant transfer from *one* of the input waves to the sum (Figure 6.17). We assume that ω_2 is intense and is not depleted so that $d\mathscr{E}_2/dz \approx 0$, but that ω_1 is weak and is affected by the interaction (the jargon often refers to ω_1 as the *signal* beam and ω_2 as the *pump* beam). Using Eqs. (6.31) we now obtain two nonzero equations:

$$\frac{d\mathscr{E}_{1i}}{dz} = \frac{-i\omega_1}{2} \left(\frac{\mu_0 \mu_r}{\varepsilon_0 \varepsilon_r}\right)^{1/2} d_{ijk} \mathscr{E}_{3j} \mathscr{E}_{2k}^*$$

$$\frac{d\mathscr{E}_{3j}}{dz} = \frac{-i\omega_3}{2} \left(\frac{\mu_0 \mu_r}{\varepsilon_0 \varepsilon_r}\right)^{1/2} d_{jik} \mathscr{E}_{1i} \mathscr{E}_{2k}$$

where it is again assumed that $\Delta k = 0$. We can also assume that the relevant d_{eff} has been determined. For convenience we define the parameter l by:

$$l = \left[\frac{d_{\mathrm{eff}}}{2\pi} \left(\frac{\omega_1 \omega_3 \mu_0 \mu_r}{\varepsilon_0 \varepsilon_r}\right)^{1/2} \mathscr{E}_2\right]^{-1} \qquad (6.36)$$

This allows the two simultaneous equations to be written:

$$\frac{d}{dz} \left(\frac{\mathscr{E}_1}{\sqrt{\omega_1}}\right) = \frac{-i\pi}{l} \left(\frac{\mathscr{E}_3}{\sqrt{\omega_3}}\right)$$

(a)

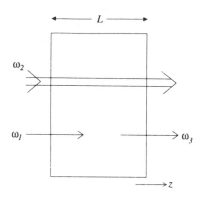

(b)

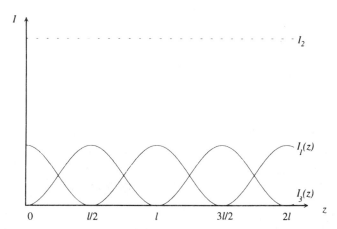

Figure 6.17 (a) Schematic representation of the parametric upconversion process. (b) Power growth and decay in frequency conversion.

$$\frac{d}{dz}\left(\frac{\mathscr{E}_3}{\sqrt{\omega_3}}\right) = \frac{-i\pi}{l}\left(\frac{\mathscr{E}_1}{\sqrt{\omega_1}}\right)$$

Assuming that there is no initial sum frequency, that is, that $\mathscr{E}_3(0) = 0$, then these equations have the solutions:

$$\mathscr{E}_1(z) = \mathscr{E}_1(0)\cos\left(\frac{\pi z}{l}\right)$$

(a)

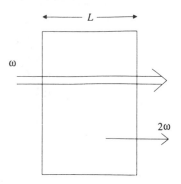

(b)

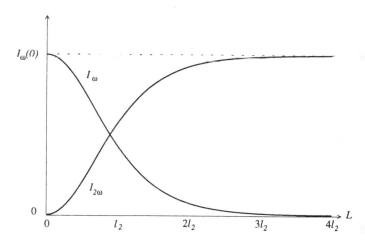

Figure 6.18 Second harmonic generation (SHG). (a) A schematic representation of the process; (b) power growth and decay in SHG.

$$\mathcal{E}_3(z) = -i\left(\frac{\omega_3}{\omega_1}\right)^{1/2} \mathcal{E}_1(0) \sin\left(\frac{\pi z}{l}\right)$$

Since the irradiance, $I = -\frac{1}{2}[(\varepsilon_0 \varepsilon_r)/(\mu_0 \mu_r)]^{1/2}\mathcal{E}\mathcal{E}^*$, we obtain

$$I_1(z) = I_1(0) \cos^2\left(\frac{\pi z}{l}\right) \qquad I_3(z) = \frac{\omega_3}{\omega_1} I_1(0) \sin^2\left(\frac{\pi z}{l}\right) \tag{6.37}$$

Graphs of $I_1(z)$ and $I_3(z)$ versus z are shown in Figure 6.17(b).

It can be seen that the ω_3 wave grows until all the power in the ω_1 wave has been depleted (that is, fully upconverted), and then the reverse process occurs

($\omega_3 - \omega_2 \rightarrow \omega_1$; this is difference generation). A full cycle is completed after a distance, l, which is known as the *interaction length* (note that I_3 in Eq. (6.37) reduces to (6.34') when $\pi z/l$ is small).

(c) Second harmonic generation (SHG)

In this case, a single intense beam generates a sum frequency with itself, i.e. ω generates 2ω. This corresponds to dividing the input into two parts, where $I_1 = I_2 = I_\omega/2$ (Figure 6.18). Assuming that $\Delta k = 0$, we can substitute these relations into Eq. (6.34) along with $\omega_1 = \omega_2 = \omega$ and $\omega_3 = 2\omega$, to obtain:

$$I_{2\omega} = \frac{\omega^2 L^2}{2} \left(\frac{\mu_0 \mu_r}{\varepsilon_0 \varepsilon_r} \right)^{3/2} d^2 I_\omega^2 \tag{6.38}$$

Note that $I_{2\omega}$ is proportional to the square of the fundamental, I_ω (for low conversion).

In contrast with the parametric upconversion case, depletion of the fundamental corresponds to the ω_1 and ω_2 waves being reduced at an equal rate; consequently, after full conversion, the reverse (difference) process cannot occur (assuming $I_{2\omega}$ is not so intense as to induce strong parametric amplification; see below).

Again dividing the fundamental into two halves, we can write two of the coupled equations as:

$$\frac{d\mathscr{E}_\omega}{dz} = \frac{-i\omega}{2} \left(\frac{\mu_0 \mu_r}{\varepsilon_0 \varepsilon_t} \right)^{1/2} d_{\text{eff}} \mathscr{E}_{2\omega} \mathscr{E}_\omega$$

$$\frac{d\mathscr{E}_{2\omega}}{dz} = \frac{-i\omega}{2} \left(\frac{\mu_0 \mu_r}{\varepsilon_0 \varepsilon_r} \right)^{1/2} d_{\text{eff}} \mathscr{E}_\omega^2$$

assuming that $\Delta k = 0$ and taking the phase of the ω_2 wave as a reference, such that $\mathscr{E}_2 = \mathscr{E}_2^*$. These have the solutions:

$$\mathscr{E}_{2\omega} = -i\mathscr{E}_\omega(0) \tanh(z/l_2)$$

$$\mathscr{E}_\omega = \mathscr{E}_\omega(0) \operatorname{sech}(z/l_2)$$

where

$$l_2 = \left[\frac{\omega}{2} \left(\frac{\mu_0 \mu_r}{\varepsilon_0 \varepsilon_r} \right)^{1/2} d\mathscr{E}_\omega(0) \right]^{-1}$$

This gives, in terms of irradiances for the length L:

$$I_{2\omega} = I_\omega(0) \tanh^2(L/l_2); \qquad I_\omega = I_\omega(0) \operatorname{sech}^2(L/l_2) \tag{6.39}$$

As must be the case, the first of these reduces to Eq. (6.38) when $I_\omega(0)$ is small, i.e. when $\tanh(L/l_2) \approx L/l_2$. Graphs of $I_{2\omega}$ and I_ω versus L are shown in Figure 6.18(b).

Theoretically (assuming 90° phase-matching, a perfectly collimated beam and zero losses), a conversion of 93% can be achieved over a distance $L = 2l_2$, and 99% over $3l_2$. Note that, as required on energy conservation grounds, $I_{2\omega}$ asymptotes to the input irradiance $I_\omega(0)$.

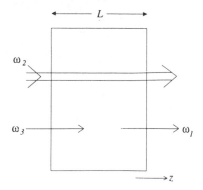

Figure 6.19 Schematic representation of a difference process.

6.3.6 Difference processes ($\omega_3 - \omega_2 \rightarrow \omega_1$ and $\omega_3 - \omega_1 \rightarrow \omega_2$)

(a) Difference generation, down-conversion, rectification
If the three waves ω_1, ω_2 and ω_3 are phase-matched then an input of ω_2 and ω_3 will yield $\omega_1 = \omega_3 - \omega_2$, and not the sum, $\omega_3 + \omega_2$ (Figure 6.19). For the low conversion efficiency interaction this proceeds in the same manner as when generating a sum output, as described by Eq. (6.34), but with I_1 and I_3 exchanged.

Parametric down-conversion: Assuming once again that ω_2 is the pump beam then down-conversion of a signal ω_3 to ω_1 is described by Eqs. (6.37), but with all the subscripts 1 and 3 exchanged. This is equivalent to shifting the $z = 0$ starting point by $l/2$.

Optical rectification: This difference process is equivalent to SHG; that is, a single input frequency gives $\omega - \omega \rightarrow 0$. Although this effect does actually occur, it is always very weak because of the problem of energy conservation.

(b) Parametric amplification
Parametric amplification is another difference process, but one in which the highest frequency, ω_3, is the intense beam (pump); see Figure 6.20. In this case we can assume that $\mathrm{d}\mathscr{E}_3/\mathrm{d}z \approx 0$ and so from Eqs. (6.31) we can write:

$$\frac{\mathrm{d}\mathscr{E}_1}{\mathrm{d}z} = \frac{-i\omega_1 d}{2}\left(\frac{\mu_0\mu_\mathrm{r}}{\varepsilon_0\varepsilon_\mathrm{r}}\right)^{1/2}\mathscr{E}_3\mathscr{E}_2^*$$

$$\frac{\mathrm{d}\mathscr{E}_2^*}{\mathrm{d}z} = \frac{-i\omega_2 d}{2}\left(\frac{\mu_0\mu_\mathrm{r}}{\varepsilon_0\varepsilon_\mathrm{r}}\right)^{1/2}\mathscr{E}_1\mathscr{E}_3^*$$

(6.40)

where we have written the complex conjugate of the second equation (again $\Delta\mathbf{k} = 0$). This time we can choose to make ω_3 the reference wave, such that $\mathscr{E}_3 = \mathscr{E}_3^*$. If we define a parameter g,

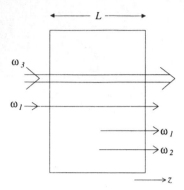

Figure 6.20 Schematic representation of parametric amplification.

$$g = d(\omega_1\omega_2)^{1/2}\left(\frac{\mu_0\mu_r}{\varepsilon_0\varepsilon_r}\right)^{1/2}\mathscr{E}_3$$

then we can rewrite Eqs. (6.40) as

$$\frac{d}{dz}\left(\frac{\mathscr{E}_1}{\sqrt{\omega_1}}\right) = \frac{-ig}{2}\left(\frac{\mathscr{E}_2^*}{\sqrt{\omega_2}}\right)$$

$$\frac{d}{dz}\left(\frac{\mathscr{E}_2^*}{\sqrt{\omega_2}}\right) = \frac{ig}{2}\left(\frac{\mathscr{E}_1}{\sqrt{\omega_1}}\right)$$

These are similar to the equations for parametric upconversion, except for a change in sign. As a result the solutions are in terms of cosh and sinh instead of cos and sin. Assuming that there is no initial wave at ω_2:

$$\mathscr{E}_1(z) = \mathscr{E}_1(0)\cosh\left(\frac{gz}{2}\right)$$

$$\mathscr{E}_2^*(z) = \left(\frac{\omega_2}{\omega_1}\right)^{1/2}\mathscr{E}_1(0)i\sinh\left(\frac{gz}{2}\right)$$

If we let $\mathscr{E}$ or z be large enough that $gz \geqslant 3$, then:

$$\cosh\left(\frac{gz}{2}\right) = \tfrac{1}{2}(e^{gz/2} + e^{-gz/2}) \approx \tfrac{1}{2}e^{gz/2}$$

and

$$\sinh\left(\frac{gz}{2}\right) = \tfrac{1}{2}(e^{gz/2} - e^{-gz/2}) \approx \tfrac{1}{2}e^{gz/2}$$

We can thus write:

$$I_1 = \tfrac{1}{4}I_1(0)e^{gz} \qquad I_2 = \tfrac{1}{4}\left(\frac{\omega_2}{\omega_1}\right)I_1(0)e^{gz} \tag{6.41}$$

We can see that a weak wave at ω_1 is amplified exponentially, with a gain factor g, whilst ω_2 is generated at the same rate. This is *parametric amplification*. The jargon refers to ω_1 as the *signal* which is said to be amplified by the *pump* ω_3, whilst an *idler* ω_2 wave is also generated. There is another useful way of seeing how parametric amplification arises and this is discussed in the section below.

The Manley–Rowe relations
We can rewrite the three main equations (6.31), assuming that $\Delta \mathbf{k} = 0$ and by substituting

$$S = \frac{d}{2}\left(\frac{\mu_0 \mu_r}{\varepsilon_0 \varepsilon_r}\right)^{1/2}$$

into them:

$$\frac{d\mathscr{E}_1}{dz} = -i\omega_1 S\mathscr{E}_3 \mathscr{E}_2^*$$

$$\frac{d\mathscr{E}_2}{dz} = -i\omega_2 S\mathscr{E}_1^* \mathscr{E}_3$$

$$\frac{d\mathscr{E}_3}{dz} = -i\omega_3 S\mathscr{E}_1 \mathscr{E}_2$$

Now consider the following derivatives, evaluated using the above relations:

$$\frac{d}{dz}\left(\frac{\mathscr{E}_1 \mathscr{E}_1^*}{\omega_1}\right) = -iS\mathscr{E}_3 \mathscr{E}_2^* \mathscr{E}_1^* + iS\mathscr{E}_3^* \mathscr{E}_2 \varepsilon_1$$

$$\frac{d}{dz}\left(\frac{\mathscr{E}_2 \mathscr{E}_2^*}{\omega_2}\right) = -iS\mathscr{E}_1^* \mathscr{E}_3 \varepsilon_2^* + iS\mathscr{E}_1 \mathscr{E}_3^* \mathscr{E}_2$$

$$\frac{d}{dz}\left(\frac{\mathscr{E}_3 \mathscr{E}_3^*}{\omega_3}\right) = +iS\mathscr{E}_1^* \mathscr{E}_2^* \mathscr{E}_3 - iS\mathscr{E}_1 \mathscr{E}_2 \mathscr{E}_3^*$$

It can be seen that all three are equal in magnitude, whilst the third is opposite in sign.
Noting that $\mathscr{E}\mathscr{E}^* \propto I$, we can write:

$$\frac{d}{dz}\left(\frac{I_1}{\omega_1}\right) = \frac{d}{dz}\left(\frac{I_2}{\omega_2}\right) = -\frac{d}{dz}\left(\frac{I_3}{\omega_3}\right) \tag{6.42}$$

Let us consider what this means:

$$I \equiv \text{energy/s/m}^2$$

$$\omega \equiv \text{photon energy}/\hbar$$

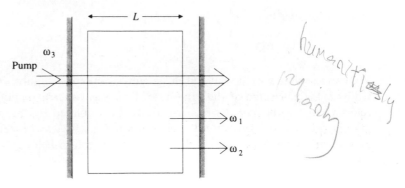

Figure 6.21 Schematic representation of an optical parametric oscillator with optical feedback provided by mirrors.

thus

$$I/\omega \propto \text{number of photons/s/m}^2 = N$$

Using N instead of I/ω in Eq. (6.42) and integrating, we obtain:

$$\Delta N_1 = \Delta N_2 = -\Delta N_3$$

These are the Manley–Rowe relations, which simply states that for all interactions between ω_1, ω_2 and ω_3 (obeying $\omega_1 + \omega_2 = \omega_3$), for every ω_1 photon that is lost, an ω_2 photon will also be lost, while an ω_3 photon will be gained, and vice versa. This is as expected on energy conservation grounds. We can now use this result to explain parametric amplification.

Beginning with the pump, ω_3 and signal, ω_1. Difference generation gives $\omega_3 - \omega_1 \rightarrow \omega_2$ (idler). If the ω_2 flux increases, so must the ω_1 and hence there is *amplification*. ω_1 'stimulates' the splitting of an ω_3 photon to ω_1 and ω_2.

(c) Optical parametric oscillator (OPO)
If we can amplify a signal, then using positive feedback, it is always possible to make an oscillator. Optical feedback can be provided by mirrors, as provided by a laser cavity (Figure 6.21). This process can be regarded as $\omega_3 \rightarrow \omega_1 + \omega_2$, except that in practice ω_1 and ω_2 build up from background 'noise' (inevitably present) by parametric amplification. The energy conservation $(\omega_1 + \omega_2 = \omega_3)$ and phase-matching $(\mathbf{k}_1 + \mathbf{k}_2 = \mathbf{k}_3)$ conditions determine the two particular frequencies that dominate.

What is very attractive about this device is that it is widely tunable. By changing the phase-matching condition, such as by varying the angle of the crystal to the cavity axis (and hence altering θ_m) or by varying the crystal temperature (and hence its refractive indices), the outputs ω_1 and ω_2 can be adjusted, within the constraint $\omega_1 + \omega_2 = \omega_3$ (Figure 6.22).

For oscillation, the gain must exceed the cavity losses. As a result a minimum pump irradiance, I_3, is required in order to reach oscillation threshold. To consider

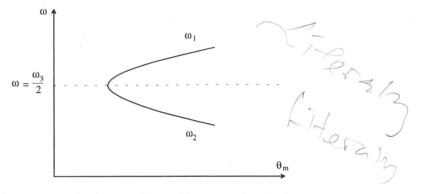

Figure 6.22 Graph of ω versus θ_m showing the tunable nature of an OPO.

this we must convert our basic equations (6.31) into the time domain as the interactions are confined spatially within the cavity. This is done by simply substituting the differential operators:

$$\frac{d}{dz} \equiv \frac{1}{c}\frac{d}{dt} = (\mu_0 \mu_r \varepsilon_0 \varepsilon_r)^{-1/2}\frac{d}{dt}$$

We must also add extra terms corresponding to the losses in the cavity, and hence write

$$\frac{d\mathscr{E}_1}{dt} = -a_1 \mathscr{E}_1 - \frac{i\omega_1 d_{\text{eff}}}{2\varepsilon_r \varepsilon_0}\mathscr{E}_2^* \mathscr{E}_3$$

$$\frac{d\mathscr{E}_2}{dt} = -a_2 \mathscr{E}_2 - \frac{i\omega_2 d_{\text{eff}}}{2\varepsilon_r \varepsilon_0}\mathscr{E}_1^* \mathscr{E}_3$$

$\mathscr{E}_1$ and $\mathscr{E}_2$ now refer to the mean fields within the cavity at any instant in time. If $d = 0$, the extra $a\mathscr{E}$ terms would simply lead to exponential decay of these fields (a corresponds to the decay rate for each field; i.e. proportional to the cavity losses for this frequency). At threshold, $d\mathscr{E}_1/dt = d\mathscr{E}_2/dt = 0$ and there is neither decay nor growth. Thus

$$a_1 \mathscr{E}_1 = -\frac{i\omega_1 d_{\text{eff}}}{2\varepsilon\varepsilon_0}\mathscr{E}_2^* \mathscr{E}_3$$

and, taking the complex conjugate of the second:

$$a_2 \mathscr{E}_2^* = \frac{i\omega_2 d_{\text{eff}}}{2\varepsilon\varepsilon_0}\mathscr{E}_1 \mathscr{E}_3^*.$$

Multiplying these together and eliminating $\mathscr{E}_2^*$ and $\mathscr{E}_1$, we obtain $a_1 a_2 = \omega_1 \omega_2 (d_{\text{eff}}/\varepsilon\varepsilon_0)^2 \mathscr{E}_3 \mathscr{E}_3^*$. The critical irradiance of the pump in order to reach threshold is thus given by:

$$I_3^{(\text{th})} \propto \frac{a_1 a_2}{\omega_1 \omega_2 d_{\text{eff}}^2} \tag{6.43}$$

We now consider the frequencies that can be made to oscillate most easily. From Eq. (6.43),

$$I_3 \propto \frac{1}{\omega_1 \omega_2} = \frac{1}{\omega_3 \omega_1 - \omega_1^2}$$

If we differentiate the denominator with respect to ω: $\mathrm{d}/\mathrm{d}\omega_1\,(\omega_3\omega_1 - \omega_1^2) = \omega_3 - 2\omega_1$, we obtain a turning point at $\omega_1 = \omega_3/2$. This corresponds to the smallest I_3 value, i.e. when $\omega_1 = \omega_3/2\,(=\omega_2)$. Conversely, by inspection $I_3 \to \infty$ as $\omega_1 \to 0\,(\omega_2 \to \omega_3)$ and as $\omega_1 \to \omega_3\,(\omega_2 \to 0)$. Clearly the degenerate OPO ($\omega_1 = \omega_2$) has the lowest pump threshold, and the more it is tuned away from degeneracy, the greater will be the pump irradiance needed. It is seen that the tuning range of an OPO is limited by the available pump power. This is typical of many optical nonlinear processes!

The application of diode-pumped solid state lasers as sources to pump OPOs is described in Chapter 8. Intracavity second harmonic generation is also exploited. This recent laser technology promises to revolutionize the practicality of these devices for which the basic mechanisms have been described in the present chapter.

References

[6.1] A. Yariv, *Quantum Electronics*, 3rd edn (Wiley, New York, 1989).
[6.2] F. Zernike and J. Midwinter, *Applied Nonlinear Optics* (Wiley, New York, 1973).

7

Fast optical modulators

The electro-optical effects described in Chapter 6 are the basis for a series of modulator devices. The microscopic effect in crystalline solids is intrinsically fast – so much so that the accompanying electronics will limit the response times. Further usable effects include acousto-optic and magneto-optic effects.

These modulators thus differ from liquid crystal modulators which usually operate on time scales of seconds to milliseconds. Here we are concerned with modulators operating over the microsecond to nanosecond timescale. Fast electro-optic, acousto-optic and magneto-optic modulators can be seen to have ever-expanding applications in laser technology, communications, information storage and laser printing. They can be used to impress information onto the phase or intensity of propagating wavefronts. They find application in Q-switched lasers, to mode lock, to switch out a short pulse, to deflect optical beams or to record information on optical disks.

7.1 Electro-optic modulators

The physical effect used in electro-optical modulators is that of an electric field causing a change in the dielectric constant of a material, as described in Eq. (6.15), through Maxwell's relation,

$$n^2 = \varepsilon_r \qquad (7.1)$$

This implies an electric field-controlled refractive index, the theory of which is described in Chapter 6 (see, for example, Eqs. (6.19)). Thus the *phase* of the optical waves can be affected, and this is used to achieve modulation of phase, amplitude, intensity or direction. In principle, fast modulators can be used in telecommunications but with the widespread use of semiconductor lasers, information can be impressed directly onto the source and hence such more direct methods are generally used.

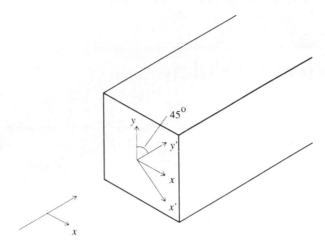

Figure 7.1 An electromagnetic wave incident upon a crystal with components along the x' and y' axes.

However, where high-power laser beams are used, it is necessary to use separate fast modulators. This is the case in *data recording*, in particular recording on optical disks where the process is known as 'disk mastering'. A relatively high-powered (several watts) argon laser is used in combination with a modulator capable of impressing information at a rate of 50 MHz.

Electro-optic retardation
The previous chapter described in appropriate three-dimensional tensor form how the refractive index components change with an applied electric field and how the equation of the optical indicatrix also changes. We now consider what happens to light that is incident on and propagates through the crystal. At $z = 0$, at the edge of the crystal, the optical field can be described by

$$E = A \exp i(\omega t - kz) \tag{7.2}$$

where the wavenumber $k = 2\pi/\lambda = \omega n/c$.

Consider a field that is incident normal to the x', y' plane with its **E** vector along the x-axis (Figure 7.1). This can be resolved at $z = 0$ into two mutually perpendicular components whose polarizations lie along the x' and y' axes.

$$E_{x'} = A \exp i(\omega t - \omega n_{x'} z/c) \tag{7.3}$$

$$E_{y'} = A \exp i(\omega t - \omega n_{y'} z/c) \tag{7.4}$$

From Chapter 6 (Eqs. (6.19)) we can substitute for $n_{x'}$ and $n_{y'}$ into the above two equations, which gives:

$$E_{x'} = A \exp i\left(\omega t - \left(\frac{\omega}{c}\right)\left(n_0 + \frac{n_0{}^3}{2} r_{63} E_z \right)z \right) \tag{7.5}$$

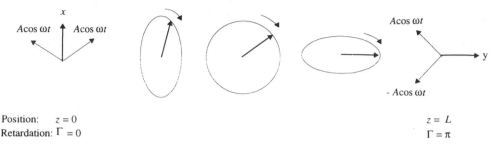

Position: $z = 0$ $z = L$
Retardation: $\Gamma = 0$ $\Gamma = \pi$

Figure 7.2 The development of the wave polarization as it propagates through the crystal.

$$E_{y'} = A \exp i\left(\omega t - \left(\frac{\omega}{c}\right)\left(n_0 - \frac{n_{0^3}}{2} r_{63} E_z \right)z \right) \tag{7.6}$$

Thus there will be a phase difference between the two polarization components at $z = L$ (the end of the crystal) that originates from the different refractive indices along the x' and y' axes. This has the effect of introducing a retardation ϕ between the two components, where

$$\Gamma = \phi_{y'} - \phi_{x'} = \frac{\omega}{c} n_0^3 r_{63} E_z L \tag{7.7}$$

with $\phi_{y'}$ and $\phi_{x'}$ the respective phases of the resolved components.

To understand the physical consequences of a retardation between the two components it is necessary to consider the development of the wave in time and space (Figure 7.2). At $z = 0$ the retardation is $\Gamma = 0$ and the wave is polarized along the x-direction. As the wave moves through the medium the retardation increases between the two components. When $\Gamma = \pi/2$ the electric field vector is circularly polarized in a clockwise sense. When $\Gamma = \pi$ the polarization is linear in the y-direction.

The retardation can also be expressed in terms of the applied voltage V_π that is required to give a retardation of π, so that

$$\Gamma = \pi \frac{V}{V_\pi} \tag{7.8}$$

where $V = E_z L$, and hence $V_\pi = c/(\omega n_0^3 r_{63})$; this is called the *half-wave voltage*.

For a crystal of potassium dihydrogen phosphate (KDP) (see Section 6.3.2), $V_\pi = 8000$ V and for the ammonium version (ADP), $V_\pi = 10\,000$ V. This effect obviously requires very high voltages but it does have the advantage of being very fast. The electro-optic effect can be used to modulate both the amplitude and the phase of a light ray propagating through a crystal.

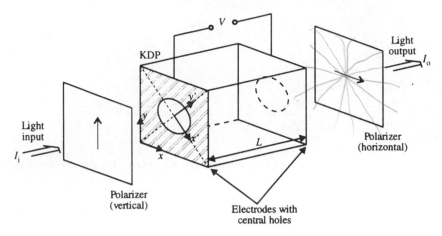

Figure 7.3 Amplitude modulation using the electro-optic effect.

7.1.1 Amplitude modulation

This is the commonly used term but in practice the amplitude of the radiation field is not altered – the term probably refers to the amplitude of the signal being modulated. It is the *phase velocity* of the individual components of the electric field which are changed and when these components are combined through a polarizer at the output of the device, intensity or irradiance modulation is achieved The modulator uses a set of polarizers on either side of the crystal (Figure 7.3). The first has its axis of polarization parallel to the x-direction, the second (output) polarizer being parallel to the y-direction. The polarizers are at 45° to the electrically induced birefringent axes. The radiation incident on the crystal at $z = 0$ is therefore polarized in the x-direction. When an electric field is applied to the crystal, the optical polarization plane is rotated with the power gradually shifting from the x to the y plane. The amplitude of the light wave components then incident on the output polarizer is dependent on the retardation, which is in turn dependent on the applied field. If the retardation is exactly π between the two components the transmitted amplitude will be equal to the incident amplitude (Figure 7.3).

The incident intensity I_{in} can be found using Eqs. (7.3) and (7.4), and is given by

$$I_{\text{in}} \propto |E_{x'}(0)|^2 + |E_{y'}(0)|^2 = 2A^2 \tag{7.9}$$

When the light leaves the crystal at $z = L$, the x' component of the amplitude is $E_{x'}(L) = A$ and the y' component $E_{y'}(L) = Ae^{-i\Gamma}$. The light then passes through the output polarizer, which is set along the y-direction, and the resulting field is found by resolving the field from the x', y' coordinates into their components in x and y. The sum of the y components of both $E_{x'}$ and $E_{y'}$ gives the total field emerging from the y polarizer:

$$(E_y)_L = \frac{A}{\sqrt{2}} (e^{i\Gamma} - 1) \tag{7.10}$$

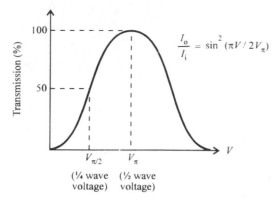

Figure 7.4 Graph showing the transmission as a function of applied voltage across the crystal.

The output intensity I_{out} is proportional to the product of the real and complex conjugate components of the electric field so that

$$I_{out} \propto (E_{yL})(E_{yL}^*) \tag{7.11}$$

Therefore,

$$I_{out} \propto \frac{A^2}{2} ((e^{-i\Gamma} - 1)(e^{i\Gamma} - 1)) = 2A^2\sin^2\frac{\Gamma}{2} \tag{7.12}$$

whence

$$\frac{I_{out}}{I_{in}} = \sin^2\frac{\Gamma}{2} = \sin^2\left(\left(\frac{\pi}{2}\right)\frac{V}{V_\pi}\right) \tag{7.13}$$

and therefore

$$I_{out} \sim \sin^2 V. \tag{7.14}$$

Considering the transmitted intensity versus applied voltage (Figure 7.4) it is clear that in order to obtain linear modification we need to bias the applied voltage to $V_{\pi/2}$. The modulator then faithfully reproduces onto the optical output, any electrical signal that is applied (provided that this signal is small compared to $V_{\pi/2}$).

Alternatively, a quarter-wave plate can also be used in order to produce linear modulation. The quarter-wave plate effectively 'biases' the cell to the linear region so that a small sinusoidal variation in the applied voltage will produce a sinusoidal transmitted output intensity around the 50% transmission point.

A number of materials are used for such modulators and these are mentioned later in this chapter in connection with available devices.

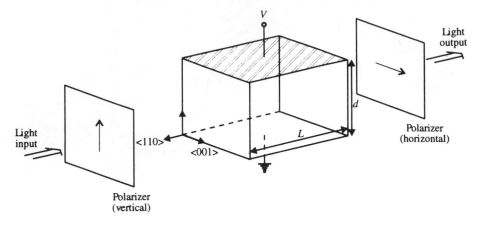

Figure 7.5 One of the geometries used to produce a transverse electro-optic modulator.

7.1.2 The transverse electro-optic modulator

Another possible geometry is one in which the electric field is applied transversely to the direction of propagation. One major disadvantage when the electric field is applied across the same ends of the crystal through which the light beam propagates is that we must use transparent electrodes and so must consider their possible interference with the beam. A transverse electric field can operate with much lower voltages and has the advantage of permitting length scaling, to increase the overall effect, without having to increase the voltage. This is particularly useful in the infrared where, because of the wavelength dependence the required fields can become quite high. Modulators of this type are often made using crystals of cubic symmetry which show no birefringence normally. For example, GaAs and CdTe with a symmetry of 43m, transmit at 10 μm (CO_2 laser wavelengths), whilst copper chloride (CuCl), which has the same symmetry, can be used in the visible region. One possible geometry for

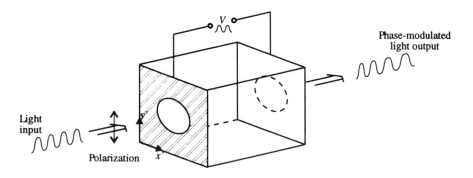

Figure 7.6 Electro-optic phase modulation.

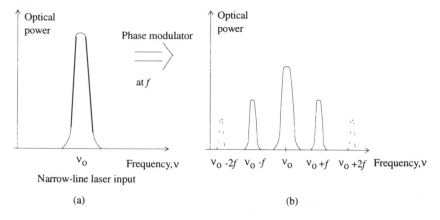

Figure 7.7 (a) The narrow-line laser input, and (b) the generation of 'side bands' due to an applied electrical signal of frequency f.

a crystal is shown in Figure 7.5. Once again the axes of the induced ellipsoid are at 45° to the input polarization and, as before, retardation of the two components of polarization can then be induced.

7.1.3 Electro-optic phase modulation

The modulators just described can also be used simply to phase modulate a beam. To do this the crystal is oriented such that one of the new dielectric axes (e.g. x') corresponds to the direction of polarization of the incident light. In this case, the applied field simply changes the refractive index that is 'seen' by the light and hence controls the optical thickness of the crystal (Figure 7.6).

If the applied electrical signal has a single frequency, f, then 'side bands' are generated at frequencies of $v_0 \pm mf$ (where m is an integer) (Figure 7.7). Assuming that KDP is being used (from Eq. (6.19)), we can write,

$$n_{x'} = n_0 + \tfrac{1}{2} r_{63} E n_0^3 \tag{7.15}$$

Again writing the phase as $\phi = \omega n z / c$ the change in phase due to a crystal of length L, due to an applied field, E, is given by

$$\delta = \frac{\omega n_0^3 r_{63} E L}{2c} = \frac{\pi n_0^3 r_{63} E L}{\lambda} \tag{7.16}$$

where δ is known as the *phase modulation index*. It can be shown that when a monochromatic beam represented by $E_{in} = A \cos \omega t$ is modulated by an electric field of $E_m \sin \omega_m t$, the electric field output has the form

$$E_{out}/A = J_0(\delta) \cos \omega t + J_1(\delta) \cos(\omega + \omega_m)t - J_1(\delta) \cos(\omega - \omega_m)t$$

$$+ J_2(\delta) \cos(\omega + 2\omega_m)t - J_2(\delta) \cos(\omega - 2\omega_m)t \tag{7.17}$$

$$+ J_3 \ldots$$

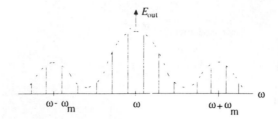

Figure 7.8 Graph showing the electric field output versus frequency when an electric field of frequency ω_m is applied longitudinally to a crystal.

The Bessel functions, J, impose an envelope function onto the side bands, as shown in Figure 7.8.

7.1.4 Electro-optic deflectors

The electro-optic effect can be used in crystals to alter the direction of propagation of a wave. This is done by varying the optical path length across the width of the crystal upon which the beam is incident. The beam then effectively 'sees' a crystal with a varying length and hence a varying delay is introduced across the beam. It is this varying delay along the wavefront (i.e. across the beam) that deflects the beam. This effect is often realized in practice by using two prisms of a 42m crystal such as KDP. The two prisms have their z-axes opposite each other, but are otherwise similarly oriented. A field applied along the z-direction then produces a change in the refractive index along the x'-direction (Figure 7.9). The refractive index 'seen' by the upper beam is given by

$$n_u = n_0 - \frac{n_0^3}{2} r_{63} E_z \tag{7.18}$$

whilst that 'seen' by the lower beam is given by

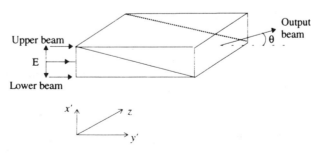

Figure 7.9 An electro-optic deflector.

(a)

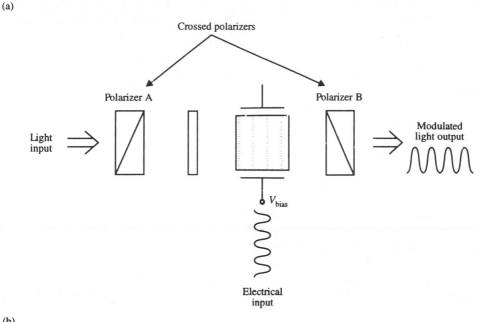

(b)

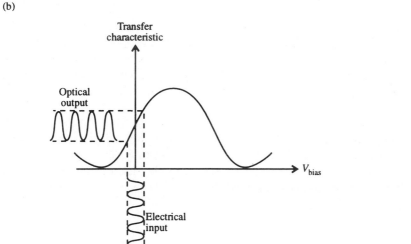

Figure 7.10 (a) Schematic diagram of the apparatus used to impress information onto an optical beam. (b) Conversion of an electrical signal into an optical signal.

$$n_1 = n_0 + \frac{n_0^3}{2} r_{63} E_z \qquad (7.19)$$

The total change in refractive index across the beam, Δn, is thus $n_0^3 r_{63} E_z$. The angular change in the output direction, θ, is a function that depends upon the change in refractive index across the beam, the distance across the beam that is affected and the length of the crystal through which the light has travelled. It can be shown that

$$\theta = \frac{L}{D}\Delta n \quad \text{for } \sin\theta \approx \theta$$

thus

$$\theta = \frac{Ln_0^3 r_{63} E_z}{D}. \tag{7.20}$$

Unfortunately, for device purposes, the effect is rather small thus requiring high voltages to give significant deflections.

7.1.5 Applications of electro-optic modulators

1. Conversion of an electrical signal into an optical signal
Figure 7.10(a) shows a schematic representation of the apparatus used to impress information onto an optical beam. The graph in Figure 7.10(b) indicates that in order to obtain a linear response we must apply a voltage within a fixed band around $V = 0$.

2. Laser Q-switching
At $V = 0$ the polarizer, Pockel's cell (see Figure 7.11), quarter-wave plate and reflector combine to stop all oscillation of light within the cavity. The energy is thus allowed to build up in the laser gain section. A sudden application of a voltage of magnitude $V_{\pi/2}$ (in a few nanoseconds) to the Pockel's cell allows oscillation to take place in the cavity and energy is dumped out in an intense, short pulse (~ 30 ns).

3. Short pulse switching
The fast switching capability of a Pockel's cell can be combined with two crossed polarizers to give a linear response to an electrical input in order to produce short output pulses (Figure 7.12). This method is used in mode-locked lasers in order to select and switch out a particular laser pulse.

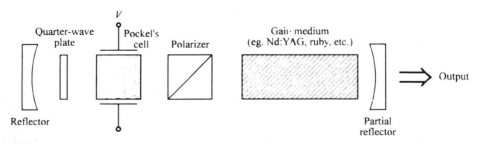

Figure 7.11 Schematic diagram showing the use of a Pockel's cell for Q-switching a laser cavity.

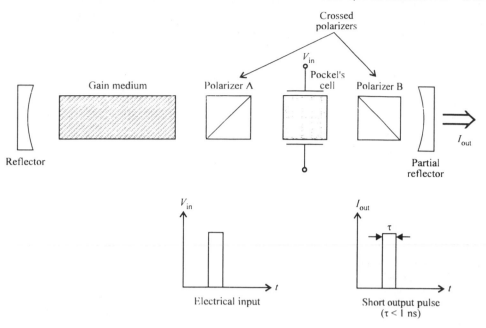

Figure 7.12 Schematic diagram showing short pulse switching.

7.1.6 Practical considerations

The choice of an electro-optic material and its geometry depends upon the specification required by an application. In general it is desirable to have:

- a large change in the refractive index per volt,
- high optical quality and transmissivity,
- a low dielectric constant (and hence low capacitance),
- a low dielectric loss tangent (this minimizes dielectric heating due to an application of high-frequency electric fields), and
- no distortions in the modulator's output due to piezoelectric resonances.

Unfortunately, all electro-optic crystals are also piezoelectric and strains within these crystals make an unwanted contribution to modulation effects. The effects can be minimized by choice of material and orientation. In this respect an optimum is provided by ADP (ammonium dihydrogen phosphate, NH_4, H_2, PO_4) in the 45° y-cut configuration, and the potassium variant, KDP, is also commonly used. Lithium niobate ($LiNbO_3$) and lithium tantalate ($LiTaO_3$) have higher electro-optical coefficients and smaller loss tangents but cannot handle as much optical power.

The details of the modulator design will also depend upon the required aperture size (a 3 mm aperture will accommodate most commercially available lasers) and the realizable levels of driver voltage output. Currently available power transistors limit the driver voltage to the order of 100 V.

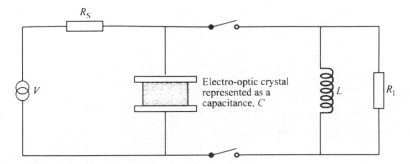

Figure 7.13 At low frequencies the crystal can be represented as a capacitance in series with the internal source resistance (switches open). At high frequencies the crystal is connected in parallel with an inductance, L, and resistance, R_L (switches closed).

The material and its dimensions determine the modulator's electrical characteristics defining thereby the design approach for an optimum electrical drive amplifier. The drive amplifier will now only depend upon the driver voltage and bandwidth that is required.

7.1.7 High-frequency considerations

The discussion so far has assumed the use of electric fields at low frequencies. The circuit used in order to apply an electric field to our crystal can be represented by that shown in Figure 7.13, where the two switches shown are open.

The frequency of the applied electric field is $\omega/2\pi$. The quantity R_S is the internal resistance of the source and C is the parallel plate capacitance of the crystal ($C = A\varepsilon/L$, where A is the cross sectional area, ε is the dielectric constant at the modulation frequency, and L is the length of the crystal along the optical path). If $R_S > 1/(\omega_0 C)$, most of the applied voltage is dropped across the internal resistance of the source, R_S, and so effectively wasted. This effect can be avoided by resonating the crystal capacitance with an inductance L at a frequency $\omega_0 = \sqrt{1/(LC)}$ and using in addition a shunt resistance R_L so that at $\omega = \omega_0$ the impedance of the parallel LCR circuit is R_L, which is chosen to be greater than the value of R_S so that most of the voltage appears across the crystal

The resonant circuit has a bandwidth over which the impedance is high,

$$\frac{\Delta\omega}{2\pi} \sim \frac{1}{2\pi R_L C}$$

which is centred on the frequency ω_0. Therefore the maximum modulator bandwidth must be less than $\Delta\omega/2\pi$, since if this is not the case then the modulated field will not follow the modulating signal with great accuracy.

As well as a suitable bandwidth of operation, as large a retardation of light entering the crystal as possible is usually required. This retardation Γ_m can be related to the maximum voltage that needs to be applied across the crystal by $\Gamma_m = (\omega n_0^3 r_{63}/C)V_m$ and hence the maximum power, $P_m = V_m^2/2R_L$.

Taking a crystal of deuterated potassium dihydrogen phosphate (DKDP) as an example, for a retardation of $\Gamma_m = \pi$, at a wavelength $\lambda = 0.5\ \mu m$, given that $A = 2\ mm^2$, $n_0 = 1.51$, $n_0^3 r_{63} = 34 \times 10^{-12}$ m/V, $\varepsilon/\varepsilon_0 = 45$ and $\Delta v = 1$ MHz the power required to produce this effect is approximately 7 W.

A typical capacitance of the crystal is given by $C = \varepsilon A/d = 4 \times 10^{-12}$ F, and this can be used to calculate the time constant of the system where $\tau = CR$. If it is assumed that the resistance R is 50 Ω, then $\tau = 200$ ps.

Transit time limitations

We know that the electro-optic retardation due to a field E can be written as,

$$\Gamma = aEL \quad \text{where} \quad a = \omega n_0^3 r_{63}/c \tag{7.21}$$

If the electric field E changes appreciably during the transit time $\tau_c = nL/c$ of the light through the crystal, we must write Γ at the output face of the crystal as

$$\Gamma(t) = a \int_0^L E(t')\mathrm{d}z = \frac{ac}{n} \int_{t-\tau_c}^t E(t')\mathrm{d}t' \tag{7.22}$$

$E(t')$ is the instantaneous low-frequency applied field. In the second integral we replace the integration over z by integration over time, realizing that the part of the wave that reaches the output face $z = L$ at time t entered the crystal at time $t - \tau_c$.

Assuming that at a certain time t, the electric field $E(t)$ has the same value throughout the crystal and that we can write $E(t') = E_m \exp(i\omega_m t)$, the expression for retardation becomes

$$\Gamma(t) = \frac{ac}{n} E_m \int_{t-\tau_c}^t \exp(i\omega_m t')\mathrm{d}t'$$

$$= \Gamma_0 \left(\frac{1 - \exp(-i\omega_m \tau_c)}{i\omega_m \tau_c} \right) \exp(i\omega_m t) \tag{7.23}$$

where $\Gamma_0 = (ac/n). \tau_c E_m$, the peak retardation. Clearly from the factor in the brackets $\Gamma(t) \to \Gamma_0$ when $\omega_m \tau_c \ll 1$, so the transit time must be small compared to the shortest modulation period. If we take the highest useful modulation frequency as that for which $\Gamma(t) = 0.9\Gamma_0$, then $\omega_m \tau_c \cong \pi/2$ and so, using $\tau_c = nL/c$, we get the highest modulation frequency to be,

$$v_{max} = \frac{c}{4Ln} \tag{7.24}$$

Typically if $L = 1$ cm and $n = 1.5$ then the maximum frequency of modulation $v_{max} \sim 5 \times 10^9$ Hz $= 5$ GHz.

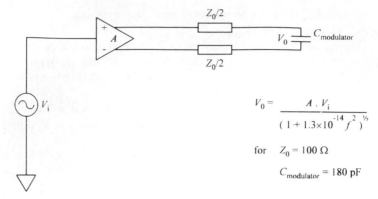

$$V_0 = \frac{A \cdot V_i}{(1 + 1.3 \times 10^{-14} f^2)^{\frac{1}{2}}}$$

for $Z_0 = 100 \ \Omega$

$C_{modulator} = 180 \ pF$

Figure 7.14 The equivalent circuit for a low-frequency modulator (<10 MHz). The crystal is modelled as a single, lumped 180 pF load.

The lumped modulator

Under certain circumstances the crystal arrangement can be treated as a 'lumped' capacitor. It is considered to have a resistance and a capacitance 'lumped' together and transmission line theory can be applied in order to understand what happens when an electric field is applied. This approximation can be used when the electric field is constant, over the length of the crystal and during the time that the light takes to pass through the crystal, i.e. during $\tau_c = nL/c$. In addition, $L \ll 2\pi c/2\pi\omega_m\sqrt{\varepsilon}$ and $L \ll 2\pi c/2\pi\omega_m n$, so that $\sqrt{\varepsilon} \gg n$.

Consider a rod of cross-section d^2 that has lumped capacitance $C = \varepsilon_0\varepsilon_r L$ and parallel conductance $G = \omega_m C \tan \delta$. The parallel capacitance C_a and conductance are

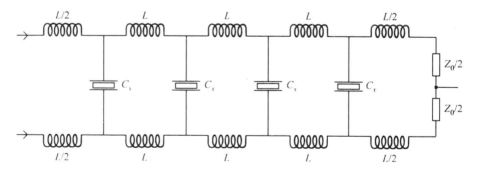

Detector rise time $\tau_c = C_t \cdot Z_0$

$C_t = n \cdot C_r$

$Z_0 = \sqrt{(L/C)}$

Figure 7.15 The equivalent circuit for a high-frequency modulator (>10 MHz), where the modulator is configured as a balanced transmission line.

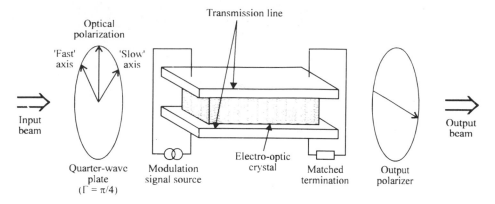

Figure 7.16 The travelling wave modulator.

unavoidable. The voltage from a generator V_g with impedance $R_g V / V_g$ is at a maximum when the impedance R_g is matched to the load specified by an ideal transformer circuit.

In practice, the range DC 10 MHz can be handled in this way. A push–pull common emitter stage with a balanced output impedance of 100 Ω performs satisfactorily and rise times of 35 ns are possible (Figure 7.14).

For a bandwidth greater than 10 MHz modulators are configured as balanced transmission lines (Figure 7.15). The electrical cutoff frequency may be extended dramatically and bandwidth improved by a factor of the order of 2.5 with no increase in power. Beyond 50 MHz, the propagation time for the modulating signal must be considered.

The travelling wave modulator

At high frequency a solution to the 'transit-time' limitation is provided by the travelling wave modulator. In this case the phase velocity of the optical field and modulating field are made identical and the modulating field is applied in the form of a travelling wave parallel to the optical field's direction of propagation. As a result, a specified part of the optical wavefront will undergo the same instantaneous modulating electric field for all positions within the crystal (Figure 7.16). As before, the retardation that is experienced by an element of the optical wavefront can be written as

$$\Gamma(t) = \frac{ac}{n} \int_{t}^{t + \tau_c} E(t', z(t')) \mathrm{d}t' \tag{7.25}$$

where $E(t', z(t'))$ is the instantaneous modulation field as 'seen' by an observer travelling with the phase front. Taking the travelling modulation field as

$$E(z, t') = E_m \exp \mathrm{i}(\omega_m t' - k_m z)$$

$$= E_m \exp \mathrm{i}(\omega_m t' - k_m(c/n)(t' - t))$$

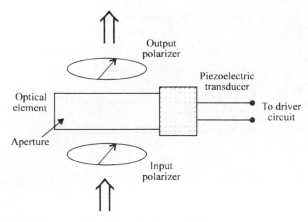

Figure 7.17 Schematic diagram of a photoelastic modulator.

since

$$z(t') = c/n(t' - t)$$

and recalling that $k_m = \omega_m/c_m$, where c_m is a phase velocity, we can integrate Eq. (7.25) to obtain

$$\Gamma(t) = \Gamma_0 \exp i\omega_m t \left[\frac{\exp i\omega_m \tau_d (1 - c/nc_m) - 1}{i\omega_m \tau_d (1 - c/nc_m)} \right] \tag{7.26}$$

where $\Gamma_0 = aLE_m = a(c/n)\tau_c E_m$ is the maximum retardation from a DC field equal to E_m. Clearly, if the two phase velocities are matched so that $c/n = c_m$, then $\Gamma(t) = \Gamma_0$ is the maximum retardation, regardless of crystal length. The maximum useful modulation frequency is that for which $\omega_m \tau_d (1 - c/nc_m) = \pi/2$, yielding

$$(v_m)_{max} = \frac{c}{4nL(1 - c/nc_m)} \tag{7.27}$$

Clearly, the closer the phase velocity of the light, c/n, to the velocity of the modulating frequency, c_m, the larger the maximum modulating frequency that can be used.

Given the potentially very fast operation of the intrinsic electro-optic effect these associated electrical considerations illustrate important basics of optoelectronic devices.

7.2 The photoelastic effect

If a sample of optically transparent material is placed under stress by being compressed or stretched, the material becomes *birefringent*. Different polarization directions exhibit different refractive indices in a manner generally similar to that described in Chapter 6. The effects therefore produce a range of device possibilities, some of which are similar to the direct effects of electric fields already described. Piezoelectric transducers are sometimes attached to the material to apply the stress. The frequency at which such stress can be applied varies widely but is often in the acoustic range.

In its simplest form, a photoelastic modulator consists of a rectangular bar of suitably transparent material (e.g. fused silica) attached to a piezoelectric transducer (Figure 7.17). The bar will vibrate at a frequency determined by its length and the speed of the longitudinal sound waves within the material. The piezoelectric drive transducer is thus tuned to this frequency and controls the amplitude of vibration. An oscillating birefringence is set up with its maximum at the centre of the bar. The size, shape and type of material used can be varied in order to design modulators to a variety of specifications.

7.3 Acousto-optical modulators

Acousto-optical modulators use changes in the refractive index across a material, which are induced by the change in density found across a sound wave to either diffract or deflect a light beam. Diffraction is the fundamental mechanism and can be used to alter the intensity of the straight-through beam. Deflection is a derived effect which can direct a light beam along an angle that depends upon the acoustic wavelength. This latter effect can be used in optical displays and interconnects. We now take a closer look at both of these effects.

7.3.1 Diffraction

A piezoelectric transducer can be bonded to an optically transparent medium in order to generate an acoustic travelling wave (or standing wave) within the medium. The photoelastic effect then sets up a periodic variation in refractive index across the medium, which is characterized by the acoustic angular frequency ω_a and wavevector $\mathbf{k}_a$ of magnitude $2\pi/\lambda$. Note that the velocity of sound v_S is given by $\omega_a/|\mathbf{k}_a|$.

By launching an ultrasonic wave into a transparent medium it is possible to form what is effectively a diffraction grating. This can then be used to diffract a laser beam, incident on the grating. By modulating the grating spacing, information can be imparted to the diffracted beam and its spatial intensity or direction can be varied.

The input acoustic wave propagates through the transparent medium. If a laser beam is also incident on the same medium it is diffracted. The diffraction is at a maximum when the diffracted beam emerges at exactly twice the Bragg angle relative to the straight-through position. This angle depends upon the frequency of the acoustic wave and hence can be controlled. Simultaneously with the diffraction, the frequency of the light wave is also shifted slightly, by an amount equal to the frequency of the acoustic wave (either a positive or a negative effect). This is a Doppler effect, since the light wave is incident on a moving 'acoustic' lattice that has a velocity v_S.

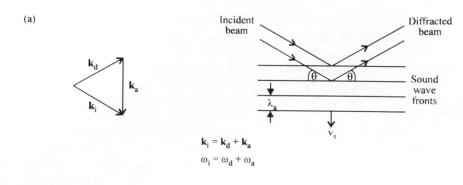

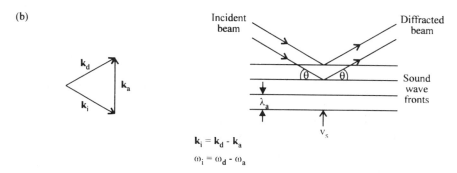

Figure 7.18 The Bragg vector diagram and corresponding physical configuration for the diffraction of light from (a) a retreating sound wave and (b) an oncoming sound wave.

The basic interaction involves three parameters. The two optical fields of frequencies ω_i (incident) and ω_d (diffracted) and the acoustic field of frequency ω_a. The relation between these quantities is expressed as

$$\omega_i = \omega_a \pm \omega_d \tag{7.28}$$

Now consider the case of a single photon incident upon the sound wave. The photon has energy $\hbar\omega_i$ and momentum $\hbar k_i$; the sound wave has a frequency ω_a and wave momentum $\hbar k_a$. Assuming that the incident photon is annihilated and a new photon of frequency ω_d is created, we can write

$$\mathbf{k}_i = \mathbf{k}_a + \mathbf{k}_d \qquad \text{conservation of momentum}$$

$$\omega_i = \omega_a + \omega_d \qquad \text{conservation of energy.}$$

Figures 7.18(a) and (b) show the vector diagrams and corresponding physical configurations for the diffraction of light from a retreating sound wave and an oncoming sound wave. Since the acoustic frequency is small compared to the optical frequency (i.e. $\omega_a \ll \omega_i$), we can make the assumption that $|k_i| \approx |k_d| = |k|$. The conservation conditions are equivalent to interference conditions, and using the above assumption we can write

$$k_a = 2k \sin \theta. \tag{7.29}$$

The magnitude of the **k** vectors can be written as

$$|\mathbf{k}_a| = \frac{2\pi}{\lambda_a} \quad \text{and} \quad |\mathbf{k}| = \frac{2\pi n}{\lambda}$$

where n is the refractive index of the transparent medium, λ is the wavelength of the incident beam in air, λ/n is the wavelength of the light in the medium, and λ_a is the wavelength of the acoustic wave, and hence also the grating spacing. By substituting the definitions for the magnitude of the k values into Eq. (7.29) we obtain

$$\frac{\lambda}{n} = 2\lambda_a \sin \theta_B$$

where θ_B is the Bragg angle. All the considerations so far have been for light propagating within the medium. This expression can be rewritten to take into account the change in refractive index the light encounters upon leaving the cell in terms of the measured angle θ_m between the first-order and zero-order beams (i.e. $\sin \theta_m/2 = n \sin \theta_B$). Hence:

$$\lambda = \lambda_m 2 \sin \frac{\theta_m}{2}.$$

The diffraction efficiency is a function of the acoustic power that synthesizes the diffraction grating. It is quoted by Yariv [7.1] as being

$$\frac{I_{\text{diffracted}}}{I_{\text{incident}}} = \sin^2 \left(\frac{\pi L}{\lambda \sqrt{2}} \sqrt{\frac{n^6 p^2 I_{\text{acoustic}}}{\rho v^3}} \right)$$

where p is the photoelastic (tensor) coefficient, ρ is the density of the medium, v is the velocity of sound in the medium, and I_{acoustic} is the acoustic intensity in watts/m^2.

The terms in front of the acoustic intensity can be used as a figure of merit and are written as $M = n^6 p^2 / \rho v^3$. The diffraction efficiency is a measure of the fraction of power that is transferred from the straight-through beam to the diffracted beam. This efficiency can be altered by varying the acoustic intensity.

7.3.2 Beam deflection

A beam of light and an acoustic cell are set up in exactly the same configuration as used earlier in order to diffract a beam. This produces a straight-through beam and a first-order diffracted beam. It is the first-order diffracted beam that is effectively deflected by varying the frequency of sound within the cell (Figure 7.19). The change

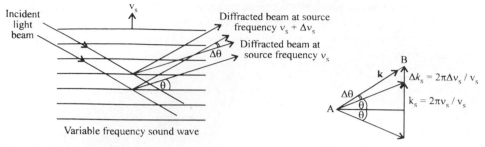

Figure 7.19 A change in the frequency of the sound wave and hence its wavevector causes a change $\Delta\theta$ in the direction of the first-order diffracted beam.

of frequency from v_s to $v_s + \Delta v_s$ causes a change in the sound wave vector of $\Delta k_s = 2\pi(\Delta v_s)/v_s$, which rotates the first-order diffracted beam by $\Delta\theta$ along the direction that least violates momentum conservation.

This is not a very efficient method of diffracting the light, since momentum is not strictly conserved. It acts like destructive interference in that it reduces the amount of energy contained in the diffracted beam. The maximum intensity of the diffracted beam then occurs in the direction that minimizes the momentum mismatch (AB in Figure 7.19). Since the angle of deflection $\Delta\theta$ and angle of diffraction θ are both small, we can write

$$\Delta\theta \approx \frac{\Delta k_s}{k_s} = \frac{\lambda}{nv_s}\,\Delta v_s$$

If the acoustic cell is to be used as an optical display device an important property to consider would be the number of resolvable spots. This is the factor by which $\Delta\theta$ exceeds the beam divergence angle $\theta_{\mathrm{div}} \sim \lambda/Dn$ (where D is the beam diameter). The number of resolvable spots is thus given by

$$N = \frac{\Delta\theta}{\theta_{\mathrm{div}}} = \left(\frac{\lambda\Delta v_s}{nv_s}\right)\left(\frac{nD}{\lambda}\right) = \frac{D}{v_s}\,\Delta v_s$$

Using this procedure in glass, the sound beam can usually be varied in frequency from 80 to 120 MHz. The optical beam diameter D is about 1 cm and $v_s = 3.1 \times 10^5$ cm s^{-1}. This gives 130 resolvable spots. The ratio of diffracted intensity to incident intensity can also be calculated using Eq. (7.19). For PbMoO$_4$, using 0.6328 μm light with an acoustic power of 1 W, beam cross-section of 1×1 mm, optical path length in the acoustic beam of 1 mm and a figure of merit M of 0.22, the ratio of diffracted to incident intensity is about 37%.

7.3.3 Practical devices

The typical spectral range for acousto-optic modulators lies between 0.3 and 10.6 μm. They have bandwidths up to 50 MHz and apertures of several mm. Tables

7.1–7.3 list some of the device types that are commercially available, together with their manufacturers, and the typical specifications to which they are made.

Table 7.1 Electro-optic modulators

Some manufacturers	New Focus Inc., Electro-optic Developments Ltd, Conoptics Inc.	
Typical specifications	Conoptics model 352	Conoptics model 363
Material	KD*P	LTA
$V_{1/2}$ at 633 nm:	586 V	34 V
Wavelength range:	300–1100 nm	450–4500 nm
Aperture diameter:	3.3 mm	2.7 mm
Piezoelectric resonance:	Yes	Yes
Capacitance:	50 pF	110 pF
Differential transit time at 50 Ω	N/A	2.9 ns
Contrast ratio		
at 633 nm:	500:1	50:1
at 1064 nm:	700:1	100:1
Length:	67 mm	258 mm

Table 7.2 Acousto-optic modulators

Some manufacturers	Optilas Ltd, Coherent Associates, Laser Lines Ltd	
Typical specifications	Laser Lines Ltd AA MTS–1200	Laser Lines Ltd AA MT–08
Material:	TeO_2	TeO_2
Mode:	shear	longitudinal
Acoustic velocity:	620–670 m s^{-1}	4200 m s^{-1}
Transducer:	$LiNbO_3$	$LiNbO_3$
Wavelength:	360–1060 nm	360–633 nm
Polarization:	Linear	Linear
Aperture:	3 × 4 mm^2	0.5 × 4 mm^2
Transmission:	98%	95%
Diffraction efficiency:	90%	90%
Carrier frequency:	Adapted to used wavelength	200 MHz
Diffraction angle:	Adapted to frequency	30 mrad at 633 nm
Modulation bandwidth at −3dB:	1 MHz	40 MHz
Rise time for 1 mm beam		
diameter:	1200 ns	160 ns
First-order extraction ratio:	2000:1	2000:1
Drive power:	Adapted to used wavelength	0–2.5 W
Input impedance:	50 Ω	50 Ω

Table 7.3 Acousto-optic deflectors and magneto-optic devices

Some manufacturers	Optilas Ltd	Optics for Research (OFR)	
Typical specifications	Optilas Ltd LS110–XY	OFR model IO–P–JR1	OFR model IO–5–UVS
Material: Wavelength:	TeO$_2$ 440 nm–1.06 μm	Input aperture: 1 mm Wavelength band: 1300 nm	Aperture: 4.8 mm Wavelength band: 380–470 nm
Time–bandwidth product (resolution):	750 × 750	Insertion loss: 1.0–2.0 dB	Reverse isolation ratio: (3000 : 1–50 000 : 1) (35–47) dB
Time aperture:	15 μs	Reverse isolation ratio: 33–46 dB	Forward transmittance: 90–95%
Sweep bandwidth:	50 MHz		

7.4 Magneto-optic devices: Faraday isolators

The application of a magnetic field to optical materials can affect their absorption, refraction and the symmetry properties of their optical constants, as discussed in Section 2.2.3. Magneto-optic effects are thus used in modulators (including spatial light modulators). Two effects are in current use: the Faraday effect is used to rotate the plane of polarization of incident light and hence form an *optical isolator*, and a laser-induced polarization change is utilized in read/write erasable memory disks [7.2] The optical isolator is a unique application that is distinctive of magneto-optical effects in that the polarization rotation effect does not reverse with the direction of the light beam. The origin of all magneto-optical effects is the change in motion of the charge carriers in the presence of a magnetic field. Consider an electron, free to move, under the influence of an electric field from an electromagnetic wave $\mathbf{E} = \mathbf{E}_0 \exp \mathrm{i}(kz - \omega t)$. In the absence of an applied magnetic field $\mathbf{B}$ the electron moves along a particular line (Figure 20a), but with an applied magnetic field the total force on the electron is given by Lorentz's expression $\mathbf{F} = e\mathbf{v} \times \mathbf{B} + e\mathbf{E}$ which produces an extra magnetic force perpendicular to the force exerted by the electric field (Figure 20b). This sends the electron into a circular orbit that is characteristic of *cyclotron resonance*. The angular frequency, $\omega_c = eB/m$, is now imposed upon the electron's motion. A free electron would give rise to free carrier magneto-optical effects which affect both the absorption and refraction of light. If the carrier is bound, the angular frequency, ω_c imposed, will *modify* any natural frequencies associated with the restoring forces in the medium. The analogies of this process with quantum mechanics have been discussed in Chapter 2; a relevant treatment is given in Section 2.2.3.

The circular motion of electrons has an important effect on any circular polarization of a light beam. A linearly polarized plane wave can be decomposed into two circular components rotating in opposite senses but with the same phase (Figure 7.21).

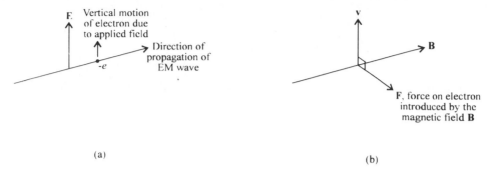

(a)   (b)

Figure 7.20 (a) The motion of an electron in an applied electric field **E**. (b) The effect of introducing a magnetic field perpendicular to the applied electric field on electron motion.

From the geometry of Figure 7.20, it can be seen that *one* of these senses will be 'excited' by the electron's resonating motion, while the other sense will be opposed (i.e. one sense follows the direction of rotation of the electron motion, while the other opposes it). If the electron is bound, with natural frequency, ω_0, the effect produces two magnetic resonances at frequencies, $\omega_0 \pm \omega_c$. From the relation between absorption and dispersion through the Kramers–Kronig relation two different refractive indices n_+ and n_- will affect the right circularly polarized (RCP) and left circularly polarized (LCP) waves, respectively. The different refractive indices change the phase velocity of the RCP and LCP waves, as shown in Figure 7.22.

On leaving the sample at $z = L$, the rotation of the two components is no longer symmetrical and the new direction of polarization, θ, is given by vectorially combining the two components at $z = L$ (Figure 7.23). The angle by which the beam leaving the crystal has rotated, θ, is given by

$$\theta = \frac{\omega L}{2c} (n_+ - n_-) \tag{7.30}$$

where n_+ and n_- represent the different refractive indices that affect the two circular components.

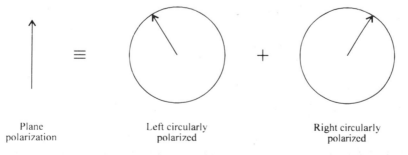

Plane   Left circularly   Right circularly
polarization   polarized   polarized

Figure 7.21 Decomposition of a plane wave into two circularly polarized components.

Figure 7.22 The effect of the differing refractive indices within the material on both the LCP and RCP components of a plane wave. In this case the RCP component is rotating in the same sense as the electrons, and hence rotates at a faster angular velocity than the LCP component.

This effect (known as Faraday rotation) was discovered by Michael Faraday in 1842 whilst transmitting light through glass in the presence of a magnetic field. The simple explanation given above applies to any material. The relative size of the effect is usually expressed by rewriting Eq. (7.30) as

$$\theta = VBL$$

where V is the Verdet constant. The explanation above also provides an understanding of a unique optical property of the Faraday rotation, which is that it *does not reverse with a reversal in the direction of the incident light beam*, unlike the electro-optic effects.

Consider Figure 7.20, with an electromagnetic wave propagating in the *opposite* direction to that shown. The electric field, **E**, still produces motion along the same line and, **B**, the magnetic field is still at right angles to this motion so that the magnetic component of the Lorentz force, **v** × **B**, continues to turn the electron in the same way. As a result the refractive index terms have the effect along the opposite beam direction that the Faraday rotation continues in the same sense with respect to the original beam. This is made use of in the optical isolator shown in Figure 7.24. It can be seen that the initial polarization direction determined by an input polarizer,

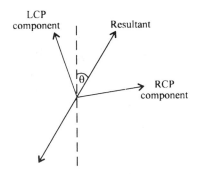

Figure 7.23 The vector combination of LCP and RCP components at $z = L$ produces a linearly polarized output wave with the plane of polarization rotated by an angle θ.

(a) Forward mode of propagation

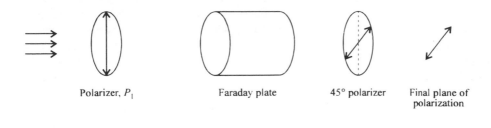

Polarizer, P_1 Faraday plate   45° polarizer   Final plane of polarization

(b) Reversed mode of propagation

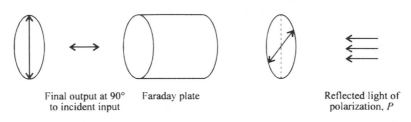

Final output at 90° to incident input   Faraday plate   Reflected light of polarization, P

Figure 7.24 Propagation of a linearly polarized electromagnetic wave, P_1 (a) one way through a Faraday plate system, and (b) on its return journey back through the same system.

P_1, is rotated by 45° after passing through the Faraday plate. Any reflected radiation passes back through the Faraday plate, continuing to have its plane of polarization rotated further until it has been rotated by 90° compared with the initial plane of polarization.

At the input polarizer P_1 this polarization is now at right angles to the setting of P_1 and so is completely attenuated. This arrangement allows the isolation of a source from any reflection and is known as an *optical isolator*. In practice, materials are used that depend on the Faraday effect of bound carriers rather than free carriers. However, the effects are essentially the same as far as the isolator function is concerned.

Applications range from very high-power laser systems with high gain amplifier stages, where stability is at stake, to laser ring-giro systems where a small amount of feedback can affect frequency stability. Application to the unidirectional propagation of optically encoded information has also grown in importance. Practical devices are offered by manufacturers and can be used with lasers from 380 nm to 1.6 μm. Suitable materials of high Verdet constant (such as Hoya Faraday rotator glass FR5, with $V \cong 26$ min/G/cm) are configured with a permanent rare-earth magnet in packages a few centimetres long and allowing optical apertures of up to 15 mm in diameter.

An exact rotation of 45° is required within the isolator and hence the length of the rotator element can be varied in order to take account of different temperatures and wavelengths. The quality of the polarizers determines the degree of isolation provided. Isolation ratios of 1000:1 (30 dB) to 4000:1 (36 dB) are typically specified, but a double series can be used for more severe isolation requirements where a ratio of up to 50 000:1 (87 dB) can be provided. Yttrium iron garnet (YIG) and gallium-doped YIG are suitable for the visible and near-infrared regions. For the far-infrared region, some *semiconductors* such as InSb have large effects (although they do require cryogenic cooling in order to keep unwanted absorption low). The theory of *refractive* Faraday rotation can be connected to magneto-absorption via band theory (see Chapter 2) using the Kramers–Kronig relation. This relation gives the Faraday rotation angle as

$$\theta = \frac{1}{2\pi n} P \int_0^\infty \frac{\omega'(n_+ \alpha_+ - n_- \alpha_-)}{\omega'^2 - \omega^2} \, d\omega'$$

where P is the principal part of the integral; α_+ and α_- are the absorption coefficients for right and left circularly polarized light, respectively; n_+ and n_- are the refractive indices for right and left circularly polarized light, respectively; ω is the resonance frequency of the material; and ω' is the incident frequency.

References

[7.1] A. Yariv, *Quantum Electronics*, 3rd edn. (Wiley, New York, 1989), p. 333.

[7.2] G.E. Thomas, 'Future trends in optical recording', *Philips Tech. Rev.* **44**, 51–57 (1988).

8

Diode-laser-pumped solid-state lasers

8.1   Introduction
8.2   High-power diode lasers
8.3   Solid-state laser media
8.4   Pumping geometries
8.5   Application of holosteric lasers to nonlinear optics
        References

8.1 Introduction

With the notable exception of diode lasers, every laser source in common use until the mid-1980s relied upon a gas discharge for its power source. In carbon dioxide lasers, for example, the discharge excites the lasing medium directly, and in 'solid-state' lasers such as Nd:YAG a gas discharge in a flashlamp optically pumps the lasing ions in the crystalline rod. While these lasers have been highly developed in many directions, their reliance on gas discharge technology limits their performance with respect to maintenance-free lifetime and efficiency.

The advent of semiconductor diode lasers of sufficient power and appropriate wavelength to excite solid-state lasers such as Nd:YAG has allowed the development of entirely solid-state laser systems, or *holosteric* lasers (from the Greek *holo*, wholly, and *steric*, solid). The benefits of this new pumping scheme are many, including high efficiency, long lifetime, high beam quality, and compact systems.

In this chapter we discuss the technology and physics of holosteric lasers, and further show how they can provide useful and reliable sources. The high beam quality of these systems helps nonlinear conversion efficiency, such that there is currently great interest in combining these all-solid-state lasers with frequency upconversion and optical parametric oscillators to produce all-solid-state coherent sources tunable throughout the visible and near infrared.

An example of one holosteric laser which helps illustrate the advantages of this technology is shown in Figure 8.1. A Nd:YAG laser is formed by depositing multilayer dielectric mirrors directly onto the ends of the solid-state laser medium, which has one curved mirror chosen to define an intracavity TEM_{00} mode radius of a few tens of microns. Rather than using a flashlamp to excite all the rod, the partially coherent 809 nm radiation from a 0.5 W diode laser is collected and focused by a pair of lenses

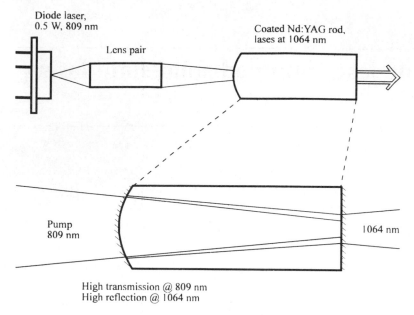

Figure 8.1 Example of a diode-laser-pumped monolithic Nd:YAG laser.

to excite only the Nd ions within the TEM_{00} mode volume. This good spatial overlap between the population inversion and the TEM_{00} mode volume ensures easy TEM_{00} operation. The Nd:YAG laser can also be very efficient: only those ions that can generate light within the TEM_{00} mode are excited, there is close to a unity quantum efficiency between the number of exciting photons and the number of photons generated by stimulated emission at 1064 nm, and the ratio of the pump and lasing photon energies is close to 1. Indeed, optical to optical slope efficiencies as high as 62% have been achieved. The high efficiency reduces problems of heating and stress in the laser rod, and the lack of UV pump radiation avoids the problem often associated with flashlamps, of photochemical degradation of the solid-state laser material.

The diode laser is remarkably efficient at converting electrical energy to light output, and many commercial devices have efficiencies of between 25 and 50%. As the spectral and spatial properties of the diode laser output can be matched so well to the solid-state laser, overall 'wallplug' efficiencies (electrical energy from a mains socket to solid-state laser output) for a holosteric laser system can be as high as 18% [8.1]. The holosteric laser system can also have a much longer maintenance-free lifetime than its flashlamp-pumped equivalent; the diode lasers have lifetimes expected to be of the order of tens of thousands of hours, rather than a couple of hundred hours for flashlamps.

One might ask why one should bother with the solid-state laser material at all. The reason is that the diode-laser-pumped solid-state laser material has many properties that are superior to those of the diode laser alone. Many diode lasers, particularly those with high powers, have a relatively poor beam quality; the beams are often highly divergent and not diffraction-limited, they are astigmatic and of large spectral bandwidth. Although this can reduce their applicability, the beams are sufficiently good that they can be efficiently collected and refocused into the TEM_{00} mode volume of a solid-state laser. The conventional cavity of the solid state laser allows the generation of beams of good spatial quality, and, with access to the intracavity flux, techniques such as single frequency selection and stabilization, second harmonic generation, and mode locking can be used. To some extent, these same techniques can be used with a diode laser in an external cavity. However, there are two areas where the diode-pumped laser concept does undoubtedly have the advantage: multiplex pumping and energy storage. The first area concerns the possibility of using many diode lasers to pump one solid-state laser to produce high output powers. This multiplex pumping can be done very simply and effectively in both coaxial and transverse [8.2] pumping geometries, and will be covered later in this chapter. The second area is of importance for the generation of laser pulses of nanosecond duration with high energy. The long upper-state lifetime of many solid-state lasers allows Q-switching to be a useful technique. A high fraction of the total energy provided by a diode-laser in a pulse of hundreds of microseconds can be stored in the population inversion in the solid-state laser, and then be released in a short pulse with high peak power. This mode of operation can be useful for nonlinear optics, LIDAR, range-finding, material processing, etc. It is difficult to conceive of a diode-only system that could provide the same peak powers in a TEM_{00} beam.

The all-solid-state nature of these systems brings them many advantages over their gas-discharge-based predecessors. In the remainder of this chapter we discuss the operation of high-power laser diodes, some of the types of solid-state laser materials that are currently of interest, and the various ways that diodes and solid-state laser media can be brought together to form useful systems. In the final sections we discuss how the holosteric laser can be combined with nonlinear optical effects to produce fixed and tunable outputs at new wavelengths.

8.2 High-power diode lasers

The initial development of holosteric lasers was stimulated by the availability of diode lasers with significant power to act as pump sources. Since the initial demonstrations, developments of high-power diode lasers and diode-pumped lasers have enjoyed a symbiotic interdependence. The first diode lasers to be used were AlGaAs lasers operating at a few milliwatts [8.3, 8.4]. These matched well the 809 nm absorption peak in Nd:YAG, and had small emitting areas that were suitable for direct imaging onto the solid-state laser rod for pumping. Since then, diode lasers have been developed to very much higher powers (20 W continuous-wave (cw) devices are commercially available at the time of writing [8.5]), and to a wider range of wavelengths suitable for pumping different materials.

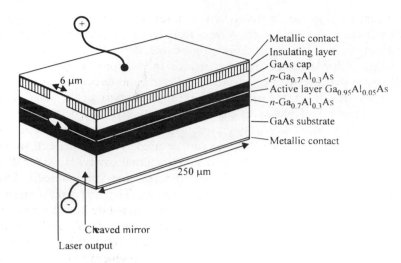

Metallic contact
Insulating layer
GaAs cap
p-Ga$_{0.7}$Al$_{0.3}$As
Active layer Ga$_{0.95}$Al$_{0.05}$As
n-Ga$_{0.7}$Al$_{0.3}$As
GaAs substrate
Metallic contact

6 μm

250 μm

Cleaved mirror
Laser output

Figure 8.2 Single-stripe, gain-guided diode laser.

The type of diode laser used for the lowest power applications is not dissimilar to that used in compact disc players, as sketched in Figure 8.2. A forward-biased pn junction provides the optical gain, as discussed in Chapter 3 and by Yariv [8.6]. In this case layers of AlGaAs are grown on top of a GaAs substrate by metal organic vapour deposition (MOCVD). The various doping levels and aluminium concentrations produce a structure with confinement for both charge carriers and the generated light (see Sections 3.43–3.46), resulting in an efficient device with low thresholds. Single and multiple quantum wells within the active region are used to increase the efficiency and contribute to the selection of the appropriate lasing wavelength [8.7]. The ends of the device are cleaved to produce mirrors with reflectivities of the order of 33%; additional coatings may be applied to increase the reflectivity of the rear mirror and to decrease the reflectivity of the front mirror.

A single stripe is usually defined in the structure of the laser. The laser in Figure 8.2 is drawn as a gain-guided laser, where proton bombardment has been used to render much of the top of the laser insulating, and thus leaving only a small stripe through which current flows to the substrate. This results in gain being localized, and an output is obtained from a region about 6×1 μm (alternatively, more complicated structures may be grown in which both the carriers and light are confined in the horizontal direction by changes in the composition of the semiconductor material; these devices are referred to as index-guided lasers). The small dimensions of the emitting region result in a large divergence of the output beam due to diffraction (see Section 3.4.2). In practice, the divergence of many laser structures is significantly greater than the value expected from diffraction alone.

There are two main limitations on the power output of these devices: heat removal from the active region as a whole, and thermal runaway at the facets. The first

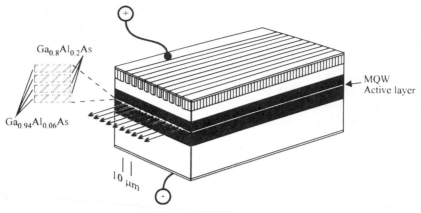

$Ga_{0.8}Al_{0.2}As$

$Ga_{0.94}Al_{0.06}As$

MQW
Active layer

$10 \, \mu m$

Figure 8.3 Gain-guided 10-stripe diode laser array.

limitation is reduced by bonding the laser *p*-side down, so that the active region is close to the cooled heatsink. The second limitation is due to bandgap changes near the surface causing the semiconductor to become partially absorbing at the laser wavelength. The absorption of the circulating laser radiation can cause additional heating, which leads to an increased absorption. This produces a thermal runaway in which the facet can boil off in microseconds. Thus, for diodes in which thermal runaway is the predominant limitation, pulse lengths of greater than a microsecond are restricted to the same peak power as true cw operation. The critical power density for the thermal runaway process to occur in AlGaAs is about 1 MW cm^{-2}, although this can be reduced somewhat by keeping the gain away from the ends of the stripe and thus defining 'non-absorbing mirrors' on the laser.

It might be thought that the way to a greater total output power would be simply to increase the width of the stripe, and thus keep below the critical power density for runaway. While this can work, there is a tendency for the current flow through a larger stripe to be nonuniform, and thus produce local regions of high gain and concomitant high power density at the facet. A solution adopted by some manufacturers is to define a number of stripes on the same piece of semiconductor, as shown in Figure 8.3. These individual stripes are electrically in parallel, and each produce their own beam. If there is optical coupling between adjacent stripes there may be a degree of phase-locking across the array, with the output beam from the array being much less divergent in the plane of the stripes than would the beam from any one stripe. Unfortunately, in many laser structures the lowest loss in the array occurs when adjacent stripes are out of phase; the resultant interference pattern then takes the form of a dual lobed beam. Thus many such arrays are manufactured with little or no coupling amongst the stripes, leading to the output divergence being essentially the same as that of each stripe (significant improvements can be made by forcing the stripes to lock together in phase through injection locking using either an external master laser, or through selective feedback of part of the output of the array). At the time of writing, one of the most popular commercial devices of this type is the Spectra Diode Labs SDL-2460, which emits 1.0 W cw from 20 stripes 5 μm wide across a 200 m aperture. The emission is into a cone of full angles 40° × 10° (FWHM), and the specified efficiency is 26%.

These watt-level devices can be seen to be attractive pump sources for solid-state lasers. The small emitting area allows simple optics to be used for re-imaging the active region onto the solid-state medium for end pumping. Lifetimes of greater than 10 000 h are predicted, and efficiencies of conversion from electrical to optical power of between 25 and 50% are achieved. Varying the relative abundance of aluminium in the $Al_xGa_{1-x}As$ composition allows the production of lasers with wavelengths of 770–900 nm. Fine tuning of the wavelength of emission is achieved by changing the temperature of the active region; the emission wavelength of the AlGaAs devices changes by about 0.3 nm °C^{-1}. While this allows diode lasers of slightly different wavelengths to be temperature tuned to match the wavelength of the absorption peak in a solid-state material, it also means that the temperature of the diode must be controlled to maintain optimum pumping conditions.

For reliable cw power levels greater than 3 W, the total heat load must be spread out over a somewhat larger area. This is achieved by defining a large number of stripes on one much wider bar of semiconductor material. Continuous-wave devices often have a set of arrays of the above type spaced out from each other by a great enough distance to ease the average thermal dissipation processes. The current technology consists of ten stripe arrays set out along a 1 cm width bar to produce 20 W of power, as shown in Figure 8.4.

The other major type of bar is optimized for operation at pulse lengths of 200–400 μs. These are the pulse lengths required for efficient pumping of a Q-switched Nd:YAG or Nd:YLF laser, which have upper-state lifetimes of 230 and 480 μs, respectively. As these arrays are commonly pulsed at up to 100 Hz, the average power limitations are less than with the cw devices, and so the fill factor of stripes on the bar can be increased. The (Spectra Diode Labs 2460) SDL-2460 contains 1000 stripes across a 1 cm wide bar, and emits 60 W for 200 μs, i.e. 12 mJ. These 'quasi-cw' bars are etched at various points across their width to reduce the possibility of lasing across the stripes. The next step to increased energy is to expand in the other direction; two-dimensional arrays are fabricated by a 'rack and stack' technique, in which several bars are stacked on top of each other to form a set of emitters. One such 2-D array that is now commercially available produces 300 mJ pulses from a 1 cm^2 active area, while still maintaining a lifetime of greater than 10^9 shots. Koechner presents a useful review of this field from an engineering viewpoint [8.8].

A possible alternative to this rack and stack geometry is the use of *surface emitting lasers*, which can take two main forms. In one, many lasers are fabricated on a substrate, and mirrors or gratings on the same substrate direct the radiation out at 90° to the laser stripes, as shown in Figure 8.5(a). In the other form, lasing is made to occur perpendicular to the semiconductor layers [8.9], as indicated in Figure 8.5(b). These types of laser are likely to become increasingly significant in the future, but

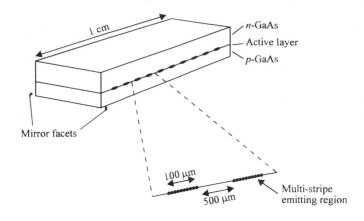

Figure 8.4 Arrangement of the Spectra Diode Labs 20 W cw diode laser bar, and a picture of the device.

currently most high-power holosteric lasers are still powered by rack and stack 2-D arrays.

The discussion above has concentrated on diodes operating at around 810 nm for pumping Nd. The same ideas are used for diodes at other wavelengths. Some of the more important combinations of laser diode and solid-state material are summarized in Table 8.1.

Table 8.1 A selection of some of the more important combinations of diode laser and solid-state laser media

Wavelength of diode (nm)	Diode material	Laser material	Wavelength of solid-state laser (nm)
670	InGaAlP	Cr:LiSAF	760–920
		Cr:LiCAF	720–840
785	AlGaAs	Tm:YAG	2010
		Tm:Ho:YAG	2100
795		Nd:YLF	1047 and 1053
801		Nd:glass	1053
809		Nd:YAG	1064
940	InGaAs	Yb:YAG	1020
980	InGaAs	Er:YLF	1600, 2800

(a)

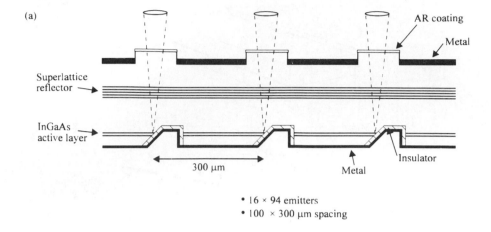

- 16 × 94 emitters
- 100 × 300 μm spacing

(b)

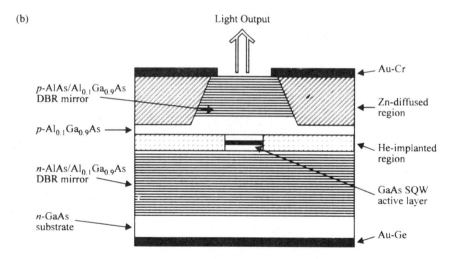

Figure 8.5 (a) Schematic diagram of a 2-D surface emitting monolithic diode array. (b) Continuous wave single transverse mode vertical cavity surface-emitting lasers fabricated by helium implantation and zinc diffusion.

8.3 Solid-state laser media

The first, and arguably still the most important, of the modern diode-pumped materials is neodymium-doped yttrium aluminium garnet (YAG), which is pumped by the well developed AlGaAs diode-laser array [8.8]. The availability of diodes of the correct wavelength to pump this material has been one reason for the initial strong interest in this material, though Nd:YAG also has many of the properties that one would hope for in a 'good' laser system. The laser ion exhibits a sharp fluorescent line on

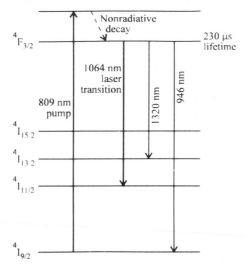

Figure 8.6 Energy level diagram of Nd:YAG. The 809 nm pump transition and strong 1064 nm transition are shown. The two other important laser transitions are to states within the $^4I_{13/2}$ and $^4I_{9/2}$ manifolds at 1320 and 946 nm, respectively.

the laser transition, it has a relatively long upper laser level lifetime, and a lower laser level that rapidly empties and is not thermally populated at near room temperature (Figure 8.6). These are common features of the transition metals, rare earths, and actinides, which have a complete outer electron shell shielding their optically active electrons from the electric fields associated with the host material. In addition, there are strong absorption bands at the pump wavelength, and the ratio of the photon energies of the pump to laser wavelengths are also very close.

The choice of host for the laser ion is also important. The host should have good optical, mechanical, thermal, and chemical properties, and should allow an appropriate density of laser ions to be incorporated into its structure without causing growth problems, strain, or variations in refractive index. YAG is one such material, in which the Nd^{3+} ion replaces the Y^{3+} ion. However, the 3% size difference between the two ions does introduce some strain into the crystal, and this limits the useful doping density to about 1.5% atomic. Several other host materials are used for containing Nd^{3+}, some of which are shown in Table 8.2. The different interactions between host and the laser ion cause the wavelengths of operation to shift slightly, the pump and laser emission bandwidths to change, and the peak stimulated emission cross-section to vary. The different thermal and mechanical properties of the different hosts also strongly affect the choice of host for a particular application. For example, very large blocks of Nd-doped glass can be fabricated, but as the thermal conductivity of the glass is relatively low, these systems tend to be limited in average power-handling capability. Glasses can be used to contain a somewhat higher dopant density of Nd ions than most crystals. In this case the maximum useful dopant density is

Table 8.2 Selected spectroscopic and thermal properties of some important solid-state laser materials. λ, wavelength of laser line; σ, stimulated emission cross-section; n, refractive index; τ_2, upper laser level lifetime; λ_{abs}, main pump wavelength that is diode-accessible; $\Delta\lambda_{abs}$, width of pump band; α_p, typical absorption coefficient of pump band; dn/dt, rate of change of refractive index with temperature; K, thermal conductivity; α_{exp}, expansion coefficient

Ion and host	λ (nm)	σ (10^{-19} cm²)	n	τ_2 (μs)	λ_{abs} (nm)	$\Delta\lambda_{abs}$ (nm)	α_p (cm^{-1})	dn/dt (10^{-6} K^{-1})	K (W cm^{-1} K^{-1})	α_{exp} (10^{-6} K^{-1})
Nd:YAG	1064	7.5	1.83	230	807.5	4	5.2	7.3	0.13	7.5
Nd π:YLF o	1047	3.7	1.45	480	792.5	6		−4.3	0.06	10.5
	1053	2.6	1.47		797.5	3				
Nd:glass (LH5)	1054	0.41	1.54	290	801.5	5	9.0	8.6	0.012	8.6
								4.6?		
LNP c	1048	3.2	1.58	1.20	801		260	−4		7–13
Ti:sapphire	700	3.5	1.76	3.2	490	150	5	13	0.35	9
	−1100									
Cr:LiSaF	780	0.5	1.41	67	670	100	50	−4∥ to c	0.03	−10∥ to c
	−1010							−2.5⊥ to c		19⊥ to c

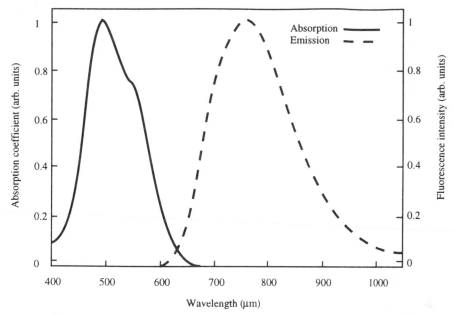

Figure 8.7 Absorption and emission spectra for the Ti:sapphire laser at room temperature.

limited by the reduction in the effective upper-state lifetime of the laser ion due to interactions between neighbouring Nd ions; this is called concentration quenching, and limits the useful dopant density in glass to a few percent. Another important class of Nd-containing material is the stoichiometric materials such as lithium neodymium tetraphosphate (LiNd P_4O_{12}), in which the Nd ion is incorporated into the basic building block of the crystal matrix. In these materials concentration quenching is less of a problem, since the regular arrangement of the Nd ions in the crystal keeps all the Nd ions a fixed distance apart.

Several other laser ions are receiving much attention, some of which were listed in Table 8.1. Holmium, thulium, and erbium can all be made to lase in the 2–3 μm range [8.10]. These wavelengths can be strongly absorbed by water, which makes these lasers useful for cutting and vaporizing in medical applications. For the same reason, these lasers are relatively 'eye-safe', as the incident radiation is absorbed near the surface of the eye, rather than being focused to a small spot on the retina. This has led to interest in these systems to develop 'eye-safe' LIDAR applications. Erbium also has a laser transition at 1.5 μm, which has led to its widespread use as a dopant in optical fibers for amplification of signals in optical fiber communication systems (see Chapter 10). The relative merits of some of the different ion and host combinations will now be described.

In contrast with the Nd lasers, certain types of new solid-state lasers use ions in which there is a strong electron–phonon interaction between the optically active

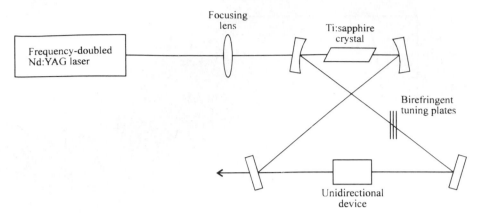

Figure 8.8 A diode-laser-pumped cw Ti : sapphire laser.

electron and the crystalline lattice. This can result in a large broadening of the emission line, which tends to reduce the peak stimulated emission cross-section. However, if there is enough pump power available to bring these systems above threshold, then there is the possibility of tuning these lasers over substantial ranges. The first spectacularly successful tunable solid-state laser was the Ti:sapphire laser. The interaction of the Ti ion with the Al_2O_3 host produces the absorption and emission spectra shown in Figure 8.7. The incorporation of the Ti:sapphire crystal into a cavity with tuning elements can allow tuning of the Ti:sapphire laser from 670–1100 nm

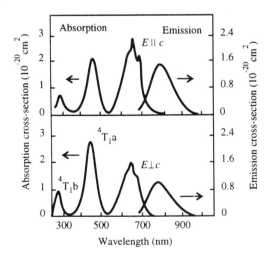

Figure 8.9 Absorption and emission spectra for Cr : LiCAF for polarization parallel and perpendicular to the *c*-axis.

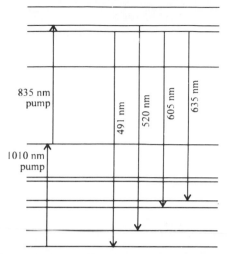

Figure 8.10 Energy level scheme for Pr^{3+}-doped ZBLANP glass, showing the two diode-accessible wavelengths used for the two-step excitation process, and the four reported lasing wavelengths.

[8.11]. However, the absorption bands of this material do not fall in the range of any high-power diode types, so the route to an all-solid-state version of this tunable laser system is by generating the second harmonic of a diode-laser-pumped Nd laser, and using this to pump coaxially the Ti:sapphire laser, as shown in Figure 8.8.

More recent interest in tunable holosteric lasers has focused on the chromium-doped materials [8.12] Cr^{3+}-doped $LiCaAlF_6$ and Cr^{3+}-doped $LiSrAlF_6$, known as LiCAF and LiSAF, respectively. LiCAF has the absorption and emission spectra shown in Figure 8.9. The advantage of these materials over Ti:sapphire is that they are directly diode-pumpable with 670 nm diodes, and still have very substantial tuning ranges in the near-infrared, 720–840 nm for LiCAF, and 760–920 nm for LiSAF.

All the laser systems mentioned above are diode-pumpable, and the output photons have a lower energy than those of the pump. Substantial efforts have been made to produce solid-state lasers in which the output photons have a higher frequency than that of the pump. In such cases, several low-energy photons must be absorbed for every high-energy photon that is emitted in lasing. These incoherent upconversion processes can take a number of forms, as reviewed by Lenth and MacFarlane [8.13]. All the processes rely on obtaining a sufficiently large population in an intermediate excited state such that there is a good chance of a further energy absorption or transfer to excite the ion further. As such, high photon densities are used, and mechanisms that would depopulate the intermediate state must be reduced. Initial work concentrated on these processes at low temperatures. Some recent exciting work has made use of the properties of doped optical fibers made from fluoride glass, which can maintain high photon fluxes over a long region, and have low phonon energies.

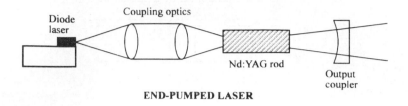

END-PUMPED LASER

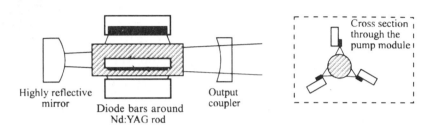

Figure 8.11 Examples of end-pumped and side-pumped holosteric lasers.

As such, cw room-temperature upconversion lasing has been demonstrated in praseodymium-doped fibers. The fiber was pumped with the (diode-accessible) wavelengths of 1010 and 835 nm, as indicated in Figure 8.10. A two-step absorption process then creates population inversions that can drive lasing at 491, 520, 605 and 635 nm.

8.4 Pumping geometries

At the start of this chapter we reviewed an end-pumped laser system, where the diode laser radiation was collected and focused to be coaxial with the Nd:YAG laser mode. This type of pumping is particularly well suited to low-threshold, TEM_{00} operation of Nd:YAG and similar lasers with the use of small pump diodes.

An alternative pumping scheme involves pumping into the side of the active region; this is particularly useful for higher-power devices where the power of many different diode arrays requires to be coupled into the active volume of the solid-state laser. In this case diode laser arrays are used much more like the flashlamps that they are replacing, with the arrays exciting large volumes of laser material transversely to the solid-state laser axis. These two schemes are outlined in Figure 8.11.

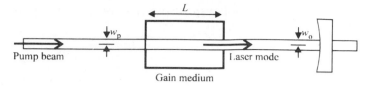

Figure 8.12 Model for the end-pumped laser. The pump beam and laser mode are both assumed to be circularly symmetric.

8.4.1 Longitudinal pumping schemes

In this section we consider the pumping powers required in diode-laser-pumped solid-state lasers in order to reach oscillation thresholds in different gain media. We start by considering a simple derivation of the pump power required to reach oscillation threshold in an idealized end-pumped geometry, and we then explore how the simple expression so derived is modified to account for departures from the idealized geometry as encountered in practical systems. Figure 8.12 illustrates the essential features of the idealized geometry. The pump power, P, is assumed to be in a cylindrical beam of radius w_p which is incident on the active medium in such a way as to be collinear with the cavity mode. The mode is configured to have the same cross-sectional area as the pump beam. The active medium is chosen to be long enough (length L) that all of the pump power is absorbed within it. Under these conditions the average excitation rate, R, per unit volume through optical pumping is given by

$$R(\text{vol}^{-1}\ \text{time}^{-1}) \sim \frac{P}{\pi w_p^2 L h v_p} \tag{8.1}$$

If we assume the fraction of this excitation that deposits an atom or ion in the upper level is η_p (pump quantum efficiency), and that the lifetime of the upper laser level is τ_2 then the steady-state population per unit volume in this state is:

$$N_2(\text{vol}^{-1}) \sim \frac{\eta_p P \tau_2}{\pi w_p^2 h v_p} \tag{8.2}$$

If the population of lower laser level can be neglected, N_2 is also the population inversion. The gain–length product associated with the laser transition is hence given by

$$\alpha L \sim N_2 \sigma_e L \tag{8.3}$$

where σ_e is the stimulated emission cross-section. Hence we obtain

$$\alpha L \sim \frac{\sigma_e P \eta_p G \tau_2}{\tau w_p^2 h v_p} \tag{8.4}$$

At threshold the gain (double pass) equals the round trip losses, which are made up of the useful output coupling loss (fractional round-trip loss T) and the parasitic loss (fractional round-trip loss δ_f). We hence obtain, on equating the round-trip gain–length product $(2\alpha L)$ to these losses, the following expression for the pump power required to reach oscillation threshold:

$$P_{th} = \frac{\pi h v_p}{2\sigma_e \eta_p \tau_2} (T + \delta_f) w_p^2 \tag{8.5}$$

A more comprehensive analysis of end-pumping has been presented by Fan and Byer [8.14] in which variations in the spatial distribution of the pump and laser cavity modes are taken into account through the use of overlap integrals. For the case where pumping is by a coherent pump beam of Gaussian beam radius w_p and the Gaussian beam radius of the cavity mode is w_0 (w_0 not necessarily equal to w_p), then Eq. (8.5) becomes,

$$P_{th} = \frac{\pi h v_p}{2\sigma_e \eta_p \tau_2} \left(\frac{T + \delta_f}{2} + \alpha_i L + \alpha_l L \right) (w_0^2 + w_p^2)[1 - \exp(-\alpha_p L)]^{-1} \tag{8.6}$$

where $(\alpha_i L)$ describes losses (single pass) in the gain medium due to bulk absorption and scattering, $(\alpha_l L)$ describes lower laser level absorption losses, and α_p is the absorption coefficient for the pump radiation in the gain medium. The term $[1 - \exp(-\alpha_p L)]$ is the fraction of the incident pump power that is deposited in the gain medium. The lower laser level absorption loss coefficient α_l is given by

$$\alpha_l = \sigma N_0 f/a \tag{8.7}$$

where σ is the absorption cross-section for the laser transition (comparable, but not necessarily equal to σ_e, the stimulated emission cross-section for the transition), N_0 is the active species doping density, and f/a is the fraction of this density excited to the lower laser level due, for example, to thermal effects.

From Eq. (8.6) it can be seen that two requirements must be met in order to achieve low-threshold operation of an end-pumped device: (i) the gain medium must be long enough to absorb a significant fraction of the pump light; and (ii) the cross-sectional area of the pump beam must be less than or equal to the cross-sectional area of the laser mode within the gain medium. The cross-sectional area of the laser mode is chosen in the first instance to be that at which adequate gain is obtained (compared to expected losses) with the pump power available.

So far, consideration has been limited to the case of a coherent pump beam (i.e. a Gaussian beam). When scaling to higher pump power is required it becomes necessary to extend the analysis to include the case of incoherent (or partially coherent) pump sources. This extension has been considered by Fan and Sanchez [8.15], the main aspects of whose analysis we now review. The laser mode radius is assumed to be constant over the length of the gain medium (i.e. the confocal parameter of the laser mode is assumed to be greater than the length of the gain medium), and further, in order to avoid having to use mode overlap integrals, the pump power is assumed to be uniform over any laser mode cross-section within the gain medium. The pump beam is treated as an incoherent beam and its focusing in the gain medium is described in the geometrical optics limit by the relation:

$$w_p(z) = w_{p_0} + \theta_p |z| \tag{8.8}$$

where $w_p(z)$ is the pump beam radius at a general point z along the axis of the gain medium, w_{p_0} is the pump beam waist, assumed to be located at the centre of the gain medium (where $z = 0$), and θ_p is the far-field divergence (half-angle) of the pump beam. The far-field divergence and waist size are related through the Helmholtz–Lagrange invariant for geometrical imaging, namely:

$$n \cdot w_{p_0} \cdot \theta_p = \text{constant, } C \tag{8.9}$$

where n is the refractive index of the medium and the constant C depends on the number of times the beam quality associated with the source lies above the diffraction limit (in one dimension). For the case of a uniform, rectangular source, the constant C can be expressed as

$$C = N_D(\lambda_p/\pi) \tag{8.10}$$

where (λ_p/π) is the diffraction-limited value of the product and N_D is the number of times the beam quality associated with the source lies above the diffraction limit. The problem now becomes one of optimizing the focusing, through altering θ_p, to achieve the highest single-pass gain, while subject to the constraint imposed by (8.10) above. Fan and Sanchez adopt the criterion that optimum focusing corresponds to the condition that $w_p(z = \pm L/2)$, i.e. the pump beam size at the edges of the gain medium, be minimized subject to the constraint imposed by (8.10). This gives the beam size at the edges of the gain medium as

$$w_p(z = \pm L/2)_{min} = (2CL/n)^{1/2} = (2N_D\lambda_p L/\pi n)^{1/2} \tag{8.11}$$

with a corresponding beam size at the centre of the gain medium of

$$w_{p_0} = \tfrac{1}{2}w_p(z = \pm L/2) \tag{8.12}$$

The pump beam must lie within the mode cross-sectional area within the gain medium, so that from (8.11) above we obtain the following condition:

$$2N_D\lambda_p L \leqslant n(\pi w_0^2) \tag{8.13}$$

If we combine Eq. (8.11) with Eq. (8.5) we obtain the following expression for the pump power required to reach oscillation threshold:

$$P_{th} = hc(L/\sigma_e n\eta_p \tau_2)N_D(T + \delta_f) \tag{8.14}$$

The benefit of obtaining close to diffraction-limited beams in order to minimize the threshold pump power is apparent from the above. The analysis applies to the case where pump beams are cylindrically symmetrical, in particular with regard to how closely they approach diffraction limited behaviour. Generally, properties of the beams from diode lasers differ markedly in the two directions orthogonal to the beam propagation direction, but the above analysis can readily be extended to incorporate such effects, and allows us to compare the very different threshold pump requirements for Nd:YAG and Cr:LiSAF lasers.

(a)

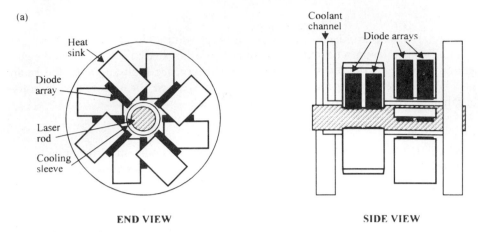

END VIEW   **SIDE VIEW**

Figure 8.13 (a) Schematic diagram showing a pump-module developed by Fibertek. A collection of 2-D diode arrays are arranged around a cylindrical Nd:YAG rod. (b) Schematic diagram of the St. Andrews 10 mJ laser, and a view of the 1064 nm fluorescence observed looking into the end of the pumped rod with a CCD camera. (The rod is in fact circular, the distortion coming from the display device.)

From the above analysis it is evident that the emitting dimensions of the pump, and their associated effects on the pump beam are very important. While $100 \times 1 \ \mu m$ diode laser arrays are good candidates for longitudinal pumping, the diode laser bars, which have a far from diffraction-limited beam emitted from a 1 cm $\times$ 1 μm area, were at first thought to be very poorly suited to end-pumping Nd:YAG lasers. However, if some cylindrical optics are included in both the pump beam and the laser cavity the dimensions of each can be brought closer to each other [8.16].

8.4.2 Transverse pumping schemes

The longitudinal or end-pumping schemes addressed above are particularly good choices for low threshold, TEM_{00} operation of solid-state lasers. To some extent, however, they are limited in the amount of pump power that can be conveniently coupled longitudinally into a small volume. Alternative techniques in terms of transverse pumping, which are more akin to the ideas involved in flashlamp pumping, can however be used to multiplex many diode lasers to pump one solid-state laser. The requirements for focusing the diode radiation to a small spot are greatly relaxed, and remarkably high total pump powers can be used. In one demonstration [8.17] over 1 kW of average power at 1 μm was generated by a Nd:YAG laser pumped by 160 diode array modules, at a 28% optical to optical efficiency. Although the excellent mode overlap present in the longitudinal pumping schemes is lost, transverse diode pumping still keeps many of the intrinsic advantages of the all-solid-state approach. Much of the novel design work in this field has been to design transverse pumping

(b)

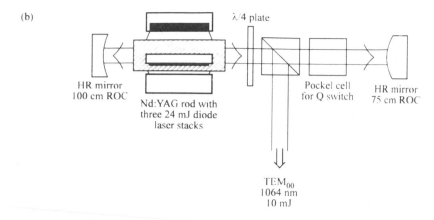

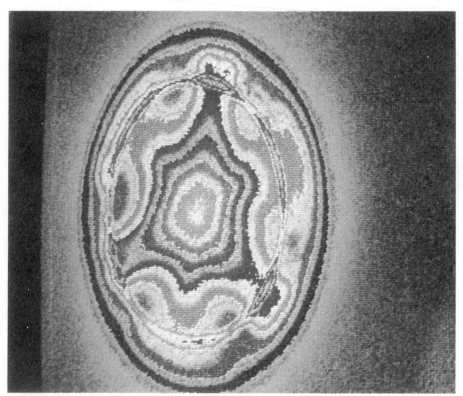

Figure 8.13 (b) Continued.

geometries in which there is still a good overlap between the population inversion distribution and the lasing mode. Two designs will be described briefly here, the Nd rod pumped by a number of diode laser bars arranged around its circumference, and the Nd slab with multiple reflections of the laser mode, at which the pump power is absorbed.

If a laser rod is transversely pumped by only one diode laser bar, then it is very difficult to get an efficient extraction of the deposited energy, since the maximum population inversion occurs where the laser mode is at its weakest. However, if the symmetry of the sytem is increased by adding more diodes around the rod, coupled with reflectors on the side of the rod opposite the diode pumps, much better overlaps between the lasing mode and the population inversion can be obtained. This is the approach taken by Burnham and co-workers [8.18] in their side-pumped gain modules. Sixteen five-bar diode arrays were arranged symmetrically around a 7.6 mm diameter Nd:YAG rod. The 200 μs long pump pulses added up to a total of 900 mJ at 806 nm, and 350 mJ of multi-transverse mode Nd:YAG laser radiation was produced. With one of these modules set up as a TEM_{00} oscillator, and four of the modules as double-pass amplifiers, their system produced 1 J of single-mode output in a repetitively pumped system. The electrical efficiency exceeded that of a flashlamp-pumped device by a factor of five, and the thermal lensing and depolarization in the amplifier modules were three times less severe than in a flashlamp-pumped equivalent.

At a much lower total power level, Q-switched Nd:YAG lasers powered by three 24 mJ diode laser arrays have been demonstrated [8.19]. These were arranged symmetrically around the 3 mm diameter rod, and the curved surface of the rod itself helps reduce the divergence of the pump radiation. A camera with a large depth of field was used to view the 1.06 μm fluorescence along the length of the rod, as shown in Figure 8.13(b). It can be seen that the fluorescence (and correspondingly the population inversion) is strongly peaked at the centre of the rod, where the circulating Nd:YAG TEM_{00} mode will be at its strongest. A Pockel's cell Q-switch and polarization output coupling were used as shown in Figure 8.13(b) to produce TEM_{00}, 10 mJ, 17 ns pulses.

The zig-zag slab is another geometry which can ensure that the circulating field overlaps well with the population inversion distribution. This geometry, as shown in Figure 8.14, spreads the heat loading over a large part of the slab. The thermally induced distortions also tend to be averaged out across the beam to leave a relatively clean beam.

The question of what are the limits of power handling capability in holosteric lasers is addressed by Basu and Byer [8.20], who propose a number of novel schemes to increase the maximum powers achievable.

8.5 Application of holosteric lasers to nonlinear optics

The lasers discussed above are very useful in their own right, but can also be combined with nonlinear optics for frequency conversion (Section 6.3). The intrinsically high

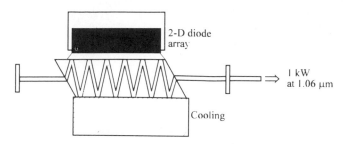

250 mJ, 200µs pulses, or
150 mJ, ns Q-switched pulses

Figure 8.14 An example of a diode array pumped zig-zag slab laser produced by McDonnel-Douglas.

quality of many of the holosteric lasers makes efficient nonlinear optics an achievable objective, even at the relatively modest powers that are attainable with these systems; the quality of the lasers can be exploited to produce remarkably high circulating intensities within the nonlinear crystals [8.21, 8.22].

Harmonic generation of continuous wave systems can be enhanced by resonating the fundamental wave. This is achieved by placing the nonlinear crystal within the laser cavity itself, or by constructing a separate 'build-up' resonant cavity. The first case is typified by the STAR laser developed by Coherent [8.23] and depicted in Figure 8.15. The circulating field within the cavity is very much higher than the field outside the cavity, and so more efficient second harmonic generation can occur. However, the laser geometry and nonlinear crystal type must be chosen with some care. If more than one longitudinal mode is operating, the green output from the laser can have a very unstable power, as sum-frequency generation between modes couples them in a nonlinear fashion, which leads to periodic switching on and off of different modes. The unidirectional ring chosen by Coherent eliminates this problem, as the laser is operating on a single frequency. The second query arises with the type of phase-matching used in the nonlinear crystal. Type I phase-matching (see Chapter 6) (o + o → e) is straightforward, as the polarization of the circulating fundamental

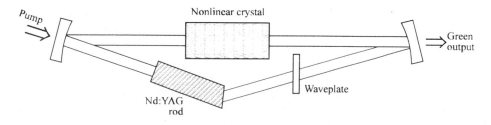

Figure 8.15 Schematic diagram of the Coherent STAR laser.

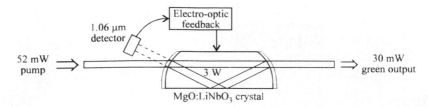

Figure 8.16 Stanford University's diode-laser-pumped nonplanar ring oscillator (frequency doubled in an external resonant ring cavity).

light is parallel to one of the axes of the nonlinear crystal, and no wave-plating effects are induced. In type II phase-matching (o + e → e), the circulating fundamental light does not propagate along the crystal axis, and the polarization state of the fundamental light can be significantly changed on propagation through the crystal. Thus type I phase-matched crystals, such as lithium triborate (LBO), are preferred for intracavity second harmonic generation of Nd:YAG [8.28]. In this particular case, the LBO is heated to 175°C to obtain noncritical phase-matching.

In principle, it is possible to decouple the nonlinear crystal from affecting the laser but still maintain the circulating field intensity by use of an external resonant cavity. This is realistic in the case of holosteric lasers due to the possibility of efficient generation of single-frequency radiation, and the intrinsically narrow linewidth of the fundamental laser which makes the stabilization of the resonant cavity onto the frequency of the fundamental cavity realistic. This scheme was impressively demonstrated by Byer and co-workers, who reported 56% conversion efficiency into the green from a 52.6 mW cw input beam [8.24]. They used $MgO:LiNbO_3$ as the nonlinear crystal, and as the substrate for the mirrors of the monolithic doubler. The resonant frequency of the cavity was maintained the same as the laser frequency by monitoring the power in the reflected beam shown in Figure 8.16, and keeping this at a minimum. A servo circuit altered the length of the resonant cavity appropriately to maintain resonance using the electro-optic effect in $LiNbO_3$. A similar device is now marketed by Lightwave Electronics; 310 mW of single-frequency radiation at 1064 nm is resonated to a power level of 14 W within the same type of resonant cavity, to produce a single-frequency green output of 200 mW. This power level is then sufficient, for example, to pump a low-threshold Ti:sapphire laser.

High peak power Q-switched Nd:YAG lasers present less of a problem in obtaining efficient harmonic generation. The high powers available in a good Gaussian beam allow very high instantaneous intensities to be generated in the nonlinear crystals. The newer nonlinear crystals have sufficiently high nonlinear coefficients and damage thresholds to allow efficient harmonic generation.

Second harmonic generation is best achieved in KTP (potassium titanyl phosphate) in a type II phase-matching scheme. For example, even with the modest energy of 12 mJ in 10 ns available from a holosteric Nd:YAG laser, a conversion efficiency to green, of up to 60% in a single pass, can be obtained. A subsequent crystal of LBO

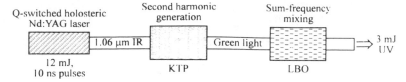

Figure 8.17 Third harmonic generation of the 1064 nm laser by generating the second harmonic of the radiation in KTP, and sum-frequency mixing the second harmonic and the residual fundamental in LBO (lithium triborate).

(lithium triborate) allows the sum-frequency mixing of the fundamental and the second harmonic to obtain up to 3 mJ of ultraviolet radiation, as shown in Figure 8.17. This scheme consists only of the two nonlinear crystals in the converging output beam of the laser depicted in Figure 8.13(b) and the result is a very significant extension of the capabilities of the laser system. In the final section of this chapter we address the addition of yet one more stage of nonlinearity, the optical parametric oscillator.

8.5.1 Optical parametric oscillators

Holosteric lasers have many properties that make them attractive to new and existing applications. One aspect which they fulfil well is to provide pump sources for optical parametric oscillators (OPOs). The basic theory of OPOs is covered in Chapter 6 and in the book by Yariv [8.11]. Essentially, three light waves in a nonlinear crystal interact through the second-order nonlinearity. The same nonlinear interactions that permit sum-frequency mixing can occur in reverse to allow a high-frequency wave to produce two waves whose frequencies sum to the input frequency. In the same way that phase-matching is necessary for sum-frequency mixing to occur efficiently, phase-matching is required for parametric amplification. When the nonlinear crystal is placed inside a resonator, the parametric gain and the feedback of one or both of the generated waves can produce an oscillator based on this nonlinear gain. Changing the phase-matching conditions (through angle or temperature tuning) changes the frequencies of the generated waves.

The OPO can have astoundingly large tuning ranges, and in the 1960s was proposed as an exciting new coherent source [8.25]. At that time, however, with the lasers and nonlinear materials that were available, the threshold for oscillation of these devices was similar to the threshold for crystal damage and hence the devices were not conspicuously successful. In the late 1980s and early 1990s, however, two technological advances came together to make OPOs practicable. The first was the invention and production of new nonlinear materials with high damage thresholds and high nonlinearities, such as urea, KTP, beta barium borate, and lithium triborate. The second was the development of high-quality lasers, such as the holosteric laser. These two developments have allowed the OPO to become a major area of interest worldwide. We illustrate this by considering two particular examples of holosteric visible OPOs.

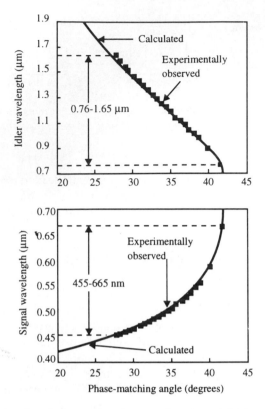

Figure 8.18 The calculated and observed tuning as a function of OPO crystal phase-matching angle for the broadly tunable visible holosteric OPO system.

To generate tunable visible light an ultraviolet pump source must be used. The frequency tripled Nd:YAG laser described in the last section is a useful example, and, indeed, was used to produce the world's first all-solid-state visible OPO system [8.26]. This operated using a type I phase-matched LBO crystal noncritically phase-matched to generate a temperature tunable output around 482 nm and 1.35 m. The noncritical phase-matching geometry allows a very low threshold for oscillation of 0.42 mJ, as there is no 'walk-off' among the different waves. When the OPO is pumped six times above threshold 50% of the incoming pump radiation is converted into a tunable output.

Somewhat greater tunability is obtained using a type I critically phase-matched geometry in LBO [8.27]. In this case, changing the angle of the crystal to change the phase-matching conditions allows tunability from 457–666 nm on one wave, with a corresponding change on the other wave from 1600–760 nm. The crystal angle required for any given pair of wavelengths can be worked out from the phase-matching relationships and the Sellmeier equations; the calculated and observed tuning curves

Plate 8.1 Photograph of a 10 mJ holosteric Nd:YAG laser.

for the 355 nm pumped LBO OPO are shown in Figure 8.18. It can be seen that the holosteric OPO has demonstrated very widely tunable operation, all in a very compact and relatively efficient package.

In the same way that beam waists can be optimized for laser gain, similar relationships can be derived for OPOs [8.28]. The angular acceptance and walk-off parameters must be taken into account, along with the parametric gain and the time required for oscillation to build up from noise. Conversion efficiencies of over 30% can be obtained with the systems described above, which clearly means that even with relatively modest pump sources the holosteric OPO is very efficient.

The early 1990s saw much effort expended worldwide on all solid-state pulsed OPOs, which were soon developed so that they were broadly tunable, reliable and efficient devices (Plate 8.1). A large number of commercial manufacturers launched new products based on OPOs in 1993, whilst research laboratories continued to try to extend their successful ideas to the operation of continuous-wave holosteric visible OPOs.

References

[8.1]　*World News Breaks: Laser Focus*, April 1989, p. 9.

[8.2]　R.L. Burnham, 'High-power transverse diode-pumped solid-state lasers', *Opt. Photon. News*, August 1990, p. 4.

[8.3]　D.L. Sipes, 'Highly efficient neodymium:yttrium aluminium garnet laser end pumped by a semiconductor laser array', *Appl. Phys. Lett.* **47**, 74–76 (1985).

[8.4]　B. Zhou, T.J. Kane, G.J. Dixon and R.L. Byer, 'Efficient, frequency stable laser-diode-pumped Nd:YAG laser', *Opt. Lett.* **10**, 62–64 (1985).

[8.5]　Spectra Diode Laboratories model SDL-3490-S.

[8.6]　A. Yaviv, *Quantum Electronics*, 3rd edn., Wiley, New York, 1989.

[8.7]　W. Streifer *et al.*, 'Advances in diode laser pumps', *IEEE J. Quantum Electron.* **24**, 883–94 (1988).

[8.8]　W. Koecherner, *Solid State Engineering*, 3rd edn., Springer Series in Optical Sciences, Vol. 1.

[8.9]　R.E. Slusher, 'Semiconductor microlasers and their applications', *Opt. Photon. News*, February 1993, p. 8.

[8.10]　R.C. Stoneman and L. Esterowitz, 'Diode-pumped mid-infrared solid-state lasers', *Opt. Photon. News*, August 1990, p. 10.

[8.11]　P.F. Moulton, 'Ti:sapphire lasers: Out of the lab and back in again', *Opt. Photon. News*, August 1990, p. 20.

[8.12]　L.L. Chase and S.A. Payne, 'New tunable solid-state lasers $Cr^{3+}:LiSrAlF_6$', *Opt. Photon. News*, August 1990, p. 16.

[8.13]　W. Lentth and R.M. MacFarlane, 'Upconversion lasers', *Opt Photon. News*, March 1992, p. 9.

[8.14]　P.J. Chandler *et al.*, 'Ion-implanted Nd:YAG planar waveguide laser', *Electron. Lett.* **25**, 985 (1989).

[8.15]　T.Y. Fan and R.L. Byer, *IEEE J. Quant. Electron.* **24**, 895–912 (1988).

[8.16]　T.Y. Fan and A. Sanchez, *IEEE J. Quant. Electron.* **26**, 311–16 (1990).

[8.17]　*Laser Focus World*, August 1992, p. 9.

[8.18]　R.L. Burnham, 'High-power transverse diode-pumped solid-state lasers', *Opt. Photon. News*, August 1990, p. 4.

[8.19]　Y. Cui, M.H. Dunn, C.J. Norrie, W. Sibbett, B.D. Sinclair, Y. Tang, and J.A.C. Terr, 'All-solid-state optical parametric oscillator for the visible', *Opt. Lett.* **17**, 646 (1992).

[8.20]　S. Basu and R.L. Byer, 'Average power limits of diode-laser-pumped solid state lasers', *Appl. Opt.* **29**, 1765 (1990).

[8.21]　R.L. Byer, 'Nonlinear frequency conversion enhances diode-pumped lasers', *Laser Focus World*, March 1989, p. 77.

[8.22]　J.T. Lin, 'Progress Report: Diode pumping and frequency conversion', *Lasers and Optronics*, July 1989, p. 61.

[8.23]　Coherent STAR laser, manufacturer's data sheets.

[8.24]　W. Kozlovsky, C.D. Navors and R.L. Byer, 'Efficient second harmonic generation of a diode-laser-pumped cw Nd:YAG laser using monolithic $MgO:LiNbO_3$ external laser cavities', *IEEE J. Quantum Electron.* **24**, 913 (1988).

[8.25]　S. Harris, *Proc. IEEE* **57**, 2096 (1969), Tunable Optical Parametric Oscillators.

[8.26]　Y. Cui, M.H. Dunn, C.J. Norrie, W. Sibbett, B.D. Sinclair, Y. Tang and J.A.C. Terry, 'All-solid-state optical parametric oscillator for the visible', *Opt. Lett.* **17**, 646 (1992).

[8.27] Y. Cui, D.E. Withers, C.F. Rae, C.J. Norrie, Y. Tang, B.D. Sinclair, W. Sibbett and M.H. Dunn, 'Widely tunable all-solid-state optical parametric oscillator for the visible and near infrared', *Opt. Lett.* **18**, 122 (1993).

[8.28] D.E. Withers, G. Robertson, A.J. Henderson, Y. Tang, Y. Cui, W. Sibbett, B.D. Sinclair and M.H. Dunn, 'Comparison of lithium triborate and beta-barium borate as nonlinear media for optical parametric oscillators', *J. Opt. Soc. Am.*, **10**, 1737 (1993).

9

Detectors

9.1 Introduction

In previous chapters we have discussed most of the components of optoelectronic systems apart from the variety of detectors that can be used. The main function of all detectors of radiation is to convert radiation absorbed over a variety of wavelengths (the power of the optical signal) into a usable electrical output (an electrical signal). The electrical signal can be either a voltage or a current, and for measurement purposes it is desirable that it be proportional to the power of the optical signal. Figue 9.1 indicates how the various components of an optoelectronic system are linked together. The importance of detectors is obvious; they determine how well the system performs its task and provide the link to the outside world.

The performance of a detector will depend upon the spectral and spatial characteristics of the incident radiation, so that for comparison purposes these characteristics

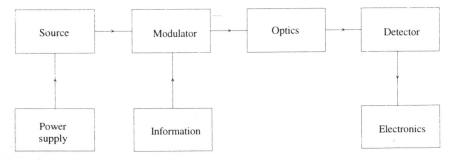

Figure 9.1 Schematic diagram showing the components of a typical optoelectronic system and how they are linked together.

must be specified. Detectors also have a range of time responses; since modulation or 'chopping' is commonly used in conjunction with a detector, modulation frequency must also be specified since it affects performance.

9.2 Parameters determining a detector's response

Let us assume for the moment that the detector responds to the input signal by producing a voltage change ΔV. ΔV will be a function of several quantities, as shown below,

$$\Delta V = \Delta V(b, f, \lambda, P, A_d)$$

where b is the applied electrical bias, f is the frequency of modulation, λ is the wavelength of the incident radiation, P the irradiance (power per unit area, i.e $W\ cm^{-2}$) and A_d is the detector area.

The spectral variation of this output signal can vary from being constant with wavelength for a thermal detector that responds to the heat produced by incident radiation, to strongly selective for those using the electronic properties of materials. As a result, calibrating sources must have their spectral form specified: for measurement of the total radiation response black bodies are often used with the known Planck distribution function of spectral power density.

We now look more closely at some of the important parameters that determine a detector's performance.

9.2.1 Origins of 'noise'

The output signal from all detection systems will include unwanted fluctuations which 'obscure' the 'true' signal. These unwanted fluctuations appear at random and are called *noise*. The overall noise associated with a detector system can be subdivided into three categories: photon noise, detector-generated noise and post-detector electronic noise (Figure 9.2) [9.1].

9.2.2 Photon noise

Photon noise represents the ultimate limit of an otherwise perfect system.

Noise due to the signal radiation
Although a source of radiation is specified by the average number of photons emitted per unit time, in practice the actual number of photons emitted in any time interval can deviate from this average. This is due to the statistical nature of optical sources and the fluctuations with time are schematically illustrated in the graph of the number of photons emitted per unit time against time (Figure 9.3).

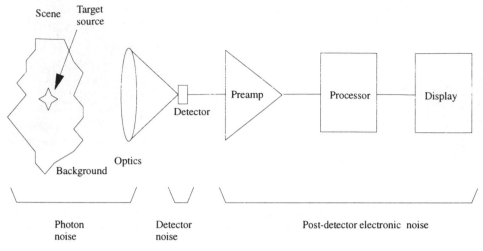

Figure 9.2 Detector system noise classification.

Noise due to background radiation

Clearly some of the background radiation will also be picked up by the optics within the system and hence produce additional noise, since it will also be of similar character to that shown in Figure 9.3.

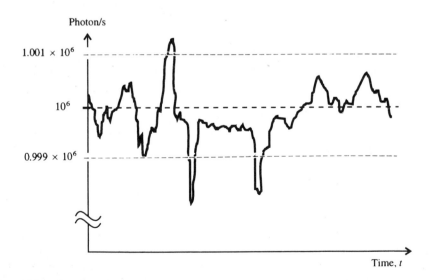

Figure 9.3 Variability of instantaneous photon flux.

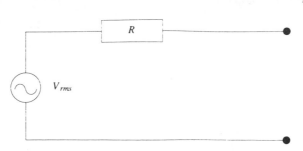

Figure 9.4 Representation of a noisy resistor by a random noise voltage generator in series with an ideal (quiet) resistor.

9.2.3 Detector-generated noise and amplifiers

Johnson noise

Johnson or Nyquist noise is caused by the thermal motion of charged particles (thermal current fluctuations) in a resistive element. The Johnson rms noise voltage, V_{rms}, is the series voltage generated in a Thevenin equivalent circuit representing the noisy resistor (Figure 9.4). It is given by

$$V_{\text{rms}} = \sqrt{4kTR\Delta}$$

where k is Boltzmann's constant, T is the temperature, R is the resistance of the detector element and Δf is the effective bandwidth of the circuit, which is a measure of the band of modulating frequencies that can be applied to the detector without significant loss in its responsivity. (Δf may be reduced by signal conditioning circuits subsequent to this stage.)

Clearly, if Δf is large the Johnson noise is also large. In order to restrict the value of Δf the radiation from a source is often 'chopped' and the detector's amplifier is switched on and off at the same frequency as the chopping frequency (Figure 9.5). This is equivalent to introducing a narrow-band electrical filter centred on the chopping frequency which reduces Δf (Section 1.3.3). We must note that Johnson noise is also produced within the post-detection electronics. In Figure 9.6 the electronics has been identified as having a resistance R_L in parallel with the detector's resistance R. The amplifier must be designed to contribute only a small amount of noise in its first stage compared to that from the detector. Different amplifier designs are required to interface with different effective detector impedances (e.g. low impedance from a thermopile or high from a pyroelectric detector).

Shot noise

Shot noise is only produced in photon detectors and is due to the discrete nature of photoelectron generation. The random arrival of photons produces a random flux of photoelectrons within the detector which in turn produce shot noise.

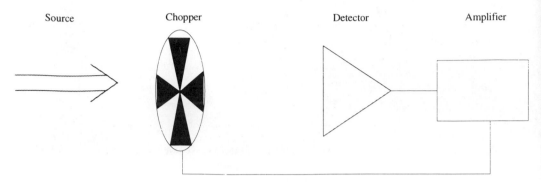

| Source | Chopper | Detector | Amplifier |

Electrical connection to chopper allows amplifier to be switched at chopper frequency

Figure 9.5 Reducing the Johnson noise by chopping the source radiation and switching the detector's amplifier at the same rate as the chopper.

Generation–recombination noise

Generation–recombination noise is observed in photoconductors and is caused by a fluctuation in current carrier generation which is due to the random arrival of photons. Thermal generation and recombination of carriers contribute noise, particularly for infrared detectors. Recently, ingenious structures have been invented in order to minimize such effects; these structures are discussed in Section 9.4.5.

1/f noise

This noise source is perhaps one that has been most actively studied but is the least well understood. The power spectrum for $1/f$ noise falls off as the frequency increases, and it is known that this is partially due to a lack of ohmic contact at the detector electrodes and to surface-state traps or dislocations.

Temperature fluctuations

In thermal detection systems any fluctuation in temperature not due to the signal can produce unwanted noise.

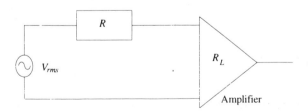

Figure 9.6 Schematic circuit showing the connection between detector and amplifier stages.

Microphonics
Microphonic noise is due to the mechanical displacement of wiring and components when a system is subjected to vibration or shock.

Post-detector electronic noise
Amplifier stages need to have a noise performance sufficiently good so as not to reduce the system performance below that of the detector itself. The most common sources of such noise are similar to detector noise and include: Johnson noise, shot noise, generation–recombination noise, $1/f$ noise, temperature fluctuations, popcorn noise and microphonics.

9.2.4 Figures of merit

Figures of merit are used to compare the measured performance of one detector against another in the same or different class, or to compare the actual performance with ideal performance. Care must be taken in using figures of merit to compare detectors since any one particular figure may not allow a true comparison (for example, the figure may not take into account the area of a detector and so if this varies between two detectors the true performance of each will not be correctly represented by the figure). We shall now define a number of useful figures of merit.

Responsivity
Responsivity is the ratio of the electrical output of a detector (measured in either volts or ampères) to the radiant input power (in watts). For example, the spectral voltage responsivity $R_v(\lambda, f)$ is the ratio of the measured output voltage, $V_s(f)$, at modulation frequency f, to the spectral radiant power $\phi_e(f, \lambda)$ around the wavelength λ and modulation frequency, f:

$$R_v(\lambda, f) = \frac{V_s(f)}{\phi_e(f, \lambda)}.$$

The 'black-body' voltage responsivity, $R_v(R, f)$, is the ratio of the detector's output voltage modulated at a frequency f, $V_s(f)$, to the total incident radiant power modulated at a frequency f from a black-body source at temperature T, $\phi_e(f, T)$. In fact, we can write this voltage responsivity as

$$R_v(T, f) = \frac{V_s(f)}{\displaystyle\int_0^\infty \phi_e(\lambda) \, d\lambda}.$$

Noise Equivalent Power (NEP)
The noise equivalent power (NEP) is the rms modulated power required to produce a signal output equal to the rms noise output. Alternatively NEP can be defined as the signal level that produces a signal to noise ratio of unity. The signal to noise ratio in terms of a current signal is defined as

$$\frac{S}{N} = \frac{i_{\text{signal out}}}{i_{\text{rms}}}$$

and in terms of a voltage signal as

$$\frac{S}{N} = \frac{V_{\text{signal out}}}{V_{\text{rms}}}.$$

Since $i_{\text{signal out}} = R_i \phi_e$, where R_i is the current responsivity we can write

$$\frac{S}{N} = \frac{R_i \phi_e}{i_{\text{rms}}}.$$

The NEP is the incident radiant power ϕ_e that produces a signal to noise ratio of one. Thus if

$$\frac{S}{N} = 1 = \frac{R_i \cdot \text{NEP}}{i_{\text{rms}}}$$

then

$$\text{NEP} = \frac{i_{\text{rms}}}{R_i}.$$

Note that we can use either the spectral responsivity $R_i(\lambda, f)$ or the black-body responsivity $R_i(T, f)$ to define two different NEPs. The units of NEP are watts.

The NEP is a useful parameter than can be used to compare similar detectors operating under similar conditions. It should not be used to compare dissimilar detectors. This is because it fails to take account of quantities such as the bandwidth Δf and detector area A_d, which may vary from detector to detector. The larger the bandwidth Δf and detector area A_d, the larger the NEP. Below we describe a parameter called the normalized detectivity D^*, which takes account of both the bandwidth Δf and the detector area A_d.

Detectivity, D

The detectivity, D, is defined as

$$D = \frac{1}{\text{NEP}}.$$

The only difference between NEP and detectivity is that for a more sensitive detector (i.e. one that detects a smaller signal) the NEP decreases while the detectivity increases, so that there is a positive correlation between detectivity and sensitivity.

Normalized detectivity, D*

·The normalized detectivity, D^*, is a more useful figure of merit and allows comparison between different detectors as it takes into account the detector area, A_d, and the bandwidth (it is measured in W^{-1} cm $\text{Hz}^{1/2}$):

$$D^* = D\sqrt{A_\mathrm{d} \cdot \Delta f} = \frac{\sqrt{A_\mathrm{d} \Delta f}}{\mathrm{NEP}}.$$

This relationship with respect to area illustrates the fundamental principle, but in practice can be modified by edge effects. Note that we can use spectral or black-body NEP to define spectral or black-body detectivity $D^*(\lambda, f)$ or $D^*(T, f)$. The normalized detectivity D^* can be expressed in a number of alternative forms, as shown below:

$$D^* = \frac{\sqrt{A_\mathrm{d}\Delta f}}{v_\mathrm{rms}} R_v = \frac{\sqrt{A_\mathrm{d}\Delta f}}{i_\mathrm{rms}} R_i = \frac{\sqrt{A_\mathrm{d}\Delta f}}{\phi_\mathrm{e}} \left(\frac{S}{N}\right)$$

For a more detailed discussion on figures of merit, see Ref [9.1].

9.3 Classes of detectors

There are two important types of detector: *thermal detectors* and *photon detectors*.

Thermal detectors rely on temperature increases of the device caused by the absorption of radiation. This temperature change causes a further change in a temperature-dependent parameter, such as the resistivity of the device which is then measured. The response time will depend upon the thermal capacity and heat sinking of the total volume of the device. The measured signal is usually proportional to the energy absorbed per unit time and, provided that the absorption efficiency is the same for all wavelengths, the output signal will be independent of wavelength.

Photon detectors operate in a different manner. When a photon is absorbed it causes a specific quantum effect. This could be, for example, photoelectric emission of electrons from a surface. These specific events can be 'counted' by the detection system which provides a measure of the rate of absorption of the light quanta. Photon detectors are not suitable for use over the entire spectrum since they rely on processes that require a minimum photon energy to initiate them. This means that they have a long wavelength limit beyond which they will not operate. However, this limitation has a fundamental effect upon noise levels such that photon detectors can have considerably less noise (since the noise from a large part of the spectrum of radiation incident upon the detector is effectively 'cut out'). Photon detectors thus have considerably greater normalized detectivities and faster response times compared with thermal detectors.

9.3.1 Thermal detectors

In this class of detectors we find devices such as Golay cells, Crooke's radiometer, bimetallic strips and liquid-in-glass thermometers, which rely upon radiation heating to cause mechanical displacements. The bolometer, thermocouple and pyroelectric devices are all electrical devices that depend on radiation heating to cause a change in one of their electrical properties such as resistance.

A suitable 'black' absorbing layer can make these detectors respond to a wide band of wavelengths. For this reason, thermal detectors have a long history in the infrared spectral region where it has been difficult to obtain long-wavelength photon detectors without cryogenic cooling. Thermal detectors do tend to be slow with a response time dependent on the heat capacity of the element and often with a frequency response of only a few hertz. For device use some thermocouples are convenient and cost effective as are some pyroelectric detectors. Table 9.1 provides a summary of some of the key parameters of some thermal detectors.

Table 9.1 Thermal detectors

Type	Wavelengths absorbed	Minimum detectable power (W)	Response time (s)	General comments
Thermocouple	All	10^{-9}	0.1	Thin-film form: cheap. Output seen as a voltage change.
Bolometer	All	10^{-8}	0.1	Change of resistance provides output measurement.
Golay cell	All	10^{-9}	0.1	Pneumatic device amplified using an optical amplifier.
Pyroelectric detectors	All	10^{-7}–10^{-8}	0.01–0.1	Cheap and can be arranged as arrays

The thermocouple

In its simplest form the thermocouple consists of two dissimilar metals connected in series. The junction formed is then covered with a black absorbing material. Radiation incident upon this material causes a temperature increase. This temperature increase then affects both of the dissimilar metals and in each metal leads to an increase in the number of electrons available for conduction around the Fermi energy, E_F. Since the exact numbers available will vary across the junction (since in each metal there are differing numbers of conduction electrons around E_F) electrons will move across the junction to form a thermoelectric emf. This thermoelectric emf will be directly proportional to the temperature of the absorbing junction, T_1 (Figure 9.7).

Note that in most cases the thermocouple is tied in series to a twin so that a reference temperature junction can be established. In the configuration shown in Figure 9.8 the voltage reading will be proportional to the temperature difference between the measuring junction and reference junction (sometimes at 0°C). The complete device is often placed into a vacuum or gas-filled enclosure. A practical

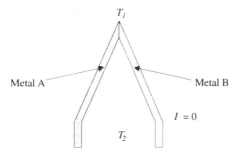

Figure 9.7 The thermoelectric effect with two dissimilar metals.

limitation to the performance of this detector is that it has a very slow response. Chopping frequencies above 20 Hz cannot normally be used. The spectral response of the thermocouple depends upon the response of the absorber and the window material if present. Normally the absorber is black and spectrally flat out to wavelengths of 40 μm.

Single thermocouples are not very practical as detectors. In fact, the responsivity can be increased by a factor of n by placing n thermocouples in series. Such a device is called a thermopile. Thermopiles are commonly made by either linking a series of wires as suggested above or by a thin-film evaporation approach, and are commonly used as infrared detectors.

The bolometer
The bolometer is mostly used for infrared detection. It need not require cooling, has a flat spectral response (limited only by the filter used) and can withstand rough conditions. When cooled it can approach a photon-limited performance level. A temperature change produced by the absorption of radiation causes a change in the electrical resistance of the bolometer material. It is this change in resistance that is used to quantify the level of radiation.

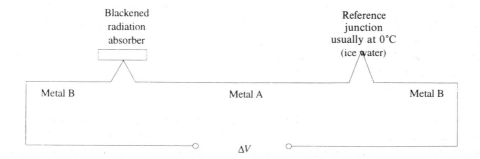

Figure 9.8 Two dissimilar metals connected in series.

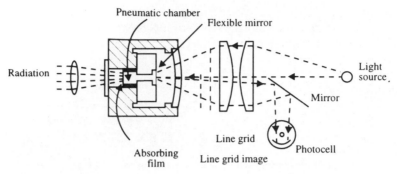

Figure 9.9 The Golay cell.

Bolometers can be divided into five main groups: the metal, thermistor, semiconductor, composite and superconducting material types. The first three are commonly used. The composite bolometer can provide low-noise and low-temperature infrared detection. The superconducting bolometer has not been widely used as it requires a very low operating temperature.

The Golay cell
The Golay cell detector is used for infrared work and is shown in Figure 9.9. It consists of a gas-filled cell (the pneumatic chamber), the ends of which are closed by very thin membranes. The membrane that receives the incident radiation is an absorbing layer of aluminium deposited on a collodion base. The other membrane is another collodion film onto which antimony has been evaporated to make it reflecting. When radiation is incident upon the receiving membrane, the gas in the cell is heated, setting up a differential pressure that distorts the second membrane. This distortion is 'observed' by an optical arrangement and converted into an electrical signal. The distortions are detected by reflecting some of the light from a lamp off the reflecting membrane onto a photocell. Two grids consisting of equally spaced fine lines are arranged so that normally the image of the lines of one grid falls onto the spaces of the other, thus allowing little light to reach the photocell until the membrane is distorted. This arrangement allows very small displacements to be measured and has an ultimate sensitivity comparable to that of a thermocouple. The detector is now largely of historical interest but illustrates how various physical effects can be utilized for detector purposes.

Pyroelectric detectors
The basis of operation of a pyroelectric detector is the spontaneous polarization that its molecules possess below a fixed temperature such as the Curie temperature T_c. Any radiation incident upon the device causes a change in temperature of the device. In turn, this causes an expansion of the crystal lattice spacing which affects the spontaneous polarization of the molecules. This change in polarization causes charge

to flow if connected to an external circuit. Note that above the Curie temperature T_c the material loses its spontaneous polarization and so cannot be used as a detector. Table 9.2 lists some of the important parameters of a few materials that are commonly used as pyroelectric detectors.

Table 9.2 Pyroelectric materials

Material	Pyroelectric coefficient ($C\ cm^{-2}\ K^{-1}$)	Dielectric constant	Specific heat ($J\ g^{-1}\ K^{-1}$)	Thermal conductivity ($W\ cm^{-1}\ K^{-1}$)	Density ($g\ cm^{-3}$)
Turmaline	4×10^{-10}				
BaTiO$_3$	2×10^{-8}	160 ($\parallel$ polar axis) 4100 ($\perp$ polar axis)	0.5	9×10^{-3}	6.0
TGS	$(2-3.5) \times 10^{-8}$	25–50	0.97	6.8×10^{-3}	1.69
Li$_2$So$_4$H$_2$O	1.0×10^{-8}	10	~ 0.4	17×10^{-3}	2.05
LiNbO$_3$	4×10^{-9}	30 ($\parallel$) 75 ($\perp$)			4.65
LiTaO$_3$	6×10^{-9}	58			
SbSI	2.6×10^{-7}	10^4	0.29		8.2
NaNO$_3$	1.2×10^{-8}	8.0	0.96		2.1

The desirable characteristics of a pyroelectric detector are high Curie temperature, large pyroelectric coefficient, small dielectric constant, large resistance, low loss tangents and a small heat capacity. The detectors can be operated fast but with loss of responsivity. The most important pyroelectric detectors consist of materials such as triglycerine sulphite (TGS), lithium tantalate and strontium barium niobate. A disadvantage of these detectors is that the pyroelectric effect is sensitive to vibration.

The detector is very useful for infrared detection where cooling of the detector is not possible and where high sensitivities are not required. The pyroelectric detector may be used in infra image detectors, or vidicons, where an array of detectors are placed side by side. The signal-to-noise ratio of such a device is better than that of a single detector with scanning optics.

9.4 Photon detectors

These are now the most important class of detector and occur in a large number of forms. Table 9.3 provides a list of some of the most commonly used, together with some of their key performance parameters.

9.4.1 Photoconductors

Photoconductors are essentially made from semiconductor materials which have a high resistance in the absence of illumination but change resistance when charge

Table 9.3 Photon detectors

Type	Wavelengths absorbed	Minimum detectable power (W)	Response time (s)	General comments
1. Photoconductors				
Intrinsic devices	Selective, up to λ_g	10^{-9}–10^{-12}	ms	Needs a bias voltage
Extrinsic devices	Selective	10^{-9}–10^{-12}	ms	Needs a bias voltage
2. Photoemissive devices	Short wavelengths only	10^{-14}	ns	Needs a high voltage
Photomultiplier tubes (PMTs)	up to 1 μm	10^{-14}	ns	Bulky devices, but allow individual photon counts
Microchannel photomultipliers	up to 1 μm	10^{-14}	ns	Arrays possible
3. Photovoltaic (*pn* junction) detectors				
Photodiodes	Si, 0.85 μm	10^{-11}	<1 μs	Reverse-biased *pn* junction with a built in electric field
pin photodiodes	Si, GaAs up to 1 μm	10^{-10}	$\sim$ns	Eliminate the slow diffusion of carriers
Avalanche photodiodes	Si up to 1 μm	10^{-9}–10^{-10}	ns–ps	Have internal gain, counts single photons
Phototransistors	Si up to 1 μm	10^{-10}	<1 μs	Provide amplification

carriers are directly excited by the absorption of photons from a nonconducting state into a conducting state that was previously empty. There are two important cases to consider:

The intrinsic case. In this case the electrons belonging to the 'pure' atoms within the semiconductor are confined to a band of energy known as the valence band (Chapter 2). Above this band exists a forbidden energy gap, and if the electrons in the valence band are given sufficient energy they can jump across this forbidden gap into the conduction band (Figure 9.10). The energy required to jump across the

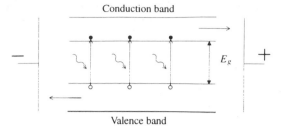

Figure 9.10 Creation of electron–hole pairs by the absorption of photons of energy greater than E_g.

bandgap E_g is provided by photons of energy $h\nu \geqslant E_g$ and once electrons are raised from the valence band to the conduction band there is a consequent contribution to the increase in conductivity from holes in the valence band as well as from the electrons in the conduction band.

The extrinsic case. This class of semiconductor extends the wavelength range that can be detected into the infrared by use of *impurity levels* within the energy gap. The

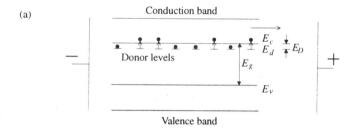

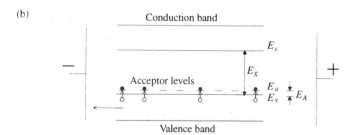

Figure 9.11 (a) Excitation of electrons from a donor band in an *n*-type material. (b) Excitation of holes from an acceptor level in *p*-type material.

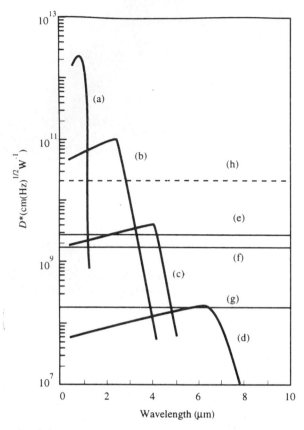

Figure 9.12 Normalized detectivity, D^*, for various detectors operated at room temperature: (a) thallium sulphide photoconductive cell; (b) PbS photoconductive cell; (c) PbSe photoconductive cell; (d) InSb photoconductive or PEM cell; (e) Hilger-Schwartz thermocouple; (f) Golay cell; (g) thermistor bolometer; (h) represents the thermal noise limit for a non-selective detector operated at 300 K.

required photon energy for carrier excitation is now $hv > E_d$ or E_a, the binding energies of donors or acceptors (Figure 9.11). With cryogenics to minimize the thermal excitation of carriers from impurity states into either the conduction band or the valence band, the shallow impurity states can extend photoconductive sensitivity into the far infrared to greater than 100 μm and provide the most sensitive detectors known for this spectral region.

In both of these cases the absorption of radiation produces an increase in conductivity, and hence these devices are called photoconductors. A voltage bias is applied across the device so that photogeneration of carriers leads to an increase in current within the detection circuit.

Photoconductors are in general capable of better detectivity and speed of response than thermal detectors. Compared with an ideal thermal detector, the photoconductor

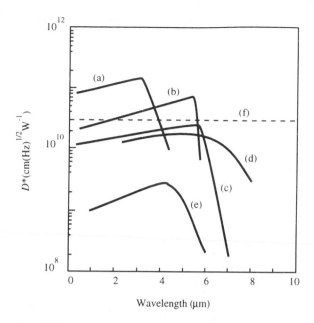

Figure 9.13 Normalized detectivity, D^*, for various photoconductive detectors operated at 77 K: (a) PbS; (b) InSb; (c) PbSe; (d) Au-doped Ge, p-type; (e) PbTe; (f) represents the thermal noise limit if the surroundings of the detector are at 300 K.

has a cutoff frequency either near its energy gap (for an intrinsic semiconductor) or at its impurity excitation energy (for an extrinsic semiconductor). This means that it 'intersects' less of the ambient radiation from its enclosure which in turn reduces the intrinsic noise level.

Figures 9.12 and 9.13 show plots of the normalized detectivity D^* against wavelength for several different types of detector. From these figures it is clear that photoconductors in general have a narrower band of wavelengths over which their detectivity is very high compared with other detectors. To obtain best performance from a photoconductor it is important to reduce all sources of noise as far as possible. Obviously it is important that the detector is sufficiently cooled so that the thermal ionization of holes and electrons is minimized. In addition, the room-temperature radiation incident upon the detector should also be minimized; this can be done by having cooled slits on the device generating the radiation (e.g. a spectrometer) and cooled windows and filters in front of the detector. Internal noise sources, such as those due to the formation of recombination centres for carriers due to impurities, can be minimized by ensuring as pure a sample as possible. The current noise is the additional noise that appears when a current is passed through a resistor. This noise mechanism is usually reduced by carefully cleaning the detector surface and washing it for long periods of time in deionized water.

A variety of semiconductors exist for infrared detection. Table 9.4 lists a few semiconductors with the principal wavelengths that they are used to detect.

Table 9.4 Semiconductors used for infrared detection

Type of semiconductor	Material	Temperature (K)	Principal wavelength detected (μm)
Intrinsic	InSb	77	5
Extrinsic	Zn-doped Ge	4	17
Extrinsic	Cu-doped Ge	4	30
Extrinsic	Sb-doped Ge	4	118
Intrinsic	HgCdTe	77	10

9.4.2 Photoemissive devices

These devices provide the most sensitive detectors for the visible and near infrared spectral regions. Photoemissive devices rely on photons incident with sufficient energy to *eject* electrons from the material surface. Once generated, the electrons are usually accelerated in a vacuum by a large electric field between the cathode (electron-emitting material) and the anode to form a current in an external circuit (Figure 9.14).

The current that is generated by the photon stream depends upon the number of photons incident per second, the fraction of these that produce electrons and the electronic charge. It is important to note that a current will only be generated if the incident photons have sufficient energy to generate free electrons. (i.e. if the photon energy $h\nu >$ the work function ϕ_0). This usually limits the spectral response of a device to around 1 μm, beyond which the photon energies are insufficient to liberate electrons from the photocathode material. Note that a *metal* photocathode will not respond to photon wavelengths longer than 300 nm, and hence beyond this wavelength the impurity levels in a semiconductor material are usually used to produce electron emission.

With the device shown in Figure 9.14 it is only possible to produce a signal current of the order of 10^{-19} A. Such a small current cannot normally be detected so that, before the detector can be used near its own limits and not the limits imposed by the electronics, electron multiplication (gain) is necessary. This is usually achieved using a photomultiplier tube, as discussed below.

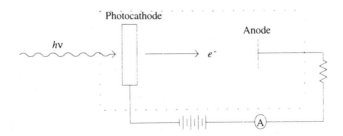

Figure 9.14 Photoemissive detection.

Vacuum tube

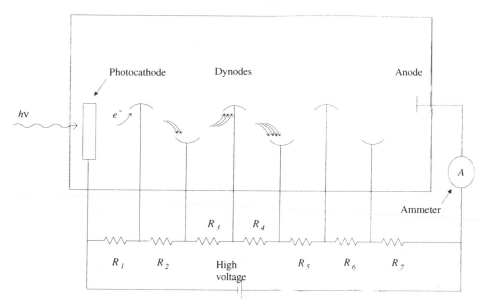

Figure 9.15 Schematic representation of photomultiplier tube.

The photoemissive device is still one of the most sensitive radiation detectors with an NEP of the order of 7×10^{-16} W and a typical response time of the order of 1–2 ns.

Photomultiplier tubes (PMTs)

As mentioned earlier, the electron current produced by a single photoemissive device is only of the order of 10^{-19} A. This is insufficient in strength to be detected by the electronics in the circuit and as a result a photomultiplier tube (PMT) is usually used. The function of such a tube is to convert the photoelectric current that is produced by the liberation of one electron by one photon into a far larger stream of electrons that produces a measurable current level.

A photomultiplier tube is shown schematically in Figure 9.15. When a photon is absorbed by the photocathode it produces one electron. This electron is accelerated towards the first dynode by a high potential difference across the photocathode and first dynode. The increased kinetic energy of the electron is enough to cause a number of electrons to be emitted from the first dynode. These are then accelerated towards the second dynode to cause yet further electron emission. In this manner a gain of the order of 10^5 can easily be achieved by about 11 dynodes and the electron current can be amplified to the order of 10^{-13} A, which can easily be detected using electronics.

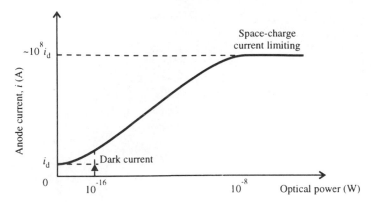

Figure 9.16 PMT linear dynamic range.

The linear dynamic range of a PMT is typically of the order of 10^8; it is limited by dark current at the lower end of the range and by space-charge effects at the high end (Figure 9.16). Space-charge limiting occurs when the electron cloud density between the final dynode and the anode is so high that it repels any incident electrons from the dynode. The dark current is due to thermionic emission (which can be dramatically reduced by cooling the tube), regenerative effects such as α, β and γ emission which cause secondary electron emission, and due to a leakage current (due to the insulator resistance not being infinite).

Microchannels
Microchannels are small tubes of glass that can behave as electron multipliers. Electrons emitted from a photocathode are incident upon one end of the tube, which has a bend in it. Once it has entered the tube, an electron undergoes a number of 'reflections' from the tube walls and at each reflection an increased number of electrons are produced (electron gain). An anode is usually placed at the end of the tube and overall gains of the order of 10^8 are possible. The dynamic range of such devices does tend to be smaller than the PMT, so that the PMT offers superior performance in most single-detector applications, but microchannels are better for constructing arrays for imaging applications. It is possible to produce an array of microchannels in a single plate of glass; these are called microchannel plates. Typical physical dimensions for such an array are diameter of 18–75 mm, thickness (channel length) of about 1 mm, channel diameter in the range 8–20 μm, with 10^4–10^7 microchannels per plate. For more details on PMTs, microchannels and microplates see Ref. [9.2].

9.4.3 Photovoltaic detectors

The photoconductive and photoemissive detectors require high applied voltages. Photovoltaic detectors can provide this internally or in some forms only need applied

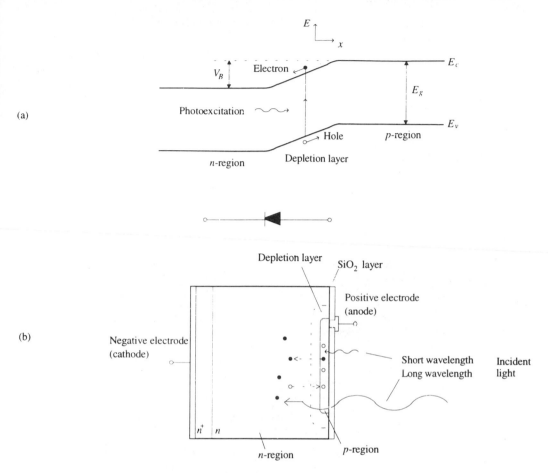

Figure 9.17 The photovoltaic effect: the *pn* junction photodiode.

voltages at logic levels. They have in consequence become the most generally used form of detector. We thus discuss this class in detail.

The photodiode

Photodiodes make use of the photovoltaic effect – the generation of a voltage across a *p–n* junction – when the semiconductor junction is exposed to light. The term 'photodiode' broadly includes such devices as solar cells but usually refers to detectors. The *p–n* junction is described in Chapter 3 in some detail. The energy level structure of such a junction is illustrated again in Figure 9.17(a), for ease of reference. A photon can excite an electron from the valence band to the conduction band in the *n*-region, the *p*-region and depletion region, and thereby create an electron–hole pair (for more details, see Section 3.2). However, both the electron and the hole are quickly swept out of the depletion layer by the built-in field and can produce either a voltage across the diode or a current through an external circuit.

For photodiode operation the incident photons must have an energy that is greater than or equal to the energy gap, E_g, and the temperature of the detector must be low enough to minimize thermal excitation of electrons across the bandgap – a condition readily satisfied for the important visible and near infrared regions.

Photodiodes can be classified by function and by construction as below:

(i)   pn photodiodes
(ii) pin photodiodes
(iii) Schottky-type photodiodes
(iv) Avalanche photodiodes

with detailed variations for individual performance specification. Table 9.5, as an example, shows those manufactured by Hamamatsu Photonics [9.3].

Table 9.5 Construction of vertices of photodiodes

Type	Construction	Features	Photodiode types
Planar diffusion type		Small dark current	Silicon photodiodes (eg. S2386. S2387 series. S1087. S1133 series) GaAsP photodiodes
Low C$_i$ planar diffusion type		Small dark current Fast response High UV sensitivity High IR sensitivity	Silicon photodiodes (S1336 series. S1337 series)
PNN· type		Small dark current High UV sensitivity Suppressed IR sensitivity	Silicon photodiodes (S1226 series, S1227 series)
PIN type		Ultra-fast response	PIN silicon photodiodes
Schottky type		High ultraviolet sensitivity	GaAsP, GaP photodiodes
Avalanche type		Internal multiplying mechanism Ultra-fast response	Silicon avalanche photodiodes

In the silicon planar diffusion type, an SiO_2 coating is applied to the *pn* junction surface resulting in a low dark current. The *low capacitance* planar diffusion type uses highly pure *n*-type material to enlarge the depletion layer, thereby decreasing the junction capacitance and consequently lowering the response time by a factor of up to 10. The *pnn$^+$ silicon* diode utilizes a low resistive n^+ layer to bring the n–n^+ boundary close to the depletion layer lowering infrared response and producing a useful shortwave performance. The *p-i-n* type, described in detail later, is for ultra-fast

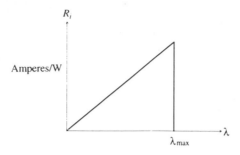

Figure 9.18 Ideal photon responsivity.

response. Gold coating sputtered on *n*-type III–V material, forming a Schottky barrier, has a small distance from the outer surface to the junction and hence good UV response. The avalanche photodiode (APD) resembles a photomultiplier by featuring internal amplification. This is also treated more fully later.

A typical configuration for a *p–n* junction device is shown in Figure 9.17. All the incident photons traverse the entire width of the junction twice, unless they are absorbed first. A reflective coating on the bottom of the detector provides this double path length for the photons which increases the probability of their absorption.

The current responsivity of such a detector, defined as $R_i = I_p/\phi_e$ where I_p is the current generated by an incident flux of ϕ_e, is (approximately in practice) proportional to wavelength λ as far as the energy gap cutoff, as is shown in Figure 9.18. By definition

$$R_i = \frac{I_p}{\phi} = \frac{\eta q \lambda}{hc}$$

Some typical spectral response characteristics are given in Figure 9.19. When the

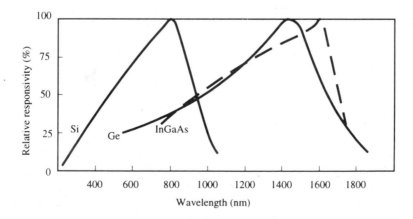

Figure 9.19 Ideal responsivity for constant quantum efficiency.

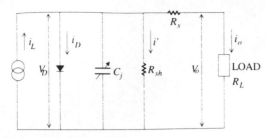

Figure 9.20 Photodiode equivalent circuit.

energy of absorbed photons falls below E_g, the response falls rapidly. The limiting wavelength is given by $\lambda = 1240/E_g$ (nm) with E_g measured in electron volts (eV).

Equivalent circuit

The photodiode equivalent circuit is shown in Figure 9.20. In the figure i_L is current generated by incident light (proportional to the amount of light), i_D is the diode current, C_j is the junction capacitance, R_{sh} is the shunt resistance, R_s is the series resistance, i' is the shunt resistance current, V_D is the voltage across the diode, i_o is the output current, and V_o is the output voltage.

Using the above equivalent circuit and solving for the output current, we have

$$i_0 = i_L - i_D - i' = i_L - i_s\left(\exp\frac{eV_D}{kT} - 1\right) - i'$$

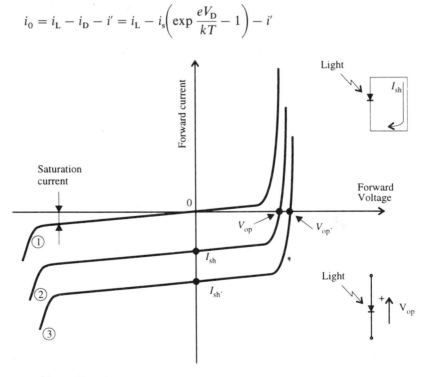

Figure 9.21 Voltage–current characteristic of a photodiode [9.3].

(a) i_{sh}

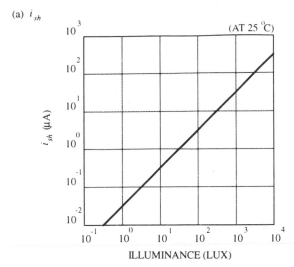

(b) V_{op}

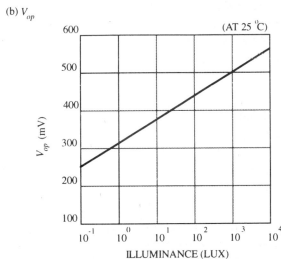

Figure 9.22 Output signals as function of irradiance [9.3].

where i_s is the photodiode reverse saturation current, e is the electron charge, k is Boltzmann's constant and T is the absolute temperature of the photodiode.

The open circuit voltage V_{0p} is the output voltage when i_0 equals 0. Therefore, we have

$$V_{0p} = \frac{kT}{e} \ln\left(\frac{i_L - i'}{i_s} + 1\right)$$

If we ignore i', since i_s increases logarithmically with respect to increasing ambient temperature, V_{0p} is inversely proportional to ambient temperature and proportional to the log of i_L. However, this relationship does not hold for very small amounts of incident light.

The short-circuit current i_{sh} is the output current when the load resistance $R_L = 0$ and $V_0 = 0$, yielding:

$$i_{sh} = i_L - i_s\left(\exp\frac{e(i_{sh} \cdot R_s)}{kT} - 1\right) - \frac{i_{sh} \cdot R_s}{R_{sh}}$$

In the above relationship, the second and third terms limit the linearity of i_{sh}. However, if R_s is a few ohms (or lower) and R_{sh} is 10^7 and 10^{11} ohms, these terms become negligible over quite a wide range. Thus it can be assumed that i_{sh} equals i_L to a good approximation.

V–i characteristics

When a voltage is applied to a photodiode in the dark state, the V–i characteristic curve observed is similar to the curve of a conventional rectifier diode, as shown in Figure 9.21. However, when the light strikes the photodiode, the curve at (1) shifts to (2) and increasing the amount of incident light shifts the characteristic curve still further to position (3). For the characteristics for (2) and (3), if the photodiode terminals are shorted, a photocurrent i_{sh} or $i_{sh'}$ proportional to the light intensity will flow in the direction from the anode to the cathode. If the circuit is open, an open circuit voltage V_{0p} or $V_{0p'}$ will be generated with the positive polarity at the anode.

The short circuit current i_{sh} is extremely linear with respect to the amount of incident light (Figure 9.22(a)). When the incident light power is within a range of 10^{-12} to 10^{-3} W, the achievable range of linearity is 6 to 8 orders of magnitude, depending on the type of photodiode and circuit in which it is used. The lower limit of this linearity is determined by the NEP, while the upper limit depends on the load resistance and the reverse bias voltage, and is given by the following equation:

$$P_{MAX} = \frac{V_{Bi} + V_R}{(R_S + R_L) \cdot R_\lambda}$$

where P_{MAX} is the input power at upper limit of linearity (W), V_{Bi} is the contact voltage (V), V_R is the reverse bias voltage (V), R_λ is the radiant sensitivity at wavelength λ, R_S is the series resistance (Ω), and R_L is the load resistance (Ω).

When laser light is condensed on an extremely small spot, however, resistance increases and the linearity deteriorates. V_{0p} varies logarithmically with respect to a change of amount of light and is greatly affected by variations in temperature, making it unsuitable for light intensity measurements (Figure 9.22(b)). Figure 9.23 illustrates the use of currents for intensity measurement. In the circuit shown in (a), the voltage $(i_L \times R_L)$ is amplified by A and the use of bias voltage V_R makes this circuit suitable for receiving short pulses of light, although the circuit has limitations with respect to linearity. This condition is shown in Figure 9.23(c). In the circuit of Figure 9.23(b), an operational amplifier is used and the characteristics of the feedback circuit are such that the equivalent input resistance is several orders of magnitude smaller than R_f enabling nearly ideal i_{sh} measurements. The value of R_f can be changed to enable i_{sh} measurements over a wide range.

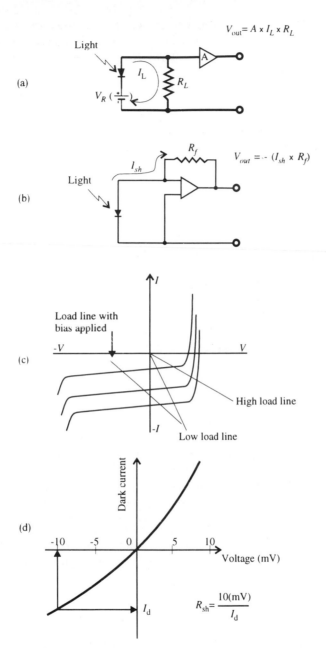

Figure 9.23 Photodiode operational circuits, characteristics and load lines [9.3].

If the zero of Figure 9.23(c) is magnified, we see, as shown in Figure 9.23(d), that the dark current is linear over a voltage range of approximately ± 10 mV. The slope in this region is termed the shunt resistance (R_{sh}) and this resistance is the cause of thermal noise currents.

The p–i–n diode

Photodiodes constructed with an intrinsic (i) region between the p and n regions of the junction are used where speed of response is important (Table 9.5). There are three main factors limiting the response time of photodiodes.

(a) Diffusion time of carriers into the depletion region

Carrier diffusion is inherently a relatively slow process. It is quantified by τ_{diff} where

$$\tau_{diff} = \frac{d^2}{2D_c}$$

where d is the distance the carrier travels and D_c, the minority carrier diffusion coefficient, may have values $\sim 10^{-3}$ m^2 s^{-1}. With d of (say) 5 μm, τ_{diff} is of the order of a few nanoseconds (i.e. $\sim 10^{-9}$ s).

(b) Drift time through the depletion region

In high electric fields (above 2000 V m^{-1}) the drift velocity of carriers saturates: we can then assume a constant velocity v_{sat}. To traverse the full depletion width, W, the transit time is

$$\tau_{drift} = W/v_{sat}$$

For a comparable dimension of 5 μm, a typical v_{sat} of 10^5 m s^{-1} occurs with a $\tau_{drift} \sim 10^{-11}$ s. Thus the drift process is much quicker.

The design of a fast response detector therefore aims to optimize the number of carriers generated in the *depletion* region – this is done by increasing its thickness by inclusion of the intrinsic layer. Carriers arriving from outside by diffusion then give rise only to a 'slow' tail to the response.

(c) Junction capacitance effects

For reasons associated with (b) photodiodes are often operated with a reverse bias giving a larger field causing faster drift. The junctions then exhibit voltage-dependent capacitance, C_j, caused by variation of stored charge at the junction. An expression for this is

$$C_j = \frac{A}{2} \left[\frac{2e\varepsilon_r\varepsilon_0}{V_0 - V} \left(\frac{N_d N_a}{N_d + N_a} \right) \right]^{1/2}$$

where A is the junction area, e is the electronic charge and external bias, V, is $\gg$ zero bias V_0. In practice, with real junctions $C_j \propto V^{-1/3}$. A cut-off frequency f_c is then given by

$$f_c = \frac{1}{2\pi R_L C_j}$$

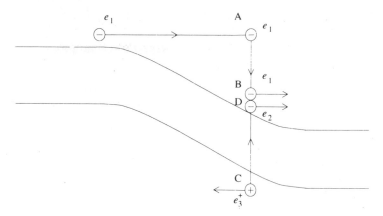

Figure 9.24 Principle of operation of an avalanche photodiode.

where R_L represents the effective shunt resistance. C_j is of the order of 10s of pF. It may be reduced by reducing the diode area and doping level N_d or increasing reverse bias voltage V. For use in optical communications *pin* diodes are always used with reverse bias.

The *pin* structure allows optimization of the junction capacitance and the drift transit time. These can be made nearly equal for best response times which can be less than 1 ps. Silicon photodiodes are the most popular type, operating between 400 and 1000 nm, while germanium can be used out to 1.8 μm. However, ternary and quaternary compound photodiodes are now extensively used in optical communications.

The avalanche photodiode

Most photodiodes used for fast response applications require considerable external amplification. Operation of a *pn* junction device with *large reverse bias* (30–50 V) causes carriers traversing the depletion region to gain sufficient energy to enable further carriers to be excited across the energy gap by impact excitation. These carriers are available for conduction and so provide *internal amplification* analogous to that in the photomultiplier. The effect of a reverse bias is to increase the difference in energy between the p and n doped regions (Figure 9.24). The generation of a photoelectron, e_1, as shown in the diagram, can now produce further electron–hole pairs at the *pn* junction since each electron e_1 has sufficient energy to induce this effect. Having reached the point A, the electron e_1 has sufficient energy above the conduction band bottom to enable it to excite an electron from the valence into the conduction band (C → D). Whilst inducing this effect the electron e_1 falls from A to B. These secondary pairs of electrons and holes then provide an avalanche of free carriers. Second-generation carriers can also lead to third and higher-order carrier generation and it is possible for a primary pair of carriers to generate 20–100 carrier pairs with a corresponding amplification factor.

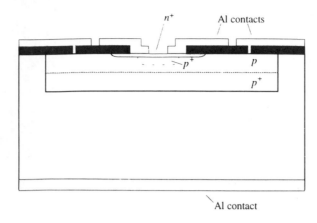

Figure 9.25 Schematic cross-section of double epitaxial silicon SPAD device.

The limit to useful amplification occurs when the avalanche process contributes excessive extra noise. Silicon is a favourable material with respect to noise performance.

Reverse-biased *pn* or *pin* photodiodes are, in general, as fast as avalanche photodiodes for applications where the signal intensity is sufficiently high. A *pn* diode of 100 μm diameter and 2 μm junction depth, for example, has a C_j of about 0.5 pF. With a 50 Ω resistance this corresponds to a capacitance-limited rise time of $\approx 2.2\,RC \approx 50$ ps. Widening the depletion layer encounters the transit time limitation of ~ 10 ps/μm and also demands small active areas.

The avalanche device is of particular interest for very weak intensity applications, particularly when single photon counting is required. A device structure for a single photon avalanche diode (SPAD) is shown in Figure 9.25.

It features double epitaxy fabrication and a thin guard ring made from a low doped *p*-region, 2.5 μm thick as well as a p^+-type 'plug' which is kept small in diameter. Most electrons generated in the neutral *p*-region are drained and any electrons generated within a few microns of the n^+p junction (0.3 μm thick) contribute to the avalanche process.

Recent determination of response time has shown that a Full Width at Half Maximum (FWHM) of less than 50 ps (implying a timing resolution of ~ 20 ps) can be obtained. An application in a time-resolved high spatial resolution luminescence microscope for lifetime measurements in semiconductors is the Edinburgh Instruments Ltd 'LIFEMAP' instrument. Due to the small detector size, combined with photon counting capability, a spatial resolution of ~ 5 μm can be combined with time resolution of ~ 20 ps for wavelengths between 400 and 1100 nm. Thus the SPAD detector enables a powerful analytical technique for quality control during device fabrication [9.4].

The detectivity of avalanche photodiodes is often compared with that of photomultiplier tubes. In 1970 McIntyre compared the two types of detector theoretically for laser pulses of 1, 10 and 100 ns duration at a wavelength of 1.06 μm. For a

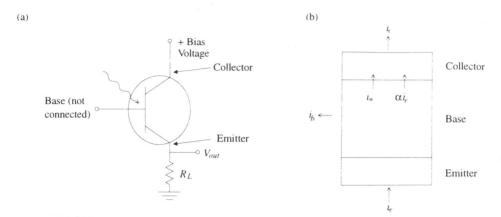

Figure 9.26 (a) The external connections that are made p–n–p phototransistor. Light incident upon the base causes an emitter current to flow, which produces a signal voltage across the load resistance R_L. (b) The currents that are set up in the phototransistor.

100 ns pulse the avalanche photodiode compared well to the photomultiplier. For shorter pulse lengths only photomultipliers with high quantum efficiencies had an advantage at medium background levels. For the 800–1100 nm region of wavelengths the quantum efficiency of photomultipliers is lower than that of photodiodes which have a better detectivity for this region.

The possibility of using avalanche photodiodes as single photon detectors was investigated by Cova *et al.* [9.5]. The detectors are cooled so that the contribution to noise from thermal sources is negligible. The detector is reverse biased with a voltage just above the breakdown voltage so that a single photon allows the formation of a large pulse. A quenching circuit also needs to be used with the diode to prevent a self-sustaining avalanche current.

Phototransistors
A bipolar phototransistor is shown in Figure 9.26. It is different from a conventional bipolar transistor in that the base–collector junction has a large area to absorb a large number of the incident photons. These photons lead to the production of electrons and holes at the *pn* junction formed by the base and collector when the base lead is floating. As a result the electrons contribute a photocurrent in the collector, while the pulse generated in the base region and those electrons swept into the base from the collector lower the base–emitter potential so that further electrons are injected across the base to the collector.

It is possible to understand the operation of the device further, by considering the external currents that are generated in the phototransistor by the incident photons (Figure 9.26) which produce the base current i_b.

By the conservation of charge we can write,

$$i_c = i_e - i_b$$

where i_c and i_e are the collector and emitter currents, respectively. The collector current has two components: (a) the normal diode reverse saturation current i_{c0}; and (b) that part of the emitter current which manages to cross to the collector. The latter term can be written as αi_e where α is less than unity and is known as the common base current gain. Therefore

$$i_{c0} + \alpha i_e = i_e - i_b$$

whence

$$i_e = \frac{(i_b + i_{c0})}{(1 - \alpha)}$$

$$= (i_b + i_{c0})\left(\frac{\alpha}{1 - \alpha} + 1\right)$$

$$= (i_b + i_{c0})(h_{fe} + 1)$$

where h_{fe} is called the common emitter gain of the transistor. When there is no light incident upon the transistor the typical value of h_{fe} is of the order of 100, and so, since $i_b = 0$, the emitter current is $i_e = i_{c0}(h_{fe} + 1)$. This is called the dark current of the device.

When the base is illuminated with radiation of wavelength λ, a base current whose magnitude depends upon this incident wavelength, i_λ, is formed. The emitter current is now $(i_\lambda + i_{c0})(h_{fe} + 1)$ which if $i_\lambda \gg i_{c0}$ is given by $i_\lambda(h_{fe} + 1)$ and it can be seen that the device has an *internal gain*.

The phototransistor does have some disadvantages, however. First, at low light levels and hence low base currents, h_{fe} drop to quite low values and hence there is low gain. Secondly the frequency response of the device is relatively poor (less than 200 kHz) due to the time it takes for carriers to diffuse across the base region.

The photon drag detector
The photon drag effect relies upon the photons in a laser beam transferring their momentum to the free carriers in materials such as p-type doped germanium. The free carriers are physically driven down the detector and hence create a voltage gradient across ring electrodes at the ends of the crystal. Photon drag detectors are insensitive compared with other detectors, but have the advantages of a fast response (less than 1 ns), a high damage threshold (greater than 20 MW/cm^2) and convenient room-temperature operation. They are usually used to measure CO_2 laser pulses in the 9–11 μm wavelength range. For a more detailed discussion of the photon drag detector see [9.6].

9.4.4 Charge-coupled devices (CCDs)

These devices are the basis of video cameras and other array detectors and produce a time-sequential electrical output. This special case of photodiode detectors is in

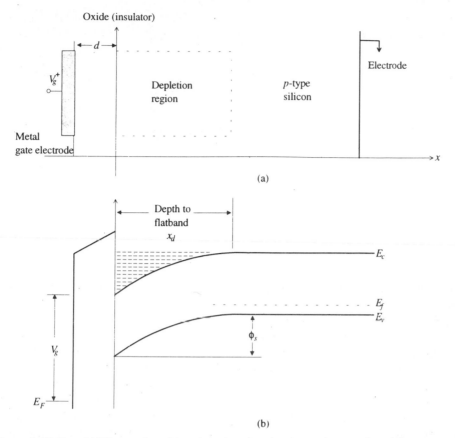

Figure 9.27 Two MOS capacitors biased so that the charge can be transferred from capacitor 1 to 2.

widespread use and is industrially important. In their simplest form, charge-coupled devices (CCDs) are merely arrays of closely spaced metal–insulator–semiconductor (MIS) capacitors. The most important type of CCD is the metal–oxide–semiconductor (MOS) capacitor, which uses silicon as the semiconductor and silicon dioxide as the insulator.

We examine the basic building block of a CCD more closely. In this section we focus on the MIS capacitor, a cross-section of which is shown in Figure 9.27(a), together with the corresponding energy-band diagram in Figure 9.27(b). A voltage V_g is applied to the metal gate in order to charge or discharge this capacitor. The energy band diagram for such an MIS capacitor shows that the Fermi levels are shifted by the applied voltage V_g. The charge in the semiconductor material under the gate will be equal and opposite to that on the gate and since the insulator in between has infinite resistance, no charge transport takes place through it. For applications in CCDs almost all MIS capacitors are metal–oxide–silicon (MOS)

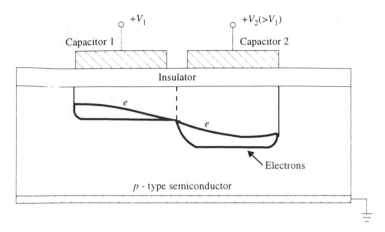

Figure 9.28 A MOSFET with eight gates.

capacitors consisting of aluminium metal, silicon dioxide insulator and silicon semiconductor.

The positive voltage applied to the gate causes the positive holes in the *p*-type semiconductor to migrate towards the ground electrode and leave behind a depletion region depleted of positive charge. In this depletion region electrons which are minority carriers can be accumulated and held against the insulator surface. If a photon with an energy greater than the energy gap is now absorbed in the depletion region, it produces an electron–hole pair. The hole is mobile and moves out towards the electrode; however, the electron is held in the depletion region against the insulating band. In this manner the photons incident on the CCD produce a charge whose magnitude is proportional to the number of photons absorbed in the depletion zone. Note that the amount of negative charge that can be collected under the gate is proportional to the gate voltage, V_g.

Forming an array of CCDs
If an array of MOS capacitors is biased by the application of a correct sequence of clock voltage pulses, the negative charge in the depletion region of each capacitor *can be transferred through the array* of capacitors. The potential energy of each well in a capacitor depends upon the voltage that is applied to the gate electrode and also the oxide thickness. As a result we can apply different gate voltages across an array of capacitors so that the electrons in each capacitor along the array are in different energy levels (Figure 9.28). Since the electronic charge in each capacitor can move to an adjacent capacitor if its electrons are in a higher energy level, the charge can be transferred along the array by varying the voltage applied to each gate along the array.

Let us now consider a metal–oxide–semiconductor field effect transistor (MOSFET) built with eight gates, as shown in Figure 9.29. A charge introduced at the source

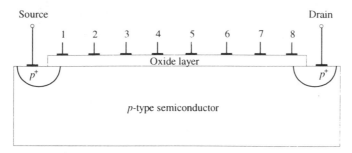

Figure 9.29 Sources of noise in a system using a CCD array.

can be transferred to the drain by passing under gates 1 to 8. This can be done by serially biasing in time gates 1 to 8. The charge flows to the potential minimum as a positive voltage is applied sequentially to the gates and eventually comes out of the drain diode.

It is possible to extend a one-dimensional array of CCDs into two dimensions to form an area array. A variety of interconnection schemes can then be used to read out the information.

Illumination of CCD arrays
It is possible to illuminate an array either from the front (i.e. the side with the gate electrodes) or from the back (the side with the *p*-type material). Front illumination suffers from the problem of a loss of detector response which is caused by interference and absorption effects at the gate electrodes. Back illumination can eliminate some of these problems, although the photon transmission through silicon must be carefully considered. For use in the infrared, back illumination allows the filtering of all visible photons. For use in the 0.3–1.1 μm regime the substrate needs to be thinned down so that the distance to the depletion region is within one diffusion length of the carrier. If ruggedness and the red response are both of importance, then a thicker substrate is desirable and the sensitivity of back illumination can be improved by using an antireflection coating.

Noise in CCD arrays
Figure 9.30 identifies some of the possible sources of noise that can exist in a system using a CCD array.

9.4.5 Non-equilibrium devices for infrared detection: minority carrier extraction and exclusion

Recent advances in heterojunction photodiodes have yielded structures that significantly improve detectivity. Narrow-gap semiconductors are used for sensitive detection in the infrared. The atmospheric window regions from 3 to 5 μm and 8 to

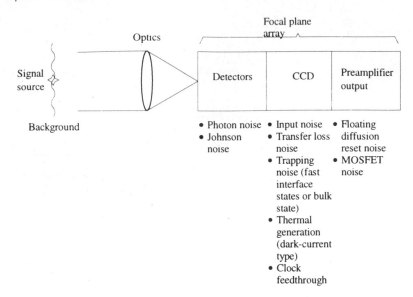

Figure 9.30 Schematic diagram of (a) homojunction excluding and (b) heterojunction extracting devices showing energy levels and carrier concentrations.

13 μm are particularly important. For the former, cadmium mercury telluride (CMT) and indium antimonide (InSb) are used but require cryogenic cooling, typically to less than 200 K. For the longer wavelength window CMT is used but requires cooling to 80 K. This region is important for thermal imaging although the requirements for vacuum encapsulation and cooling systems add to the expense and inconvenience of use of devices.

Recently new semiconductor structures have been developed by C.T. Elliott and colleagues [9.7] at the DRA Malvern, which cause the detectors to operate in a non-equilibrium mode such that the carrier densities are held below their equilibrium, near-intrinsic values. The structures use *carrier exclusion* in photoconductors and *extraction* in photodiodes. Near room-temperature performance is obtained.

As indicated in the previous discussion of noise sources, the signal-to-noise ratio of near-intrinsic long-wavelength detectors is limited by statistical fluctuations in the rates of thermal generation and recombination of electron–hole pairs. The effects of fluctuating recombination can be avoided by arranging for the process to take place in a region of the device where it has little effect. One example is near the contacts in swept-out photoconductors and another is in the neutral regions of photodiodes. The generation process itself, with its associated fluctuations, cannot however be avoided.

In a slab of semiconductor material of thickness d, which is of the order of the optical absorption length, let us assume that all the electron–hole pairs generated within it, due either to incident photons or thermal effects, are counted in some way. The detectivity of this device can then be written as

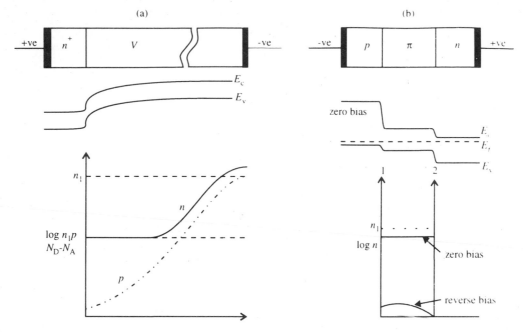

Figure 9.31 Schematic diagram of (a) homojunction excluding and (b) heterojunction extracting devices showing energy levels and carrier concentrations.

$$D_\lambda^* = \frac{\eta}{E_\lambda} \frac{1}{(2Gd)^{1/2}}$$

where η is the fraction of incident photons which are absorbed and E_λ is the energy of a photon. The quantity G is the total carrier generation in the material and is made up of several components so that we may write

$$G = G_A + G_{S-R} + G_{Rad}$$

G_{Rad} is due to photons emitted from the detector enclosure and photons received through the input optics. This background radiation generation sets the fundamental limit to the detector performance. G_{S-R}, Shockley–Read generation effects, are due to defects or impurities which give rise to energy levels in the forbidden gap, and do not impose a fundamental limit. The remaining term, G_A, the Auger generation, can arise from several possible processes in various semiconductor situations. In the case of both InSb and CMT, the mechanism is due to impact ionization by electrons and is dominant at temperatures where the materials are intrinsic or near-intrinsic. This generation rate is proportional to the *electron density*. Thus we expect that, if this density can be held below its equilibrium value in the key part of the device structure, an improved signal-to-noise ratio can be obtained.

Figure 9.31 shows the device structures for (a) the excluding contact photoconductor and (b) the extracting photodiode. In (a) the contact presents no barrier to majority carrier flow (electrons) but does not inject minority carriers. This can be achieved

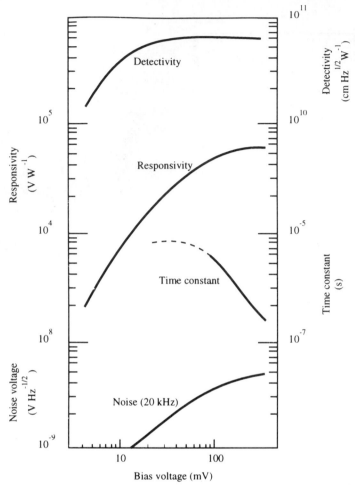

Figure 9.32 Characteristics of a 50 μm square $Cd_xHg_{1-x}Te$ detector operated at 80 K as a function of bias voltage [9.9] (Capocci 1987). The measurements were made in 30° field of view and the responsivity and detectivity values refer to the peak wavelength response at 12 μm.

with a degenerately doped n^+ region. When this contact is biased positively, both electron and hole carrier densities are reduced below their equilibrium values (i.e. the device is held in a state away from intrinsic conduction where it would normally be at room temperature) and the majority electron concentration falls to its extrinsic value $N_D - N_A$. This region extends to a length of some 100 μm in CMT. The electron concentration at 295 K is theoretically reduced by a factor of 80 and the detectivity increased by a factor of 9 to within a factor of 3 of the background limit.

Alternatively, similar Auger suppression can be achieved by carrier extraction in diode structures based on near-intrinsic p-type material. Such a structure and its carrier densities are shown in Figure 9.31(b).

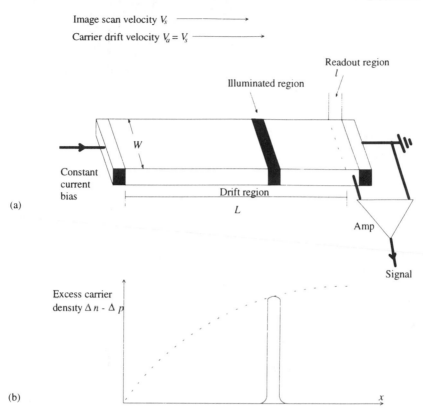

Image scan velocity V_s

Carrier drift velocity $V_d = V_s$

Readout region
l

Illuminated region

W

Constant
current
bias

Drift region

(a)

L

Amp

Signal

Excess carrier
density $\Delta n - \Delta p$

(b)

x

Figure 9.33 The operating principle of a SPRITE detector. (a) $Cd_xHg_{1-x}Te$ filament with three ohmic contacts; (b) shows the build-up of excess carrier density in the device as a pixel of the image is scanned along it.

9.4.6 Infrared array detectors

In recent years there has been significant progress in infrared array detector technology for thermal imaging. The driving force has been the military interest in night vision and the requirements for homing devices within missiles. The semiconductor $Cd_xHg_{1-x}Te$ (CMT) has been the preferred material and is widely used in the 8–14 μm region in linear arrays of up to 200 elements with individual detector sizes of 50 μm. This detector requires to be cooled to 77 K: its performance is summarized in Figure 9.32.

In 1977 Elliott [9.8] proposed the so-called SPRITE detector, in which the time delay and signal integration functions of a serial scan thermal imager are performed within a single strip of CMT, thus replacing a whole row of discrete elements of a conventional detector and the associated external amplifiers and time delay circuitry.

The operating principle of the device is illustrated in Figure 9.33. It consists of a strip of n-type CMT, typically 700 μm long, 60 μm wide and 10 μm thick with three

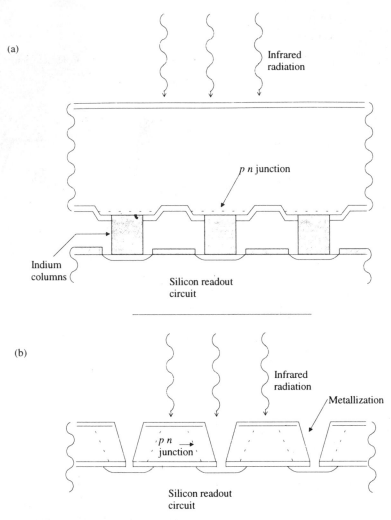

Figure 9.34 Hydridization of a 2D array of photodiodes to a silicon readout circuit. (a) Using In columns. (b) Small holes are etched in the detector material, diodes are formed around the holes and these are connected to the readout circuit with a suitable metallization.

ohmic contacts. A constant-current circuit provides bias through the two end contacts and it is important that this circuit has a good high-frequency response. The third contact is a potential probe for signal readout. The two most important points concerning the operation of the device are as follows. First, that the bias field is sufficiently high for excess minority carriers generated over a significant fraction of the filament length to reach the negative end contact before recombination occurs in the bulk, i.e. the ambipolar drift length, $V_a \tau$, is comparable to or greater than L, where V_a is the ambipolar velocity, $\mu_a E$. Second, the bias level is chosen so that the

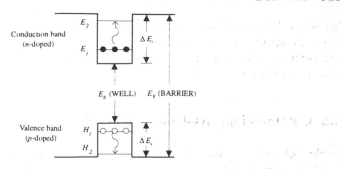

Figure 9.35 Band structure of quantum-well (depths ΔE_c and ΔE_v). Intersubband absorption between electron levels E_1 and E_2 or hole levels H_1 to H_2 is schematically shown.

ambipolar drift velocity, V_a, is equal to the imager scan velocity. In most practical situations the ambipolar velocity approximates closely to the minority-hole drift velocity.

Consider now an element of the image scanned along the filament, as illustrated in Figure 9.33(a). The excess-carrier density in the filament, at a position corresponding to the illuminated element, increases during the scan as $[1 - \exp(-x/V_a\tau)]$, where x is the distance from the positive bias contact along the scan direction and τ is the excess carrier lifetime. When the illuminated region enters the readout zone the increased conductivity modulates the voltage on the contact and provides an output signal.

Focal plane arrays

The one-dimensional arrays discussed above have been recently extended to 2-D arrays with electronic multiplexing on the focal plane. These arrays have the dual advantages of requiring no optical scanning and improving the temperature resolution of an imager by $\sqrt{n}$, where n is the number of elements. Such arrays are usually known as *staring* arrays. They are often manufactured as hybrid silicon-CMT integrated circuits and perform a function that is similar to the CCD which is used in the visible. Figure 9.34 shows an example of such a detector using CMT photodiodes.

9.4.7 Quantum-well infrared photoconductors (QWIPs)

The narrow-gap semiconductor materials such as CMT, in general, are more difficult to grow, process and fabricate in high quality than the larger gap materials, e.g. based on GaAs. Quantum-well intersubband transistors (see Section 2.2) can be utilized to provide detectors. The principles are illustrated in Figure 9.35. The excited carrier can escape from the quantum well and be collected as a photocurrent by means of doping the quantum well n or p-type as shown in Figure 9.35. Thus, although the process is essentially *intraband*, it has similarities to the extrinsic photoconductors described earlier. Since the original work, circa 1987, rapid progress has been made leading to $D^* \sim 10^9 - 10^{10}$ for detectors operating near 10 μm at room temperatures. Array detectors may be fabricated. An attractive feature is the ability to 'engineer' the subband transition to give selective detection at any desired wavelength. An extensive review is given by Levine [9.9].

9.5 Commercially available detectors

Tables 9.6 and 9.7 below provide a list of some commercially available detectors, together with their manufacturers specifications and applications. A more comprehensive list of detector manufacturers is given in [9.10].

Table 9.6 Commerically available thermal and photon detectors

(a) Thermal detectors

Device type	Manufacturers	Specifications
Thermocouple	Dexter Research Center Inc. ARi Industries Inc. Franco Corradi Wahl Instruments Inc.	Dexter dual-element thermopile detector, model DR34 Per element: Number of junctions: 40 $D^*(500\ K, DC) > 2 \times 10^8$ cm Hz$^{1/2}$ W^{-1} Responsivity: 25 V W^{-1} Resistance: 7.5 kΩ Time constant: 28 ms Package: TO-5 can. Argon filled with KBr window.
Bolometer	EDO/Barnes Engineering Infrared Labs. Inc QMC Instruments Ltd.	QMC Instruments Ltd. Ultra-sensitive hot-electron bolometer type QN1/X NEP $= 1.5 \times 10^{-13}$ W Hz$^{-1/2}$ at 270 Hz Speed: Up to 1 MHz at -3 dB Spectral response: 30–3000 GHz Hot-electron detection: Allows optimum sensitivity in mechanically and electromechanically noisy environments. Reliability: Can be cycled continuously on a daily basis or kept at operating temperature for months at a time.
Pryoelectric	Alrad Instruments Ltd. Delta Developments Philips IR Defence Components Hamamatsu Corporation	Edinburgh Instruments PVF2 Wavelength range: 2.5–>30 μm λ peak (μm): flat response Operating temperature: 233–305 K Responsivity: 1–2 μA/W Detectivity D^*: 1×10^5 cm Hz$^{1/2}$ W^{-1} Bandwidth: DC-200 Hz Diameter: 1 cm.

Table 9.6 (*continued*)
(b) Photon detectors

Device type	Manufacturers	Specifications		
Photoconductors	Edinburgh Instruments EG & G Heimann Kolmar Technologies Photonic Detectors Inc.	Edinburgh Instruments	PCRG	PCMPC12
		Wavelength range:	2–12 μm	8–13 μm
		λ peak:	8 μm	12 $\pm$ 1 μm
		Operating temperature:	273 K	77 K
		Response time:	<0.4 ns	500 ns
		Responsivity:	10–120 mV/W	5×10^3 V W^{-1}
		D^* (cm Hz$^{-1/2}$ W^{-1}):	10^6–10^7	2×10^{10}
		Bandwidth:	>100 kHz	10^3–10^6 Hz
Photoemissive (PMT tube)	Alrad Instruments Ltd. Instrumat SA Onel Corporation Thorn EMI Electron Tubes Ltd.			
Photoemissive (PMT microchannel plate)	Comstock Inc. Galileo Electro-Optics Corp. Philips Components.			

Table 9.6 *(continued)*
(b) Photon detectors cont.

Device type	Manufacturers	Specifications
Photovoltaic Photodiodes	Advanced Photonix Inc. Hamamatsu Corporation Megatronic Ltd.	Edinburgh Instruments PV MPV11 Wavelength range: 8–12 μm λ peak: >10.6 μm Operating temperature: 77 K Responsivity: 2–5 A/W D^*: 3×10^{10} cm Hz$^{1/2}$ W^{-1} Bandwidth: DC to >100 MHz
p–i–n diodes	AT&T Fibre-data Limited Hitachi NEC Electronics	Epitax ETX 75T/100T at $T = 25°C$ and $V_R = 5$ V Spectral response: 0.8–1.7 μm Responsivity at 1.3 μm: 0.80 A W^{-1} at 1.5 μm: 0.90 A W^{-1} Dark current: ~1–5 nA Capacitance: 1.0–1.5 pF Fise/fall time: 1 ns (10–90%, at 50 Ω) Max. reverse voltage: 25 V Max. reverse current: 10 mA Max. forward current: 0.5 mA Power dissipation: 100 mW Operating temperature: −40 − +85°C Storage temperature: −60 − +125°C

Table 9.6 *(continued)*
(b) Photon detectors cont.

| Avalanche photodiodes | Advanced Photonix Inc.
Devar Inc.
GCA Electronics Ltd
NEC Electronics
Hamamatsu | Hamamatsu S2383
Diameter: 1.0 mm
Effective area: 0.78 mm^2
Spectral response range: 400–1000 nm
Peak wavelength: 800 nm (at 25°C)
Breakdown voltage V_B: 150–300 V ($i_d = 100$ nA)
Dark current, I_d: 0.6–6 nA
Terminal capacitance, C_t: 7 pF
Cutoff frequency: 700 MHz
(with a load resistance of 50 Ω)
Gain at $\lambda = 830$ nm: ~100
Ambient operating temperature: $-20 \sim +60°C$
Storage temperature: $-55 \sim +100°C$ |
| Phototransistors | EG&G Vactec
Hamamatsu Corporation
NEC Electronics Inc.
Silonex Inc. | Hamamatsu S2041
Collector–emitter voltage V_{ce}: 35 V
Collector power dissipation: 75 mW
Peak wavelength: 720 nm
Photocurrent I_c: 0.6 mA
(at 1000 lux with $V_{ce} = 5$ V)
Dark current I_{ce}: 0.05 μA
(at $V_{ce} = 10$ V max)
Rise time t_r: 2 μs
(at $V_{ce} = 5$ V, $R_L = 100$ Ω)
Fall time t_F: 3 μs
(at $V_{ce} = 5$ V, $R_L = 100$ Ω) |

Table 9.6 *(continued)*
(b) Photon detectors cont.

Device type	Manufacturers	Specifications
Photon drag	Edinburgh Instruments Ltd. Hamamatsu Corporation Monolight Instruments Ltd.	Edinburgh Instruments PD2 Aperture: 10 mm Sensitivity: 180 mV/MW Uniformity: $\pm 2\%$ Max. power density: 20 MW/cm^2 Edinburgh Instruments PD3 Aperture: 20 mm Sensitivity: 45 mV/MW Uniformity: $\pm 3\%$ Max. power density 40 MW/cm^2
CCDs	General Electric, RCA Rockwell, Aerojet Electrosystems, Honeywell, Texas Instruments, Mitsubishi, Fijitsu	SBRC Architecture: Hybrid Size: 7×64 Detector Material: Si: As Spectral range: 1 25 μm Operating temperature: 8 K Device type: Photoconductor Mitsubishi monolithic interline transfer 256×256 PtSi 1 5 μm 77 K Schottky barrier

Table 9.7   Types and applications of sensors and detectors. [9.3].

Products	Major Applications
High Energy Particles	
Si Photodiodes	High Energy Physics, Nuclear Medicine, Industrial Measuring Instruments
UV	
UV Enhanced Si Photodiodes	Pollution Analyzers, Spectrophotometers, Medical Instruments, UV Detectors, Colorimeters
Schottky Type GaAsP Photodiodes	Pollution Analyzers, Spectrophotometers, Colorimeters, UV Detectors
Schottky Type GaP Photodiodes	UV Detectors
Visible	
Filtered Si Photodiodes	Camera AE, Exposure Meters, Auto-strobo, Auto-focus, Illuminometers
Diffusion Type GaAsP Photodiodes	Camera AE, Exposure Meters, Auto-strobo, Illuminometers, Flame Monitors
CdS Cells	Camera AE, Exposure Meters, Auto-dimmers, Musical Instruments, Flame Monitors, Melody Cards, Light Sensing Toys, Street Light
Near IR	
Si Photodiodes	Optical Communication, Computers, Automatic Control System, Home Appliances, Car Electronics, Photometric Instruments, Medical Instruments, High Energy Physics
Si Photo-transistors	Photoelectric Switches, Photoelectric Counters, Smoke Detectors
Photo IC Sensors	Photoelectric Switch, Tape Mark Detection, Paper Detection for Copier or Printer, Encoder
InGaAs, Ge Photodiodes	Optical Communication, IR Laser Monitors, Radiation Thermometers
IR	
PbS, PbSe Cells	Radiation Thermometers, Flame Monitors, Photoelectric Switches, Moisture Meters, Gas Analyzers, Spectrophotometers
InAs, InSb	IR Laser Detectors, Spectrophotometers, Radiation Thermometers
MCT (HgCdTe)	Thermal Imaging, Radiation Thermometers, FTIR Spectroscopy, CO_2 Laser Detection
Photon Drag Detectors	CO_2 Laser Detection
Pyroelectric Detectors	Automatic Door, Intrusion Alarm, Fire Detectors
Image Sensors — 2-D / 1-D	
MOS Linear Image Sensors	Multichannel Spectrophotometers, Spectrum Analyzers, High Energy Physics
Photodiode Arrays	Multichannel Spectrophotometers, Color Analyzers, Spectrum Analyzers, Position Detectors
Position Sensors — Discrete Type	
Discrete Detectors	Various Types are Available to Suite a Wide Application
Non-discrete Type	
PSD (Position Sensitive Detectors)	Auto-focus, Range Finders, Laser Optics, Automatic Control Systems, Proximity Switches, High Energy Physics

Sensors
- Point Sensors
  - High Energy Particles
  - UV
  - Visible
  - Near IR
  - IR
- Image Sensors
  - 2-D
  - 1-D
- Position Sensors
  - Discrete Type
  - Non-discrete Type

References

[9.1] E.L. Dereniak and D.G. Crowe, *Optical Radiation Detectors*, Chapter 2 section 2.3, Wiley, 1984.

[9.2] E.L. Dereniak and D.G. Crowe, *Optical Radiation Detectors*, Chapter 5, Wiley, 1984.

[9.3] Hamamatsu Photodiodes Catalog 1991.

[9.4] G.S. Buller, J.S. Massa and A.C. Walker, *Rev. Sci. Instrum.* **63**, 2994 (1992).

[9.5] S. Cova, A. Longoni and A. Andreoni, *Rev. Sci. Instrum.* **52**, 408, 1981.

[9.6] A.F. Gibson, M.F. Kimmitt and A.C. Walker, 'Photon drag in germanium', *Appl. Phys. Lett.* **17** (1970).

[9.7] C.T. Elliott and N.T. Gordon, Chap. 10, *Handbook on Semiconductors*, Ed. T.S. Moss, **4**, 841, Elsevier Science Publishers, 1993.

[9.8] C.T. Elliott, *Electron. Lett.* **17**, 312, 1981.

[9.9] B.F. Levine, 'Device physics in quantum well infrared photodetectors', *Semicond. Sci. Technol.* **8**, 400, 1993.

[9.10] Laser Focus World. *The Buyers Guide* (Penwell Publications), 302, 1993.

10

Fiber optic devices

10.1 Waveguiding by fibers
10.2 Fibers in communication
10.3 Optical fiber sensors
References
Further reading

10.1 Waveguiding by fibers

10.1.1 Introduction

Optical fibers are flexible waveguides for light. They provide a well-controlled optical path allowing light to propagate over potentially very long distances, since light is essentially contained within a fiber by total internal reflection. This principle was demonstrated by Tyndall in 1870, using a water jet as the transparent medium to confine the light. Theoretical studies on this effect continued through the early years of this century, but manufacture of useful fibers did not take place due to a lack of suitable technology. In 1954 cladded waveguides were first proposed as a method of transporting optical images without aberrations [10.1, 10.2]. By the mid-1960s the first serious considerations of circular dielectric waveguides for light in communications were presented by Kao and Hockham of Standard Telecommunications Laboratories [10.3]. Subsequent development has seen optical fibers become established as a mature technology.

Fibers are clearly important in *communications*, since a light signal can be modulated to carry information, and current performance of fibers in digital transmission allows long distances (hundreds to thousands of km) and high data rates (hundreds to thousands of Mbits per second) to be used. As well as being an attractive medium for communication, fibers can allow us to exploit a range of optical effects and interactions in the field of *sensing and measurement*, so that sensitive optical techniques such as interferometry can be applied in demanding situations outside the laboratory.

This chapter covers the basic theory of light propagation within fibers, followed by reviews of applications to communications and to sensing.

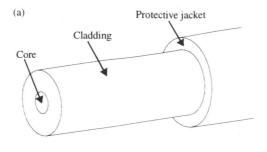

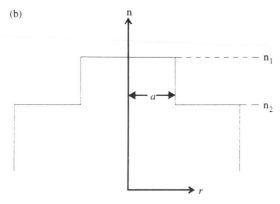

Figure 10.1 (a) Cylindrical dielectric optical fiber. (b) Graph showing the refractive index versus radial distance of a step-index fiber.

10.1.2 Light propagation within fibers

The commonest fiber is a cylindrical dielectric structure comprising an inner *core* surrounded by a concentric *cladding* (Figure 10.1a). Practical fibers are usually protected by an outer buffer or jacket layer beyond the cladding, but note the terminology: the core/cladding structure is the basic optical unit. The refractive indices of core and cladding are n_1 and n_2, respectively, with $n_1 > n_2$, and the indices are uniform within each part of the fiber. This fiber therefore has a *step-index* profile (Figure 10.1b). Other, more complicated index profiles are desirable in some applications, and will be introduced later.

(a) The ray model
The propagation of electromagnetic waves along a waveguiding structure, be it a metallic tube or dielectric fiber, is governed by solutions of the wave equation in the presence of appropriate boundary conditions. These are given in Chapter 1 and describe the 'modes'. Much can be learned, however, from simpler concepts.

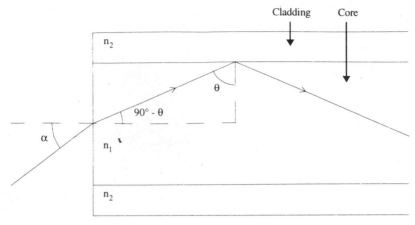

Figure 10.2 Propagation of a ray that is 'just' totally internally reflected at the core–cladding interface.

A picture of light guiding within a step-index fiber arises from the *ray model*, which is valid if the fiber core radius is large compared with the light wavelength. A light ray travelling within the core at an angle θ to the normal meets the core–cladding interface, and will be totally internally reflected, provided that $\theta > \theta_c$, where the critical angle, θ_c, is determined by the refractive indices n_1 and n_2 (Figure 10.2). Thus such a ray is trapped within the core, bouncing between the opposite faces as it propagates and the core–cladding dielectric fiber structure acts as a waveguide. Rays travelling at small enough angles can propagate in this manner, with the limiting case of $\theta = 90°$ defining the axial ray, which travels straight down the fiber. The other limiting case, at the largest angle to the axis, is shown in Figure 10.2, for which the external ray, in a medium of refractive index n_0, meets the input fiber face at an angle α_{max} given by:

$$n_0 \sin \alpha_{max} = n_1 (1 - \sin^2\theta_c)^{1/2} = (n_1^2 - n_2^2)^{1/2} \qquad (10.1)$$

The quantity in Eq. (10.1) is called the *numerical aperture* (NA) of the fiber. For a fiber in air, $n_0 \cong 1$, so the acceptance angle is given by $\sin^{-1}$ (NA). Typical values of NA are in the region 0.1–0.4, corresponding to acceptance angles of approximately 6–24°. It is useful to introduce the parameter

$$\Delta = (n_1 - n_2)/n_1 \qquad (10.2)$$

which is a normalized index difference between the core and cladding. From Eqs. (10.1) and (10.2) it follows that

$$NA = n_1 \sqrt{2\Delta} \qquad (10.3)$$

so that a small numerical aperture results if the difference between core and cladding refractive indices is small. Rays that travel in a plane containing the fiber axis are

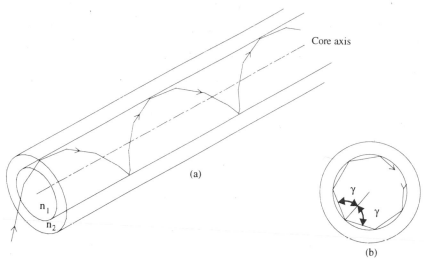

Core axis

(a)

n_1

n_2

γ

γ

(b)

Figure 10.3 The helical path taken by a skew ray in an optical fiber. (a) Skew ray path down the fiber; (b) cross-sectional view of the fiber.

termed *meridional rays*. It is also possible for a ray to enter a fiber with a 'sideways' angle, so that after refraction into the core, its path does not cross the fiber axis, but spirals around it, travelling in an annular region in the outer part of the core (Figure 10.3). Such paths are termed *skew rays*.

(b) The wave model: modes in fibers

The ray model is convenient for visualizing the guiding property of fibers, but a full description requires the wave solutions leading to the concept of *modes*. There is a correspondence between the two models for large-core fibers, however, since the lowest-order mode is analogous to the axial ray, while the higher-order modes correspond to the rays travelling at larger angles to the axis. The transition from the ray model to the wave picture is conveniently illustrated, in approximate form, by the simple geometry of the *slab waveguide* (Figure 10.4a). An infinite slab of transparent dielectric of refractive index n_1 and transverse width $2a$ is surrounded by material of index $n_2 < n_1$. The slab acts as a waveguide by confining light within the central slab by total internal reflection along the ray path shown for the approximation $2a > \lambda$, where λ is the free space wavelength. In wave terms, the ray directions correspond to wavevectors for propagation at equal angles either side of the z-axis, with plane wave fronts as shown in Figure 10.4(b). The wave pattern can only persist if the wave phase in one complete round trip is an integral multiple of 2π, equivalent to the formation of a standing wave pattern across the slab in the x-direction. The x component of the wavevector is $n_1 k \cos \theta$, so the phase-matching condition is written [10.4]:

$$2akn_1 \cos \theta + \delta = m\pi \quad (m = 0, 1, 2 \ldots) \tag{10.4}$$

(a)

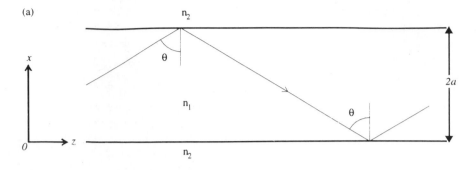

(b)

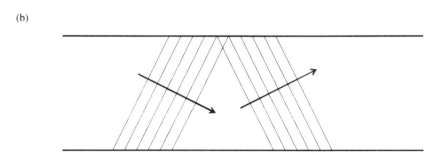

Figure 10.4 Propagation of light in a transparent dielectric medium. (a) The ray model; (b) the wave model.

where $k = 2\pi/\lambda$, where λ and δ are the phase change at reflection. Thus only particular angles θ correspond to integral values of m, so that there is a discrete set of allowed wave patterns or *modes* for wave propagation along the slab waveguide. In the limit of very large (a/λ), this wave description tends to the ray model, which allows θ to vary almost continuously. A further property of the slab waveguide solution that will be useful in interpreting waves in fibers is the wavevector component along the guide's axis z, $n_1 k \sin \theta$, conventionally written as β. The condition for total internal reflection is $\sin \theta \geqslant n_2/n_1$, from which it follows that

$$n_2 k < \beta < n_1 k \tag{10.5}$$

so that the wavevector for propagation along the z-axis lies between limits imposed by the slab and its surrounding medium.

Light within a fiber with a circular cross-section must be described by a wave solution of Maxwell's equations with the boundary conditions imposed by the dielectric properties of the core and cladding. The modes now correspond to the permitted set of waves within a circular-section waveguide. The general solution is complicated, but can be simplified by the 'weakly guiding approximation', in which the refractive index of the core is only slightly higher than that of the cladding. The parameter Δ is much less than unity for weak guiding: practical fibers have Δ in the range 0.002–0.02. A weakly guiding fiber supports a set of modes whose electric and magnetic fields can be shown to be dominated by their transverse components [10.5], with orthogonal polarizations in the x- and y-directions; these are the so-called LP (linearly polarized) modes. Selecting the x polarization, the electric fields take the form

$$E_x \propto J_v(\kappa r) \cos v\phi \quad \text{in core } r < a \tag{10.6}$$

$$E_x \propto H_v(i\gamma r) \cos v\phi \quad \text{in cladding } r > a \tag{10.7}$$

where $J_v(r)$ is the Bessel function, $H_v(r)$ the Hankel function, κ and γ are related to the wavenumber $k = 2\pi/\lambda$, and the integer v is an angular mode number. The variable ϕ is the azimuthal angle of the cylindrical coordinate system (r, ϕ, z) with its axis coincident with the fiber axis. The forms of these functions determine the patterns of the fields propagating within the fiber: in the core, the fields have a number of radial and circumferential maxima, whereas in the cladding, the fields decay smoothly outwards from the boundary $r = a$. Similar equations describe the orthogonal polarization E_y.

Since the modes are travelling wave solutions, the components all vary with time t and distance z along the fiber core as $\exp[i(\omega t - \beta z)]$, where the β is the *propagation constant* of the mode in question, and, as in Eq. (10.5), lies between the limits

$$n_2 k < \beta < n_1 k \tag{10.8}$$

(β/k) is conveniently described as an *effective refractive index*. An alternative simplified notation for fiber propagation modes is due to Gloge [10.6] and is written as $LP_{v\mu}$, where the label v describes the ϕ dependence, as above, while μ is the number of radial maxima in the Bessel function. Figure 10.5 shows three examples of LP modes, beginning with the lowest-order LP_{01}. The relation between the indices v, μ and the field distributions is clear.

The LP modes are *guided modes*; they do not lose power by radiation and their fields tend to zero at an infinite distance into the cladding. In a perfect lossless fiber, light would propagate in one or more such modes unchanged over an arbitrarily long distance. Each mode can be pictured as a series of wavevectors zig-zagging down the fiber, obeying the law of reflection at the core–cladding interface. In real fibers, imperfections such as a slightly irregular core radius, slightly nonuniform refractive index, and bends along the length all tend to induce *mode coupling*, in which power is redistributed amongst the possible guided modes. Mode coupling can be pictured in terms of the ray model, in which the geometric paths of the rays are changed slightly if the ray encounters an imperfection in the fiber's refractive index or core geometry.

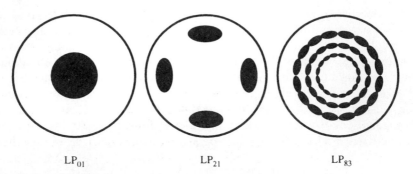

Figure 10.5 Field distributions in three different LP modes in a circular waveguide.

The wave solution (Eqs. (10.6) and (10.7)) implies that there is a finite electric field within the cladding, its magnitude decreasing exponentially at large distances from the core–cladding boundary. Guided modes do not therefore carry all their optical power completely within the core, but usually the fraction of cladding power is small. Fibers are designed with cladding sufficiently thick to ensure a negligible guided field at their outer surface.

Guided modes are not the only possible solutions. *Radiation modes* can exist, and indeed are necessary to form a complete orthogonal set of possible mode solutions [10.6]. It is possible to lose power from a fiber by bending it; in mode terms, guided modes are coupled to radiation modes by the bend, so power is lost from a bent fiber. Bending loss can be understood simply by analogy with the ray model by reference to Figure 10.6; the bend can reduce the angle of incidence at the core–cladding interface to the extent that light is no longer reflected, but is refracted into the cladding. However, this ray model is as usual an oversimplification, since complete total internal reflection does not occur at a curved dielectric interface [10.7]; radiation can tunnel through and appear as a transmitted wave in the cladding. In the real world, fibers must be bent to some degree, but it is fortunate that bending loss is very small in practice unless the radius of the bend is quite small, of the order of a few mm.

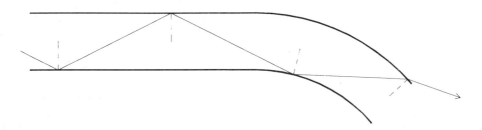

Figure 10.6 The bending loss in a fiber in terms of the ray picture.

10.1.3 Fiber types

(a) Step-index fiber

If the refractive index changes discontinuously as a function of the radial distance, the fiber is described as 'step index'.

The number of guided modes possible in a given fiber depends on the light wavelength λ, the core radius a and the index difference Δ. It is intuitively reasonable that a fiber with a core radius large compared with λ should support a larger number of modes than a very small core. A useful parameter in estimating the number of guided modes is the V number of a fiber, defined as the dimensionless quantity

$$V = \frac{2\pi}{\lambda}\, a(n_1^2 - n_2^2)^{1/2} \tag{10.9}$$

which is a normalized form of the fiber core radius a. Thus a fiber core large compared with λ would have a large V number. To a reasonable approximation, the number of guided modes N is given by [10.6]:

$$N = \tfrac{1}{2}V^2 \tag{10.10}$$

provided that V is large. For example, with $n_1 = 1.53$, $n_2 = 1.50$, $\lambda = 1$ μm and a core radius of 50 μm, $N = 4500$!

An obvious problem with such a waveguide is that the optical path lengths corresponding to the various modes (or the various ray paths) are all different, leading to different propagation times along a length of fiber. This is a problem of particular significance in communications, known as *modal dispersion*. The effect can be estimated using a ray optics model applied to a step-index fiber. We will defer the consideration of *chromatic dispersion* and group velocity (important for communications) to Section 10.1.4, and examine the geometrical effects alone. Consider the ray paths for the axial ray and the extreme meridional ray that defines the NA (Figure 10.7). The axial ray clearly travels the shortest path of length L equal to the fiber length. Therefore the light transit time $t_{min} = (n_1 L/c)$. The length of the extreme ray's path is $L/\sin\theta_c$, from which it is readily shown that the difference in light propagation time, $\delta t = (t_{max} - t_{min})$, is given by

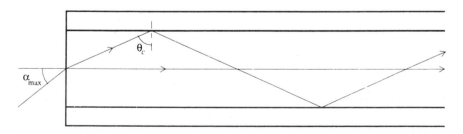

Figure 10.7 Modal dispersion: ray paths for the axial and extreme meridional ray in a step-index fiber.

Figure 10.8 Pulse broadening due to modal dispersion.

$$\delta t = \frac{Ln_1^2}{cn_2}\frac{(n_1 - n_2)}{n_1} \tag{10.11}$$

or, applying the weakly guiding fiber approximation $\Delta \ll 1$, this can be simplified to

$$\delta t = \frac{Ln_1\Delta}{c} \tag{10.12}$$

Although this estimate of modal dispersion has been derived for meridional rays in a step-index fiber, it is equally applicable to skew rays [10.8].

For a step-index fiber with $\Delta = 0.01$, $n_1 = 1.45$, the delay difference δt introduced by modal dispersion is approximately 50 ns per km of fiber. This spread in transit time is an obvious limitation if the fiber is to be used in high data rate communications, since a series of optical pulses launched into a long fiber will begin to broaden and possibly overlap at the output end. The limiting case of pulse overlap is illustrated in Figure 10.8: if input pulses of duration δt are separated by $2\delta t$, then modal dispersion will double the pulse duration. Output pulses can still just be distinguished without overlap if the input pulse frequency is less than the fiber's *bandwidth B*, where

$$B = \frac{1}{2\delta t} \tag{10.13}$$

From Eqs. (10.10) and (10.11) it follows that the bandwidth–length product is

$$BL = \frac{c}{2n_1\Delta} = \frac{n_1 c}{(NA)^2} \tag{10.14}$$

The *BL* product for the fiber quoted above is $\sim 10^{10}$ Hz m, or in more conventional units, ~ 10 MHz km.

(b) Graded-index fiber
This limitation on the performance of step-index fiber led to the development of *graded-index fiber*. The index profile of a graded-index fiber is shown in Figure 10.9. Within the core, the refractive index is a function of radius r, which for many practical designs can be written as

$$n(r) = n_1\sqrt{(1 - 2\Delta(r/a)^\alpha)} \quad r < a \tag{10.15}$$

where a is the core radius and α is a profile parameter that characterizes the shape of the index profile. For example, $\alpha = 1$ gives a triangular profile, $\alpha = 2$ is a parabolic

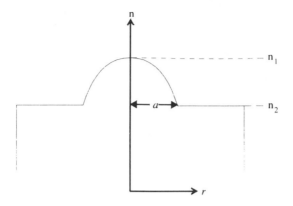

Figure 10.9 The index profile of a graded-index fiber.

profile, while $\alpha \Rightarrow \infty$ reduces to the step-index profile. Graded-index fibers still have cladding with n(r) continuous across the interface with the core and usually the index is constant in the cladding.

The ray paths are rather more complicated than those of the step-index fiber, but are such that modal dispersion effects are reduced. Figure 10.10 shows three main types of ray path: (a) the central axial ray, which is straight; (b) a meridional ray (in a plane containing the axis); and (c) a helical or skew ray (not crossing the axis). Although rays (b) and (c) clearly have longer geometrical paths than (a), they traverse the outer areas of the core where the refractive index is smaller than it is in the axial region, and so travel with greater velocity. Hence the index profile helps to compensate for modal differences in path length. It can be shown [10.8] that a parabolic index profile leads to a sinusoidal meridional ray path, and most graded-index fibers have a similar profile. The spread in propagation time due to modal dispersion in a parabolic-index fiber is smaller than that for a step-index fiber by a factor $\Delta/8$, a considerable improvement [10.9]. For this reason, graded-index fiber was widely used as the first-generation communications fiber.

(c) Multimode and monomode fibers
If the number of modes is significantly large, the fiber is said to be *multimode*. Each guided wave has its own wavefront, so that the light emerging from a multimode fiber is not spatially coherent. This limits multimode fibers to those applications in which coherence is not required, such as intensity-based light modulation, or simply as a delivery system for optical power.

For a given light wavelength λ, the number of guided modes falls if the fiber core radius is reduced. It can be shown [10.8] that the fiber will support only the lowest-order LP_{01} mode if $V < 2.405$: this value corresponds to the first zero of the Bessel function $J_0(r)$. Such a fiber is known as a *monomode* or *single-mode* fiber, and its advantage over the multimode fiber is the absence of intermodal dispersion and

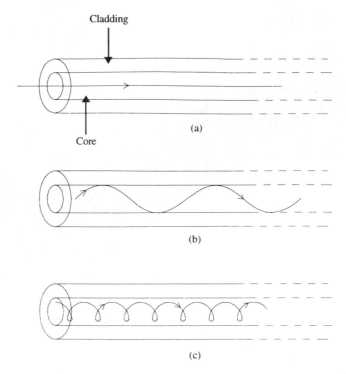

Figure 10.10 Three different ray paths in a graded-index fiber: (a) the central ray; (b) a meridional ray; and (c) a skew ray.

its preservation of spatial coherence of the guided wave. The condition on the *V*-number allows us to estimate the required core radius: for $n_1 = 1.515$, $n_2 = 1.500$ and $\lambda = 1$ μm, the core radius $a = 1.8$ μm. Monomode fibers have very small cores! In practice, the outer diameter of the cladding is usually standardized at 125 μm for both monomode and multimode fibers, so they are not readily distinguished by their appearance.

Equation (10.9) shows that the *V*-number is wavelength dependent. This means that monomode fibers are specified for use at a particular wavelength. If the light wavelength is too short, the *V*-number becomes greater and the fiber will support several modes. On the other hand, if the λ is greater than optimum, then the fiber will still be monomode, but more of the field of the guided wave will be forced further out into the cladding, so that energy will be lost and its propagation will become lossy. Monomode fibers are specified by a '*cutoff wavelength*', below which they cease to be monomode. The modal behaviour of a step-index fiber can be represented by a diagram such as Figure 10.11, which plots a normalized propagation constant *b* against *V* number for several lower-order modes. A good approximation which helps in interpreting *b* is [10.5]:

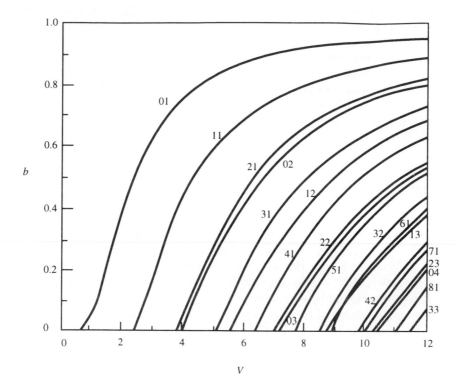

Figure 10.11 The normalized propagation constant b as a function of normalized frequency V for a number of LP modes [10.6].

$$b \cong \frac{(\beta/k) - n_2}{n_1 - n_2} \tag{10.16}$$

which, from Eq. (10.8), must lie between 0 and 1. Each mode except LP_{01} has a limiting value of V below which it cannot exist: as the wavelength is increased for a fixed core radius, the V number decreases and each mode (except the lowest) reaches cutoff in turn, this being the limit $b = 0$ for that particular mode. Eventually the condition $V < 2.405$ is reached and the fiber is single-mode, only the LP_{01} mode surviving. Although the curve for LP_{01} approaches $b = 0$ at a finite value of V, this indicates that the mode is still propagating, but with a negligible fraction of power remaining in the core once V is less than about 0.5. We shall see that monomode fibers are extremely important in both communications and in fiber sensors in which the coherence of light must be maintained through one or more fibers.

At this point, it is worth returning to the weak guiding condition, $\Delta \ll 1$. If this is not fulfilled, the LP mode theory outlined above does not apply, and the mode description becomes complicated. Weak guiding brings two practical advantages. First, for a given core radius, a smaller index difference Δ reduces the V-number and therefore reduces the number of guided modes. In communications, this is helpful in reducing the modal dispersion of a propagating pulse. The second advantage of weak guiding is that the core radius of a monomode fiber, although small, is still larger than the light wavelength. As Δ is made smaller, the corresponding core radius for monomode operation becomes larger, from the condition $V < 2.405$. Therefore the practical issues of monomode fiber alignment and connection become less demanding than for the case of a larger index differences. Of course, Δ cannot be made arbitrarily small: this would lead to lossy fiber with a high bending loss.

(d) Polarization-preserving fibers

It is worth considering briefly the *polarization* properties of fibers. Ordinary monomode fiber actually supports two modes with orthogonal polarizations. The state of polarization of the guided wave is not necessarily maintained in a real fiber with bends, stresses and thermal effects introducing birefringence. A difference between the principal refractive indices for the orthogonal polarizations occurs. It is possible to overcome this difficulty by making fibers with a high degree of built-in stress [10.10], as shown in Figure 10.12, so that the core refractive index distribution, by the photoelastic effect, is effectively elliptical. Such a fiber is described as highly birefringent ('hi-bi'), and provided the external stresses due to bending or thermal effects are smaller than the built-in stress, then the orthogonal states defined by the built-in stress axes will be preserved as 'eigenstates' of propagation. Linearly polarized light launched into an eigenstate of a hi-bi fiber remains in that state, unaffected by external effects such as temperature change or bending of the fiber.

10.1.4 Manufacture and performance of fibers

The properties of fibers described so far have not depended on the particular dielectric material of core and cladding. The major factor in enabling fibers to be used widely

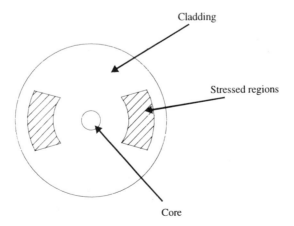

Figure 10.12 Cross-section of a 'bow-tie' highly birefringent monomode optical fiber.

has been the advance in the technology of manufacturing high-purity glasses with well-controlled refractive indices from which fibers can be drawn. The general method of manufacturing silica glass fibers [10.10, 10.11] begins with the production of a *preform* with a diameter in the region of 50 mm. The perform is heated and drawn down to produce a long continuous fiber. The technology of preform and fiber production can be separated into liquid phase and vapour phase techniques. An example of the former is the 'rod in tube' process, in which the preform comprises a glass rod (the core) surrounded by a glass tube (the cladding) of slightly lower refractive index. The preform is held vertically in a furnace so that the fiber can be drawn down from the heated preform. This method is suited to manufacture of large core step-index fibers.

Production of graded-index fibers with optimum performance demands precise control of the refractive index profile; this can be achieved by vapour phase processes, in which high-purity compounds containing silicon and dopant atoms (such as germanium) react in a controlled oxygen atmosphere to build up layers of doped silica. The refractive index of silica is well characterized by the dopant concentration. For example, germania (GeO_2) raises the refractive index, and is therefore suitable in a doped-core fiber. Vapour phase methods produce fibers with lower losses from impurities compared with fibers drawn from the liquid phase.

The outside (cladding) diameter of the fiber is controlled by the drawing speed and is monitored during the process. A cladding diameter of 125 μm is common. The buffer coating, usually a polymer material, is applied to the fiber as the final manufacturing stage to protect the fiber against breakage.

The two great advantages of fibers are their *attenuation* and *dispersion* performance compared with conventional transmission systems. We now focus on these two aspects before describing example specifications for real fibers.

(a) Attenuation

Optical attenuation is defined as the ratio of optical input power to output power, conventionally expressed in decibels (dB):

$$A \text{ (in dB)} = 10 \log_{10} \frac{P_i}{P_0} \tag{10.17}$$

and the fiber attenuation is usually given in units of dB km^{-1}. Attenuation in a multimode fiber depends on the particular distribution of optical power among the modes: if power is launched initially into low-order modes (rays at a small angle to the axis), then mode coupling will gradually increase the proportion of higher-order modes. Conversely, if the source launches light into very high-order modes, these will decay in the first few metres of fiber. In either case, the measured attenuation will not be constant along the fiber. In order to measure the intrinsic attenuation, the optical power must be in its equilibrium mode distribution. This is achieved by measuring attenuation at the far end of a long fiber length, or by using a 'mode scrambling' technique that introduces small random bends to induce rapid mode

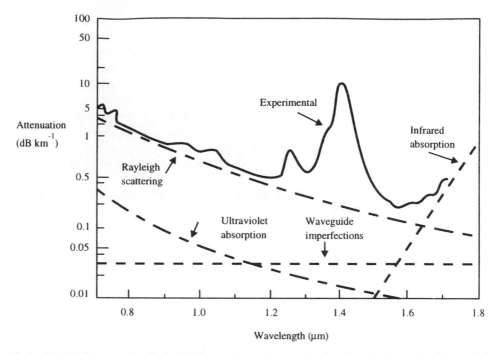

Figure 10.13 The measured attenuation spectrum for an ultralow-loss single-mode fiber (solid line) with the calculated attenuation spectra for some of the loss mechanisms contributing to the overall fiber attenuation (dashed and dotted lines) [10.26].

coupling along a relatively short fiber length. The attenuation value should then be independent of launch conditions, and characteristic of the fiber itself [10.12].

Attenuation in multimode fibers is a few dB km^{-1}, slightly less than that of copper cable systems, while the best performance is achieved in single-mode fiber, which can be as low as 0.2 dB km^{-1}. Attenuation is a function of wavelength, as can be seen from the curve in Figure 10.13 for a typical silica glass fiber. The shape of this curve is determined by the physical processes of interaction between light in the core and the glass material: some of these effects are intrinsically irreducible, being set by physical limits, while part of the attenuation is set by the chemical purity of the glass. The measured attenuation shows a minimum value around $\lambda = 1.55$ μm. The prominent peak near 1.40 μm is caused by absorption due to a molecular vibration by the hydroxyl (OH) group, present from very small amounts of water incorporated in the glass itself. The general rise in attenuation at shorter wavelengths is due to Rayleigh scattering of light within the fibre core by small random fluctuations in density and refractive index inevitably 'frozen' into the glass on cooling. Rayleigh scattering introduces an attenuation term proportional to $1/\lambda^4$, so can be minimized by operating at the longest practicable wavelength. However, in glass, a rising attenuation occurs for $\lambda > 1.5$ μm, from absorption processes associated with lattice vibrations in the glass itself at longer infrared wavelengths (see Chapter 2). It will be seen that fiber transmission is most efficient either side of the 1.4 μm peak, and these wavelengths have indeed become standard in the optical telecommunications field. This is not to imply that fibers are unimportant at shorter, visible wavelengths. There are many applications, in sensing and measurement for example, where fiber lengths of a few tens of metres are adequate and visible or near infrared wavelengths are appropriate, and the penalty of higher fiber attenuation is not an issue.

Fiber can also be manufactured from plastic (polymer) materials [10.13]; for example, a polymethyl methacrylate or polystyrene core with a polymer cladding in a step-index configuration. Because of the high elasticity of polymers, the fiber can be made in larger diameters than glass fiber, and the NA is generally larger, which facilitates coupling of LED light into the fiber and relaxes mechanical tolerances on the connectors. However, the attenuation is much larger, of the order of 200 dB km^{-1}. Therefore, plastic fiber is not competing with glass in long-distance communications (there is no plastic single-mode fiber), but has a role in short-distance, lower data rate links and for flexible, low-cost, light delivery in the automobile market, for example. A typical attenuation spectrum of polystyrene core fiber is shown in Figure 10.14; the minimum loss occurs in the visible part of the spectrum (660–670 nm), unlike silica fibers. Research on the materials and production process of polymer fiber continues: for example, graded-index polymer fiber has been produced with a bandwidth–length product about 20–40 times that of conventional step-index polymer [10.14].

(b) Dispersion

Attenuation describes the power transmission efficiency of a fiber; dispersion is a measure of its capacity to transmit rapidly changing optical pulses or, in other words, the information-handling capacity of a fiber when viewed as part of a communications system. Modal dispersion has already been described in connection with step and graded-index multimode fibers. This source of dispersion is 'geometrical' in origin. There is a second type of dispersion, *chromatic* dispersion, which arises from the microscopic dispersive property of the refractive index of glass. An optical pulse launched into a fiber will contain a finite range of wavelengths, so that the refractive index and hence group velocity is not completely constant for all components of the pulse. Thus the pulse will be broadened. The amount of this dispersion can be calculated from elementary wave theory, using the relations for group velocity v, propagation constant β and angular frequency ω as follows:

$$v = \frac{\mathrm{d}\omega}{\mathrm{d}\beta}; \qquad \omega = \frac{2\pi c}{\lambda}; \qquad \beta = \mathrm{n}(\lambda)\,\frac{2\pi}{\lambda} \tag{10.18}$$

from which, by differentiating with respect to λ,

$$v = c\left/\left(\mathrm{n} - \lambda\,\frac{\mathrm{d}n}{\mathrm{d}\lambda}\right)\right. \tag{10.19}$$

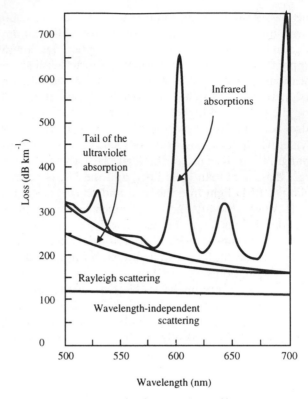

Figure 10.14 The attenuation spectrum of polystyrene core fiber.

The propagation time for an optical signal over a length L of dispersive fiber is known as the *group delay*:

$$t(\lambda) = \frac{L}{v} = \frac{L}{c}\left(\mathrm{n} - \lambda\,\frac{\mathrm{dn}}{\mathrm{d}\lambda}\right) \tag{10.20}$$

so the spread in propagation time Δt, or dispersion, for a pulse containing a wavelength spread $\Delta\lambda$ is

$$\Delta t = \frac{\mathrm{d}t}{\mathrm{d}\lambda}\,\Delta\lambda$$

$$= -\frac{L}{c}\,\lambda\,\frac{\mathrm{d}^2\mathrm{n}}{\mathrm{d}\lambda^2}\,\Delta\lambda \tag{10.21}$$

or, omitting the sign, the size of the chromatic dispersion effect depends on the second derivative of the core material refractive index with wavelength, and is usually quoted as the *material dispersion, M*:

$$M = \frac{\lambda}{c} \frac{d^2 n}{d\lambda^2}$$ (10.22)

for which the units are conventionally ps nm^{-1} km^{-1}. Thus the pulse broadening per km of fiber with a source of spectral width 10 nm launched into a fiber with $M = 100$ ps nm^{-1} km^{-1} would be 1000 ps or 1 ns. The value of M varies with wavelength, and for silica, M is zero around $\lambda = 1300$ nm [10.15]. This makes 1300 nm an attractive choice for communications, since both attenuation and dispersion are small.

Another component of chromatic dispersion is *waveguide dispersion*. This arises from changes in the waveguiding action of the fiber in response to the change in the effective refractive index seen by the range of wavelengths propagating. This effect is small for multimode fibers, but has an important application in optimizing the performance of single-mode fibers. The fiber core index profile and dimensions can be arranged to give waveguide dispersion of opposite sign to material dispersion at wavelengths above 1300 nm, so can be used to compensate for material dispersion. By altering core diameter and index difference Δ, or by moving away from a simple step-index profile to a triangular or more complicated refractive index profile, the designer can minimize the dispersion at wavelengths other than 1300 nm. This controlled-dispersion fiber is at the heart of the latest fiber telecommunications systems, as will be seen in Section 10.2.

10.1.5 Examples of fiber specifications

Finally, we consider some typical specifications of the various types of fibers described so far; these are summarized in Table 10.1. The conventional description of fiber size is given by two figures giving the core/cladding diameters (in μm), e.g. 100/140 multimode fiber.

Table 10.1 Summary of typical optical fiber specifications

	Single-mode fiber	Multimode fiber	Plastic fiber
Size (μm)	9/125	50/125	–
Buffer diameter (μm)	850	900	1000
Profile	Step	Graded	Step
NA	0.1	0.2	0.5
Attenuation (dB km^{-1})	0.5–0.6	1.5–3.0	250

10.2 Fibers in communication

10.2.1 Introduction

Information can be transmitted in either analog or digital form, that is, as a time-varying continuous signal or as a coded stream of discrete pulses. The purpose of a communications system is to deliver the information at an acceptable rate without error or distortion. From the previous section, it is clear that fibers are an excellent medium for this purpose. The low attenuation and high bandwidth–length product of single-mode fiber in particular has led to worldwide deployment of fiber in medium- and long-distance communications. A great deal of knowledge through practical experience has been accumulated over the last 20 years in using fibers for telephone and data networks. The commercial incentives to improve the technology have led to an evolution of systems, as will be outlined below. In addition to the optical fibers themselves, associated technologies of fiber connectors, fusion splicing, packaging of sources and detectors, have been developed, together with products for monitoring and fault diagnosis such as optical time domain reflectometers and optical equivalents of the electrical multimeter.

10.2.2 Outline of a simple fiber communications system

A simple system comprises an optical source (a laser or LED), a length of fiber, and a detector. The input information modulates the optical intensity in either analog or digital fashion, and the corresponding output signal is available as the detector voltage. The word 'system' implies that we are dealing with a number of components, which must function together in a compatible manner for optimum performance. The optical source must be capable of being modulated without distortion at the highest frequency desired in the transmitted information; the optical wavelength must match the fiber for efficient low-loss transmission; the wavelength spread must be small to minimize dispersion; there should be sufficient optical power at the detector to give an acceptable signal-to-noise ratio; and the detector itself should be linear with sufficient bandwidth. Therefore the development of suitable solid-state light sources (Chapter 3) and detectors (Chapter 9) has been just as important as the developments in fiber itself from a systems viewpoint.

An analog system modulates the transmitted signal as a continuous time-varying waveform, for example a speech signal or a video signal. At the receiver, the varying optical intensity is converted back to a more or less faithful reproduction of the original voltage input. In a digital system, there is an intermediate step in which the analog signal is sampled. The amplitude of each sample is then assigned a discrete value on a digital scale. This digitized value is transmitted in a coded fashion over the communications link, and the process reversed at the receiver end. A common method of digital coding, known as pulse code modulation (PCM), is illustrated in Figure 10.15. The analog sample is converted, via pulse amplitude modulation (PAM), to a binary coded value for transmission by a sequence of pulses in two states, representing 0 or 1. These quantized units of information are known as 'bits'; the more bits that are assigned to each sample, the smaller are the discrete digitizing steps, so the digitized waveform becomes a more accurate representation of the original signal. A standard code uses 8-bit digitization, leading to $2^8 = 256$ discrete steps. The advantage of such coding is that the transmission medium is carrying the information only as two levels, 0 or 1, so that noise introduced during transmission has much

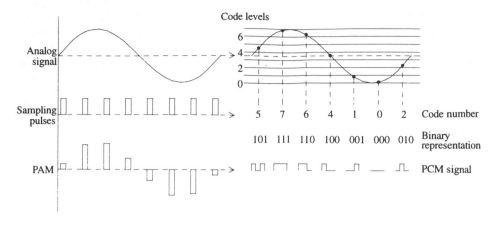

Figure 10.15 The quantization and encoding of an analog signal into PCM using a linear quantizer with eight levels [10.26].

less effect on the signal than in the analog case. The effect of noise, where a transmitted '1', for example, would be received as a '0', can be estimated as a bit error rate. Digital systems do require a higher bandwidth than analog, because 'N' bits are transmitted for each sample.

In communications parlance, we have described an intensity modulation, direct detection (IM/DD) system; more advanced systems will be introduced later. The fiber properties described in Section 10.1.4 relate to system design directly. Attenuation determines the received signal amplitude, and therefore the signal-to-noise ratio at the detector. If the attenuation is too great, the received signal will be noisy (if analog) or contain errors (if digital). To avoid this degradation in signal quality, a long link would include *repeaters* to amplify an analog signal back to an acceptable level, or alternatively to regenerate a digital signal. The attenuation per unit length thus determines the repeater spacing; hence the interest in the low attenuation of single-mode fiber for very long-distance applications such as undersea cables. Dispersion is the other important factor in that it determines the bandwidth, or the maximum rate of change of the information signal carried by the fiber. Pulse broadening by dispersion, being proportional to distance, also influences repeater spacing. For efficient use of installed systems, it is usual to carry multiple signals simultaneously – *multiplexing* – and a high bandwidth is desirable.

10.2.3 Historical development of fiber communication systems

As with many technological developments, it is difficult to establish a precise date at which fiber optics began. Short lengths, less than 1 m, of optical fibers were used in early work for transmitting optical images along fiber bundles. Hopkins and Kapany [10.2], describing such experiments in 1954, reported that the technique had been proposed by John Logie Baird in a patent dated 1927. However, 1970 is a significant date in fiber history, as this was the year in which single-mode optical fiber was manufactured by Corning Glass [10.16] with an attenuation of 20 dB km^{-1}, comparable with the performance of copper coaxial cable. Some time elapsed before fiber was rugged enough for field trials, but in the UK in 1977, graded-index multimode fiber was installed in experimental links at BT Research Laboratories near Ipswich, followed by a 9 km trial route between Stevenage and Hitchin. Similar trial systems were installed in the USA, Japan and Germany [10.17] in order to gain an understanding of the operation of fibers under realistic conditions. These first-generation systems used graded-index multimode fibers with LED sources operating at 850 nm, transmitting digital signals at pulse rates of up to 140 Mbit s^{-1}. Their performance was sufficiently good to increase confidence in fiber technology and hence by the early 1980s there was a growth in intercity (trunk) fiber links which had lengths between 100 and 300 km. Repeaters were placed every 10 km and information transfer rates were between 32 and 140 Mbit s^{-1}.

While graded-index fibers were becoming accepted as a reliable communications medium, work was in progress to exploit the benefits of operating at longer wavelengths. The low attenuation and small dispersion at 1300 nm could provide dramatic gains in performance. Early systems at 1300 nm used LEDs and multimode fibers [10.18], rather than diode lasers, which at that stage had not reached acceptable levels of cost and reliability. The low fiber loss of approximately 0.6 dB km^{-1} at 1300 nm compared with a value of about 2.5 dB km^{-1} at 850 nm, offset the lower optical power of the LED and gave similar transmission distances to those of laser systems at 850 nm wavelength with low cost and high reliability.

Once laser diode sources were available at 1300 nm, they were coupled with single-mode fiber (loss $\sim$0.4 dB km^{-1}) to give a loss-limited distance improvement of a factor of 6 over 850 nm multimode fiber systems [10.19]. As indicated in the previous section, material dispersion in fused silica tends to zero near 1300 nm, and this low dispersion provides a high bandwidth. In practical 1300 nm systems, a bandwidth–length product in excess of 10 GHz km is available.

The next major advance was made in systems based on transmission at 1550 nm, at the attenuation minimum of single-mode silica fiber. The attenuation drops from typically 0.4 to 0.25 dB km^{-1} as the wavelength is changed from 1300 to 1550 nm. However, material dispersion also increases, so to maintain high bandwidth, controlled-dispersion fiber as described in Section 10.1.4 must be used. The ultimate value of the bandwidth–length product, set by a minimum possible dispersion and the quantum limit at the receiver, is $\sim$900 Gbit s^{-1} km [10.19]. A second approach uses very narrow-linewidth sources, such as distributed feedback diode lasers which themselves operate on a single mode (Chapter 3).

Even this does not represent the limit in fiber optic communications. A number of promising techniques exist, at present in the research and development phase, such as coherent optical detection, fiber lasers, soliton transmission and fibers in materials other than silica for transmission in the mid-infrared. These are the elements that may be incorporated into future optical transmission systems, and they are reviewed in Section 10.2.4 below.

The evolution of fiber communication systems described in this section is summarized in Table 10.2.

Table 10.2 Evolution of fiber optic communication systems, based on the classification proposed by Henry *et al.* [10.19]

Generation	Wavelength (nm)	Source	Fiber	Comments
First	850	LED	Multimode, graded-index	Original systems
Second (i)	1300	LED	Multimode, graded-index	
(ii)	1300	Diode laser	Single-mode	Current systems
Third	1550	Diode laser	Single-mode, dispersion-shifted	Current systems
Fourth	1550	Diode laser (single mode)	As above	Coherent detection
Fifth	>2000?	Diode laser	Mid-IR	Future systems

10.2.4 Technological developments

(a) Controlled-dispersion fibers
As already indicated in Section 10.1.4, the total dispersion is the sum of contributions from material dispersion and waveguide dispersion. We recall that material dispersion (Eq. (10.22)) in fused silica changes sign with wavelength, crossing zero near 1300 nm and becoming negative for $\lambda > 1300$ nm. Waveguide dispersion arises from the dependence of effective refractive index, β/k, on λ. If we consider an optical pulse in a single-mode fiber, its shorter λ components experience a higher refractive index than the longer wavelengths, where more of the field has spread into the cladding. Unlike the material dispersion, the amount of waveguide dispersion can be manipulated by the fiber designer, so that the two components cancel to give zero dispersion, to first order, at a chosen wavelength. If this zero occurs at 1550 nm, as in Figure 10.16, the result is *dispersion-shifted* fiber, which has the twin benefits of very small dispersion and minimum attenuation [10.20].

Dispersion-flattened fiber, as its name suggests, has a low dispersion over a range of wavelengths, arising from two zeros in the total curve (Figure 10.16). This is achieved by changing from a simple step-index to a W-profile [10.21], as in Figure 10.17. These fibers are required for wavelength division multiplexing applications, where signals over a relatively wide range of wavelengths may be propagating simultaneously.

(b) Multiplexing
Multiplexing is the term given to simultaneous transmission of many signals by a single communcations route. It is standard practice in voice telephony, for example,

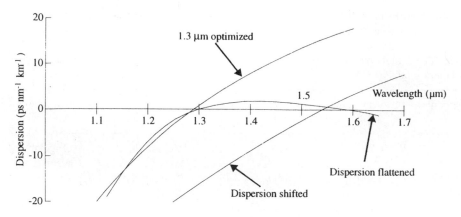

Figure 10.16 Total dispersion characteristics for various types of single-mode fiber [10.26].

to multiplex digitized speech signals by interleaving the pulses for transmission. Specifications for a standard telephony system in the US range from 24 to 233 000 voice channels multiplexed onto coaxial cable or microwave waveguide [10.22].

Optical transmission clearly allows this type of pulse interleaving, or *time-division multiplexing*, but also offers the possibility of optical multiplexing methods. One such method is *wavelength division multiplexing* (WDM), in which each channel is assigned a specific wavelength, separated from neighbouring channels by a known amount. The total wavelength spread of all the channels is designed to keep within the limits related to fiber dispersion, so that all the channels can be transmitted simultaneously along one fiber, thus increasing the system capacity (Figure 10.18). A fiber coupler, which is functionally equivalent to a beam splitter and beam combiner, can be used at the input end to multiplex wavelengths λ_1, λ_2, ..., at the fiber input, with a dispersive element such as a diffraction grating to separate the signals at the demultiplexer [10.23]. Lipson *et al.* [10.24] describe a six-channel WDM demonstra-

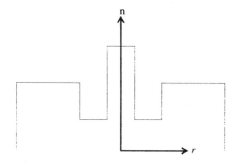

Figure 10.17 The W-profile fiber.

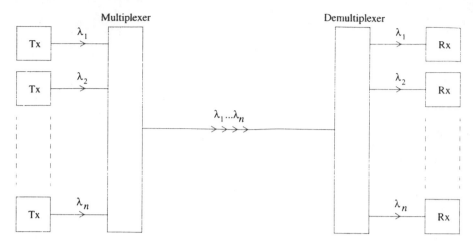

Figure 10.18 An illustration of wavelength division multiplexing.

tion, three channels between 1275 and 1335 nm, and three between 1510 and 1570 nm. Multilayer interference filters have also been used as wavelength-selective elements for WDM components [10.25].

(c) Coherent optical communications

Although it has been developed to operate at impressively high bit rates, the intensity modulation, direct detection scheme outlined in Section 10.2.2 is a relatively crude way of using light. The light signal is merely a binary 'on–off' pulse, achieved by modulating the source intensity. The light frequency and phase are not used in IM/DD. The alternative is *coherent communication*, which has been used for many years in conventional broadcasting at radio frequencies. This offers several improvements in comparison with IM/DD:

(i)   significantly less optical power is needed at the detector for the same bit error rate, leading to increased repeater spacing;
(ii) higher bit rates using existing repeater spacings;
(iii) potential usage of the vast bandwidth (thousands of GHz) available at optical wave frequencies.

A basic technique for coherent detection is the combination of the incoming (generally weak) optical signal (power P_s) with a second optical wave from a *local oscillator* laser (power P_L). The two waves mix at the photodetector to give a photocurrent signal of the form [10.26]:

$$I = P_s + P_L + 2\sqrt{P_s P_L} \cos(\omega_s t - \omega_L t + \phi) \tag{10.23}$$

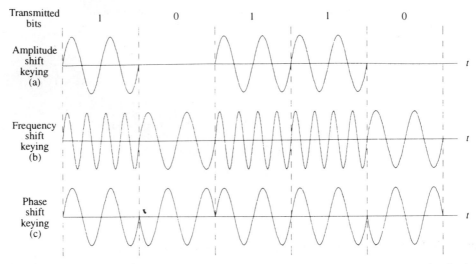

Figure 10.19 Modulated carrier waveforms used for binary data transmission: (a) amplitude shift keying (ASK); (b) frequency shift keying (FSK); (c) phase shift keying (PSK).

where ω_s and ω_L are the corresponding signal frequencies and ϕ is the phase difference between the two signals. It is clear that Eq. (10.23) contains amplitude, frequency and phase information present in the original signal. Therefore, if any one of these parameters is modulated, then the modulation can be recovered from the detector signal *I*. For example, if the signal amplitude is increased, then the amplitude of the detected cosine wave will also increase, and if the signal frequency or phase is changed, then a corresponding frequency or phase change will occur in the detected wave. For digital transmission, only two binary states are required, and the three types of modulation just described can be implemented as shown in Figure 10.19. In relation to digital transmission, they are referred to as *amplitude shift keying* (ASK), *frequency shift keying* (FSK) and *phase shift keying* (PSK), respectively.

Equation (10.23) shows that the original optical signal at frequency ω_s appears at the difference frequency ($\omega_s - \omega_L$). If the signal and local oscillator frequencies are identical, the modulation appears directly at baseband frequencies, as is the case in IM/DD. This is called *homodyne detection*. But if the local oscillator laser frequency is different from that of the transmitting laser, the process is known as *heterodyne detection*. In the frequency domain, the signal spectrum has been shifted bodily by an amount equal to the difference frequency by mixing it with the local oscillator. Thus, heterodyne detection gives the potential to multiplex signals and recover them by coherent receivers with high wavelength selectivity. The frequencies involved are much higher than with radio techniques and are therefore not easy.

(d) Fiber amplifiers

The optical fibers described so far have been passive waveguides. It is also possible to make active fibers in which the core acts as an optical gain medium, so that

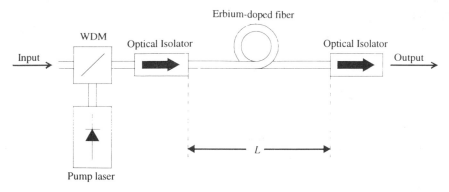

Figure 10.20 Schematic diagram of an erbium-doped fiber amplifier (EDFA).

incoming light is amplified. One class of such fibers has the core silica glass doped with rare earth ions, commonly erbium [10.27]. The optical amplification is brought about by a population inversion caused by optical pumping, in which suitably energetic photons are introduced into the fiber core from an external source, or pump laser (Figure 10.20). A wavelength multiplexer couples both the pump radiation and the incoming optical signal into the erbium-doped core. The energy level diagram for the erbium ion Er^{3+} is shown in Figure 10.21: pump radiation can be 980 or 1480 nm, and stimulated emission can occur in the region of 1550 nm wavelength.

The erbium-doped fiber amplifier (EDFA) clearly has an important application as a signal repeater in fiber communications, and has now received sufficient development to become a commercial product. Some impressive demonstration experiments of potential performance have been reported. Ryu *et al.* [10.28] describe a 195 km submarine fiber operating without a repeater, by amplifying the input signal by 8 dB using a booster EDFA after the transmitting laser. Murakami *et al.* [10.29] report data transmission at 10 Gbit s^{-1} over 6000 km of dispersion-shifted fiber at 1552 nm, employing 119 EDFAs spaced at 50 km intervals.

If the active fiber is contained within an optical resonant cavity, then the device becomes a *fiber laser*. In a remarkably early paper (1964), Koester and Snitzer [10.30] described 1 m long neodymium-doped glass fibers as 1060 nm lasers. Research in this field was active in the mid-1980s, with a number of different resonant optical cavities being investigated, such as the Fabry–Perot and the fiber ring resonator, and designs that allow wavelength tuning of the output within the important 1520–1560 nm range [10.31].

(e) Soliton transmission

So far, the repeater spacing has been determined by the attenuation of the fiber and the minimum signal at the receiver for an acceptably low bit error rate. This gives repeater spacings of the order of 100 km with single-moder fiber operating at 1550 nm. Dispersion, being dependent on transmission length, becomes a factor limiting the

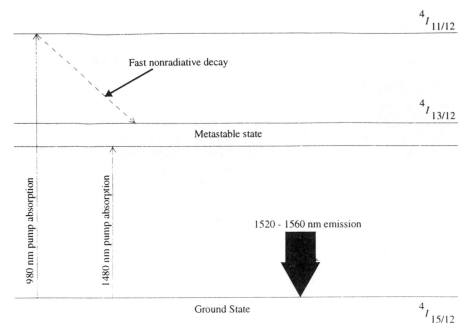

Figure 10.21 Energy levels for the Er^{3+} ion in silica glass.

repeater spacing at high data rates. However, the treatment of pulse propagation up to now has assumed that the fiber is a linear medium. In fact, the refractive index has a small nonlinear component which depends on the square of the electric field of the propagating wave:

$$n = n_0 + n'|E|^2 \tag{10.24}$$

This small nonlinearity in the presence of weak dispersion can be shown to lead to a solution of the wave equation which is a pulse of a particular shape and amplitude that travels through the fiber without dispersion – a *soliton* [10.32]. The soliton will still be subject to attenuation by absorption and scattering, and if attenuated sufficiently, it will degrade to an ordinary pulse subject to dispersion. Thus, provided solitons periodically pass through repeaters to preserve their amplitude, they will not broaden on transmission as do conventional pulses. Very short duration solitons therefore offer the prospect of extending data rates beyond the dispersion limit, and this is an active area of research for application in future systems [10.33, 10.34].

(f) Infrared ultralow-loss fibers
The attenuation minimum for silica fiber of about 0.2 dB km^{-1} at 1550 nm wavelength (Figure 10.12) is a window between the OH absorption peak near 1400 nm and the rising attenuation at longer wavelengths. The latter is due to absorption peaks in the

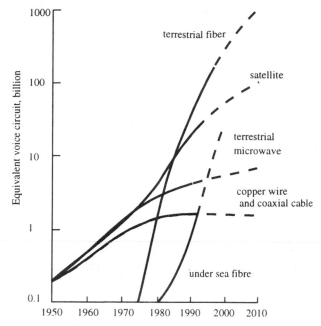

Figure 10.22 The global transmission capacity.

infrared ($\lambda > 2$ μm) caused by molecular vibrations of the glass material. This is an intrinsic property of the silica glass, and represents a fundamental limit to its use. Since Rayleigh scattering decreases as $1/\lambda^4$, operation at wavelengths longer than 1550 nm would be an attractive prospect if fibers were available in a suitable material with smaller infrared absorption. Alternative types of glasses have been studied, with the prospects of an order of magnitude smaller attenuation at wavelengths in the range 2–4 μm [10.26]. However, a great deal of development and operational experience will still be required before it is feasible to replace silica fibers.

10.2.5 Systems in use

The dramatic increase in the share of global information transmitted by fiber is evident from Figure 10.22: worldwide, fiber on land and under sea exceeds 50% of the total capacity, and the figure in the UK is about 70% of long-distance communication traffic [10.35].

As an example of the way fiber has moved into the world of communications, we consider the telephone trunk network in the UK [10.26]. From the early trials briefly described in Section 10.2.3, first-generation systems operating at 850 nm with graded-index multimode fiber were installed in the late 1970s in the trunk network with data rates of 34 and 140 Mbit s^{-1}, repeaters being spaced between 8 and 10 km apart. A 1300 nm LED source/graded-index fiber system was used in a 205 km link between London and Birmingham, followed by demonstrations of single-mode systems. Deployment of 1300 nm single-mode systems took place in the mid-1980s, with bit rates of 140 and 565 Mbit s^{-1} and repeater spacings from 30 to 60 km. Finally, 1550 nm single-mode dispersion-shifted fiber has become standard on trunk routes. The majority of long-distance telephone calls in the UK are now transmitted by fiber along part of their routes.

The coherent techniques described in Section 10.2.4(c) have undergone field trials in the UK. The world's first operational trial of such a link was reported in 1988 by BT Research Laboratories, consisting of a 176 km loop on existing fiber between Cambridge and Bedford [10.36]. Differential phase shift keying (DPSK) modulation was employed (that is, the optical phase change between successive bits contains the '0' or '1' information) at a rate of 565 Mbit s^{-1}, with heterodyne detection. There were no intermediate repeaters over the entire 176 km. This trial was followed by a BT demonstration with two optical channels over a 200 km fiber route between Edinburgh and Newcastle, with an erbium-doped fiber amplifier located at Galashiels, 70 km from Edinburgh. Both DPSK and frequency shift keying modulation was demonstrated, carrying simulated data at 622 Mbit s^{-1} [10.35].

As a dramatic demonstration of the confidence gained in the performance and reliability of optical transmission, fiber has now been installed in undersea cables in many parts of the world [10.37]. The trans-Atlantic TAT-9 cable began service in 1992, operating at 565 Mbit s^{-1} over a total cable length of 10 000 km, complementing TAT-8, the first optical trans-Atlantic cable, in service since 1988. Trans-Pacific optical cables between Japan, Hawaii and the USA also began service in 1988. These cables and associated optoelectronics are designed for a 25-year service life. Optical routes compare favourably with satellite links for international traffic, being secure, reliable and free of the round-trip delay time to geostationary orbit.

10.3 Optical fiber sensors

10.3.1 Introduction

Optical fibers can also be used in applications that are quite different from those just described, namely in *sensing and measurement*. Fibers have been incorporated into optical systems to detect or measure a wide range of quantities [10.38, 10.39] including temperature, pressure, acoustic signals, vibration, rotation, strain, electric current, magnetic fields, fluid flow rate, gas concentration [10.40] and even X-rays [10.41]. Fibers can simply be used for delivery and collection of light in otherwise conventional optical systems, or they can be used as the basis for new types of sensors, in which the measurand (the parameter to be measured) interacts with and modulates the light in a fiber.

Generally speaking, the technology of fiber sensors is not yet as mature as that of fiber communications, but has benefited from the development of communication fibers and components such as fiber directional couplers. Fiber sensors have not displaced conventional sensors on any scale in the industrial sector: they are slowly advancing into areas that are less sensitive to cost, or where a particular performance advantage exists compared with conventional methods. In particular, in the late 1970s, development of optical fiber sensors began in the defence areas of undersea hydrophones [10.42] and fiber gyroscopes [10.43, 10.44]. Fiber gyroscopes have now entered the commercial markets and are undergoing operational trials in aircraft and even in motor vehicle navigation systems. Fiber sensors are suitable for niche applications such as biomedical temperature measurement, noncontact vibration and displacement sensors in engineering metrology, and pressure switches and contact sensors in industrial control.

Fiber offers the possibility of *distributed sensors* in which the measurand can interact with a long length of fiber [10.45], and techniques such as optical time domain reflectometry can be used to locate a measurement at a given time. A single fiber could, as an example, monitor temperature in areas of an electrical substation and produce a warning of 'hot-spots' in power transformers. Distributed temperature measurements with a resolution of 1 K and spatial resolution of 10 m have been made on a 22 km sensing length of fiber [10.46].

The previous section introduced the concept of multiplexing signals on a communication link. Multiplexing the outputs of an array of sensors is another attraction of fiber-based systems: the possible multiplexing schemes [10.47] comprise time-division, frequency-division, wavelength-division and coherence multiplexing, which is based on selection of a particular interferometric sensor by its optical path imbalance. An interesting form of multiplexing has been demonstrated [10.48] in which a multimode fiber was used as a mechanical vibration sensor while transmitting an analog video signal.

It can be seen that optical fibers open up new areas for measurement systems; the all-dielectric construction gives freedom from electromagnetic interference; robust and compact optical sensors are available for specialized applications; multiplexed and distributed sensing over considerable distances is an area of current research and shows promise in the control and monitoring of large systems and structures.

10.3.2 Classification of optical fiber sensors

Because of the wide range of sensor types it is convenient to classify them according to basic principles of operation. Sensors in which the measurand acts upon light outside the fiber are known as *extrinsic sensors*. Conversely, if the light remains guided within the fiber in the measurement volume, then the sensor is of the *intrinsic* type. For example, an intrinsic force sensor, shown in Figure 10.23, operates by producing a controlled series of bends in a fiber that modulate the transmitted light in response to the applied force. In the extrinsic displacement sensor (Figure 10.24) the light intensity collected by the right-hand (output) fiber is modulated by movement of the shutter in the collimated beam.

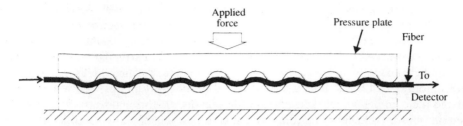

Figure 10.23 A force sensor: based on bending-induced intensity losses.

Another useful distinction can be drawn between fiber sensors based on *incoherent* or *coherent* optical techniques. Incoherent systems usually employ LED sources and multimode fibers, with modulation of the light intensity by the measurand. Both the examples given above would be operated as incoherent sensors. Coherent sensor systems clearly require laser sources with monomode fibers to preserve the relative optical phase between source and detector. Fiber interferometers are in this category, and are important in allowing this sensitive optical technique to be applied outside carefully controlled laboratory conditions.

A further classification labels sensors as *direct* or *indirect*. In direct sensors, the measurand acts on the fiber without any intermediate conversion or transmission. An example is that of the fiber optic hydrophone, in which a length of fiber is directly exposed to an acoustic field, the varying pressure inducing strains within the fiber core, consequently changing its refractive index. In the indirect sensor, there is some intermediate mechanism between the measurand and the fiber or optical interaction. For example, the shutter modulator sensor in Figure 10.24 could be configured as a pressure sensor by attaching the shutter to a diaphragm exposed to the pressure. Pressure change then indirectly modulates the light intensity.

With such a wide range of physical variables and optical techniques, a large variety of sensor designs are possible. In this chapter it is not feasible to provide a comprehensive review of all the possible solutions for all the possible measurement problems, but it is worth noting that some fundamental design principles and limits

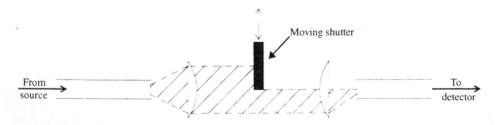

Figure 10.24 A simple fiber optic displacement sensor (the shutter provides modulation).

are set by the physics of the optical and electronic systems. For example, the smallest detectable change in the measurand is determined by the amount of *noise* on the sensor output. A ratio of signal to noise power of unity occurs at the resolution limit of the measurand. The principles are similar to those described for detectors in Chapter 9. Noise can never be eliminated, but can be minimized by suitable design principles such as maximizing the detected optical power. For an optimum design of sensor, all the components of the system (optical sources, fibers, detectors and method of interaction with the measurand) must perform to meet the desired specification in a practical way at an acceptable cost – sometimes these are conflicting requirements!

10.3.3 Multimode fiber sensors

We now consider the types of sensors that can be constructed using multimode optical fiber. The basic transduction mechanisms based on intensity and on wavelength are described, followed by some examples of sensors for two common measurands, temperature and pressure.

(a) Intensity-modulated sensors

The usual mechanism employed in intrinsic intensity modulated sensors is that of *bend loss*, which has already been referred to in the context of mode propagation, where power is radiated into the cladding at the outside of a bend and can be removed from the fiber by an absorbing coating. The force sensor already shown (Figure 10.23) produces a series of small bends, usually termed *microbends*, in a multimode fiber. The guided optical power detected at the fiber exit is a sensitive measure of the relative displacement of the two ridged deformer plates defining the microbends. The sensitivity to displacement can be optimized by matching the spatial periodicity of the array of bends to the mode propagation properties of the fiber, so that the guided modes become efficiently coupled out by the microbends. The design and performance of such devices is discussed by Lagakos *et al.* [10.49], who showed that an incoherent LED source can allow the displacement resolution to approach the shot noise limit of 0.001 nm. Measurands such as force, temperature and acceleration can be converted to displacement by suitable design of the microbend sensing head.

The shutter modulator in Figure 10.24 is a simple example of an *extrinsic* intensity sensor. A more sophisticated version employed in a prototype hydrophone is described by Spillman [10.50], consisting of a pair of parallel gratings, one fixed, the other movable in its own plane. Relative movement modulates a collimated HeNe beam which is delivered and collected by optical fibers and graded-index lenses. This was one of the first multimode sensors to demonstrate performance comparable to that of electrical devices.

A basic problem with all intensity-based sensors is that parameters other than the measurand can affect the intensity of the received signal. Such spurious signal level changes can be caused by changes in the source intensity and in the coupling between source and fiber, bending losses in the connecting fibers and contamination of the optical surfaces. Therefore to obtain reliable results from an intensity-based sensor, an intensity reference is needed. This is accomplished by including a second optical channel, or *reference*, which is not sensitive to the measurand, but is otherwise ideally identical to the signal channel. Any variations in source, fiber transmission or detection will be common to both channels, so by dividing the signal by a reference, these systematic changes can be eliminated from the final output of the sensor.

The example sensors shown in Figures 10.23 and 10.24 do not include reference channels and would therefore be susceptible to intensity drifts. Incorporation of referencing without compromising the performance or increasing the cost is a challenge to the designer of intensity-based sensors. The grating displacement sensor can be intensity referenced by using two source wavelengths with gratings that modulate one wavelength but transmit the other unchanged. Engineered prototype devices for angular position measurement suitable for aerospace use are described by Huggins *et al.* [10.51]. An optically read shaft encoder employed a second wavelength for intensity referencing, giving an output linear to 0.1% of full scale.

Not all intensity-based sensors require referencing. The vortex shedding technique to measure air flow described by Webster *et al.* [10.52] detects the bend loss caused by flow-induced vibration of an optical fiber in a wind tunnel. In this case, the measurand (air speed) is proportional to the frequency of modulation of the optical intensity transmitted by the vibrating fiber, so intensity referencing is unnecessary.

(b) Wavelength-modulated sensors

A range of wavelengths can propagate along a given fiber, depending on the optical bandwidth of the light source and on the fiber size and index properties. Broadband (white) light can be transmitted by a multimode fiber, and a monomode fiber will not necessarily remain single mode if wavelengths below cutoff are present. Wavelength is therefore another parameter available to the fiber sensor designer, and is attractive compared with intensity modulation, as the spectrum of an optical signal is much less subject to unwanted change than is the intensity. This improvement is bought at the expense of a more complicated optical system to measure the spectrum, or some spectrally dependent parameter such as the ratio of intensities at different wavelengths.

A schematic wavelength-modulated sensor is shown in Figure 10.25, comprising a broad-band source, a transduction region in which the spectrum is modified by the measurand, and a dispersing element before the detector(s). Techniques to modulate the wavelength include [10.53] mechanical movement of dispersing elements (prisms or diffraction gratings), and temperature measurement by the thermal dependence of the absorption spectrum in doped fiber [10.54].

10.3.4 Examples of multimode fiber sensors

(a) Temperature measurement

An example of an extrinsic fiber thermometer is described by Zhang *et al.* [10.55]. The sensor operates by illuminating a small (<1 mm cube) sample of fluorescent

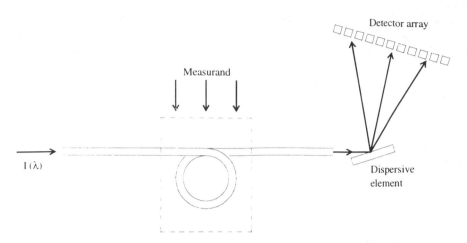

Figure 10.25 Schematic diagram of a wavelength-modulated sensor. It comprises a broadband source, a transduction region (where the spectrum is modified by the measurand), and a dispersing element before the detector array.

material (Cr:LiSAF) bonded onto the end of a 200/230 μm multimode optical fiber. When a light pulse from a 670 nm source is absorbed by the sample, it fluoresces, emitting radiation with $\lambda > 700$ nm as an exponentially decaying response to the illuminating pulse. The decay time (fluorescence lifetime) of this response decreases with increasing temperature and therefore, by signal processing, this quantity can be converted to temperature of the sample region on the end of the fiber. The signal processing dispenses with the need for a reference channel by a feedback loop to regulate the laser source modulation period to be proportional to the fluorescence lifetime. The authors quote temperature results with standard deviations better than 0.01 °C in the range 20–50 °C, a range suitable for biomedical applications, where an all-dielectric thermometer probe has advantages for monitoring tissue temperature in treatments involving microwave heating or scanners using high EM fields.

(b) Pressure measurement
Industry is unlikely to accept new technology such as optically based sensing and measurement systems until there is sufficient proof of reliable operation and a clear advantage over established methods. The optical pressure sensor described by Murphy and Jones [10.56] is an example of the approach most likely to succeed, in which the conventional electrical measurement is replaced, as a retrofit, by an optical fiber strain gauge in an otherwise standard diaphragm pressure transducer. The possibility of an all-optical pressure transducer with fiber links is attractive for use in applications with a potential risk of explosion.

The optical strain gauge exploits the strain dependence of birefringence together with a broadband technique known as chromatic detection, whereby the ratio of the

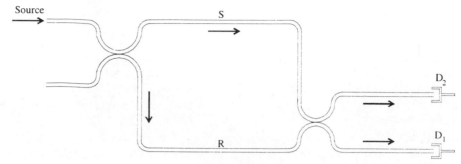

Figure 10.26 Two-beam Mach–Zehnder fiber interferometer.

outputs from two photodetectors with different but overlapping spectral responses provides immunity to intensity fluctuations in the source. Experimental trials gave a pressure range 0–413 kPa, a resolution of 0.3 kPa and hysteresis of 1.6 %

10.3.5 Monomode fiber sensors

Monomode fibers preserve the spatial coherence of a guided optical wave and may therefore be used in sensors which exploit coherent techniques, such as interferometry or polarization modulation. Interferometric sensors offer the highest potential performance and have undergone a great deal of development for applications in both the defence and aerospace sectors. The principles of optical phase measurement by interferometry are outlined below, followed by an examination of some sensors.

(a) The optical phase in a fiber
The output intensity of an optical interferometer depends upon the optical phase difference. As a simple example, consider a two-beam Mach–Zehnder fiber inter- ferometer (Figure 10.26). Light from a coherent source is launched into a monomode fiber and divided by a fiber directional coupler into two arms, termed the signal and reference arms. If the signal arm contains a sensing element in which the optical path length is a function of the measurand, while the path length of the reference arm remains constant, then when the two beams recombine in the second coupler, they interfere to give an intensity which depends on the *optical path difference* between the two arms.

The optical phase change of light of wavelength λ through a fiber of length l and effective core refractive index n is given by

$$\phi = \frac{2\pi n l}{\lambda} \tag{10.25}$$

The potential effect of a measurand X is to change both n and l for the length of fiber considered, resulting in a phase change that depends on the derivative:

$$\frac{d\phi}{dX} = \frac{2\pi}{\lambda} \left[n \frac{\partial l}{\partial X} + l \frac{\partial n}{\partial X} \right] \tag{10.26}$$

For example, if X were temperature, then the first term in the brackets would correspond to thermal expansion and the second to the thermo-optic effect. The measured temperature sensitivity of monomode fused silica fiber [10.57] is ~ 100 rad $m^{-1} K^{-1}$. Other possible measurements are strain ($\varepsilon = \delta l/l$), with sensitivity 6.5 rad m^{-1}/microstrain, and pressure.

Sensors may also be based on birefringent fiber sensing elements, in which the measurement changes the *state of polarization* of the guided wave. In Section 10.1 we discussed highly birefringent fiber that can support two orthogonal eigenstates. Any sensing mechanism that changes the birefringence of the fiber also changes the relative phase of the two eigenmodes. Denoting this phase difference by ϕ_2,

$$\phi_2 = \frac{2\pi\Delta n l}{\lambda} \tag{10.27}$$

where Δn is the refractive index difference between eigenmodes. The sensitivity of ϕ_2 to X is given by an expression analogous to Eq. (10.26). The temperature sensitivity of the polarization eigenmode phase difference in typical hi-bi fiber is smaller than that of the optical phase, ~ 5 rad $m^{-1} K^{-1}$.

Thus we can see that optical fiber itself can act as a sensitive transducer between physical variables (temperature, strain or pressure), and optical phase which can be measured by interferometry.

(b) Optical fiber interferometers

The Mach–Zehnder interferometer was introduced as a simple example of an all-fiber arrangement. It is a two-beam interferometer, with two intensity outputs observable at the detectors of the form

$$I_1 = I_0(1 + V \cos \phi) \tag{10.28}$$

$$I_2 = I_0(1 - V \cos \phi) \tag{10.29}$$

where I_0 is the mean optical power and V is a dimensionless constant between 0 and 1, called the *visibility* of the interference output. At maximum visibility ($V = 1$) the intensities and states of polarization of the signal and reference beams must be equal at the detectors, and the coherence length of the source must be long compared to the path length difference between the two arms. A more detailed treatment of the Mach–Zehnder [10.57] shows that V also depends on the splitting ratios of the two directional couplers. The output intensities I_1 and I_2 are in antiphase with respect to ϕ. This is a useful property, since by taking the difference between the two output signals, the DC offset and common-mode variations in the two channels are removed.

It can be seen that if one arm (designated the signal arm) of the Mach–Zehnder interferometer is exposed to the measurand, while the other (reference) arm is not,

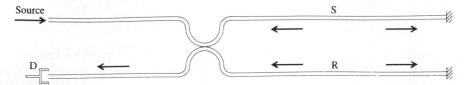

Figure 10.27 The optical fiber Michelson interferometer.

then the optical phase difference ϕ between the arms can be recovered from measurement of both outputs. Because the transfer functions (10.28) and (10.29) are periodic, signal processing is required.

The optical fiber *Michelson* interferometer is shown in Figure 10.27. The light is again divided into signal and reference arms, but is reflected from the end of each one to the same directional coupler. It can be shown that the visibility of the Michelson arrangement is independent of the coupler properties. Practical advantages are that it uses only one coupler and that the signal arm is single-ended, being used in reflection, allowing it to be configured as a sensor probe more easily than the fiber in the Mach–Zehnder case. However, it is difficult, but not impossible, to exploit the complementary output which is directed back to the source. This feedback of optical power can cause instabilities if a laser diode is used, unless isolating optics are employed.

The *Sagnac* interferometer (Figure 10.28) is another two-beam system, but in this case the two beams are separated not spatially but by their direction of propagation

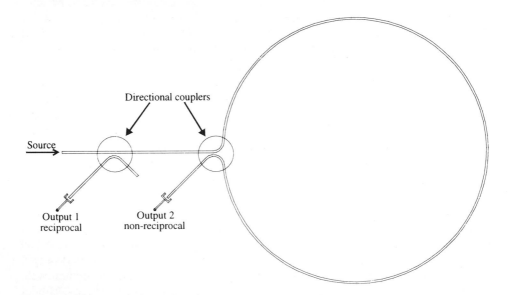

Figure 10.28 The fiber optic Sagnac interferometer.

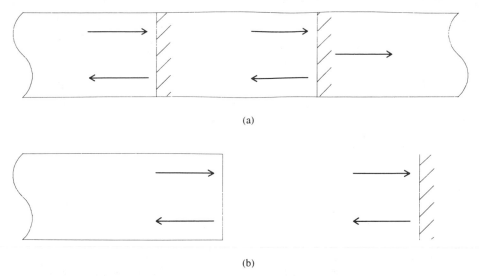

(a)

(b)

Figure 10.29 The fiber Fabry–Perot interferometer: (a) set up as an intrinsic sensor, and (b) set up as an extrinsic sensor.

in a closed loop of fiber. A change in phase can only be produced by affecting the two counter-propagating beams unequally, thus excluding static temperature or strain effects. The Sagnac loop is sensitive to rotation in its own plane, which causes the propagation times of the two beams to be unequal. This is the principle of the fiber optic gyroscope, and the Sagnac loop has been the subject of considerable research and development [10.43, 10.44].

Multiple-beam interferometers can show greater sensitivity than their two-beam counterparts. An example is the fiber Fabry–Perot interferometer shown in Figure 10.29. The guided beam is multiply reflected between the reflective coatings. If the reflectivity is high, then the transfer function is sharply peaked with respect to phase change determined by the change in optical length of the fiber between the coatings (Figure 10.30). At low reflectivities, the function approaches the cosine shape of the two-beam system. The fiber Fabry–Perot can be set up either as an intrinsic sensor, as shown, or as an extrinsic sensor in which the reflections occur at the fiber end and at an external surface.

The interferometer types described so far are based on the phase difference between two geometrically distinct optical paths. Polarization interferometers, based on highly birefringent fiber, generate an interference output using the two orthogonally polarized eigenmodes. These two modes can be made to interfere by placing a linear polarizer (analyzer) at the output, resolving the two polarization states in a given direction, so that they produce an intensity of the form

$$I = I_0(1 + V \cos \phi_2) \tag{10.30}$$

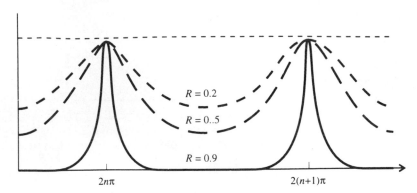

Figure 10.30 Intensity versus phase for the Fabry–Perot interferometer. The fringes increase sharpness with higher mirror reflectivity, *R*.

where ϕ_2 is the phase difference between the two eigenmodes, as defined by Eq. (10.27). The measurand, when applied to the hi-bi sensing fiber (Figure 10.31), will produce phase changes at the output. One example of this is the sensitivity of hi-bi fiber to temperature, which can be easily demonstrated by heating a length of such fiber and observing sinusoidal oscillations of the output intensity transmitted by a linear polarizer. However, it is also possible to design hi-bi fiber with a very small temperature sensitivity [10.58] ($<2.5 \times 10^{-3}$ rad m^{-1} K^{-1}) for application to polarimetric strain or pressure measurement.

(c) Signal processing for interferometers
A disadvantage of interferometric sensors lies in the periodic form of their transfer function. Suppose that the two outputs from a Mach–Zehnder interferometer are combined differentially (by subtracting Eqs. (10.28) and (10.29)), resulting in an output of the form

$$I \propto \cos(\phi_s + \phi_d) \tag{10.31}$$

where ϕ_s is the time-varying phase change due to the measurand, i.e. the desired signal, while ϕ_d represents any other component of phase difference between the interferometer arms, which may include noise and effects of environmental drifts. For small signals ϕ_s we can write

$$I \propto (\cos \phi_d - \phi_s \sin \phi_d) \tag{10.32}$$

so that the small-signal sensitivity is

$$\frac{\mathrm{d}I}{\mathrm{d}\phi_s} \propto \sin \phi_d \tag{10.33}$$

i.e. the sensitivity varies with ϕ_d and can be perturbed by environmental effects, such as temperature changes or drifts in the source wavelength. Indeed, the sensitivity can

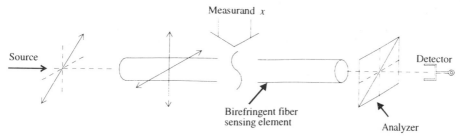

Figure 10.31 The basic polarimetric (strain or pressure) sensor.

become zero, an effect known as signal fading, corresponding to the turning points of the transfer function where the slope is horizontal.

Fortunately, at the expense of some added complication, we can constrain the interferometer to operate at its maximum sensitivity for small-signal operation. This can be achieved by including a *phase modulator* in the references arm which controls the ϕ_d term directly, as in Figure 10.32. A servo loop feeds back to maintain $I \cong 0$, or from Eq. (10.31), $\phi_s + \phi_d \cong \pi/2 + m\pi$. A suitable phase modulator consists of a length of fiber wrapped around a piezoelectric cylinder [10.59]. The servo voltage is applied to the cylinder to stretch the wrapped fiber length slightly, introducing a phase change as the optical length is changed. The measurand can be extracted from the servo loop voltage at either of two points, depending on the signal frequency

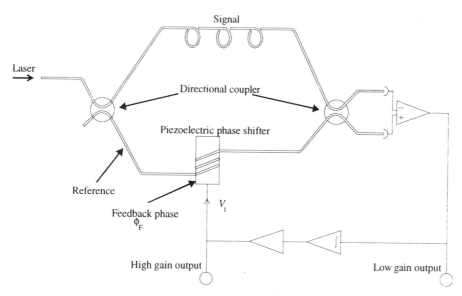

Figure 10.32 Active homodyne signal processing for a fiber interferometer.

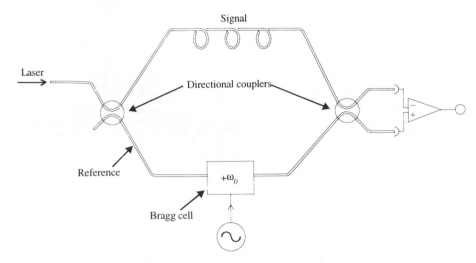

Figure 10.33 Heterodyne processing for the fiber interferometer.

compared with the servo bandwidth. High signal frequencies are present at the low gain output of Figure 10.32, while signal frequencies within the servo bandwidth can be derived from the high gain output. This method of interferometer control is known as *homodyne signal processing.*

Homodyne processing will only operate over a measurement range set by the phase modulator. An alternative approach is shown in Figure 10.33, which is *heterodyne processing.* Instead of introducing a phase shift into the reference arm, a frequency shift ω_r is imposed on the reference beam, so that the output signal is of the form

$$I \propto 1 + V \cos(\phi_s + \phi_d + \omega_r t) \tag{10.34}$$

This output is a time-dependent oscillation of the carrier frequency ω_r with a phase modulation. A number of standard communications-based techniques such as the phase-locked loop exist for the demodulation of such a signal and recovery of the phase induced by the measurand.

If the phase shift ϕ_s is large (several radians) then an alternative method based on the *phase stepping* technique [10.60] can be used. Essentially, this consists of repeatedly switching a known constant phase step in and out of the optical path, with a corresponding phase shift in the transfer function. For example, a phase step of $-\pi/2$ will convert the $(1 + V \cos \phi)$ transfer function to a $(1 + V \sin \phi)$, assuming that the value of ϕ has not changed significantly in the phase step time. The two output signals can be processed and divided to yield a value of $\tan \phi$, and hence ϕ, over the full range $0 < \phi < 2\pi$.

10.3.6 Examples of single-mode fiber sensors

The most highly developed optical fiber sensors are the hydrophone and the fiber optic gyroscope. Their optical and engineering design has evolved to a greater degree than other fiber sensor systems, because of the advantages of non-electrical systems and the potential for multiplexing [10.42].

More typical of fiber sensor design is the research prototype stage, in which a system is demonstrated in a particular application and its performance assessed against the design objectives and against conventional sensors. Prototypes are not necessarily engineered to the extent of the hydrophone or gyroscope but nevertheless may represent a considerable development of the basic interferometer types outlined above.

A suitable example of such a prototype interferometer is the Michelson system described by Hand *et al.* [10.61]. It is used in the detection of ultrasonic acoustic emission from a metal workpiece during machining, as a monitor of machine tool wear. The interferometer must detect small-amplitude (1–10 nm) surface waves in the frequency range 0.1–1 MHz on the surface of a metal block during a machining operation with the attendant acoustic noise and vibration at frequencies up to about 10 kHz. The instrument shown in Figure 10.34 is an extrinsic all-fiber Michelson interferometer in which the signal arm carries light reflected from the workpiece surface. Homodyne signal processing is implemented by a servo loop driving a piezoelectric phase modulator in the reference arm. The arrangement includes three features not shown in the simple fiber Michelson of Figure 10.27. First, the reference arm is terminated in a fiber loop reflector rather than a mirror, to allow the amplitude and polarization state to match that of the returned signal for maximum fringe visibility; second, the antiphase output is extracted from the launch arm by a polarization technique; and finally, the signal arm consists of hi-bi fiber to maintain the state of polarization of the light to and from the target surface, independent of bending or vibration of the signal arm during operation. This prototype interferometer

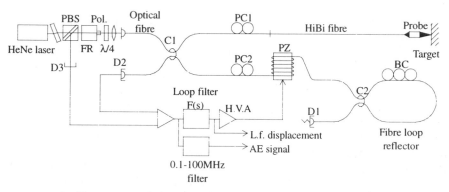

Figure 10.34 The fiber optic Michelson interferometer system used in the detection of ultrasonic acoustic emission from a metal workpiece.

has been demonstrated in repeated trials on a milling machine with the servo system successfully locking out the low-frequency vibration, achieving a 0.1 nm amplitude noise floor over the bandwidth 0.1–1 MHz.

The Fabry–Perot arrangement lends itself to either intrinsic or extrinsic sensors. Since the interference occurs between the two surfaces defining the optical cavity, its operation is inherently insensitive to mechanical or thermal disturbance of the addressing fiber. The fiber Fabry–Perot is a particularly robust interferometer, formed by a short length of fiber reflectively spliced to the addressing fiber and used in either reflection or transmission [10.62]. Intrinsic reflective 2 mm long fiber Fabry–Perot sensors have been demonstrated by Kidd *et al.* [10.63] as calorimeter gauges in transient wind tunnel experiments. Thermal energy incident on the end face of the sensor produces an optical phase change proportional to the mean temperature change of the sensing fiber. The sensors were demonstrated in experiments in a modified shock tube, in which a mean temperature change of 200 mK was measured during a 20 ms thermal transient, the noise floor being 11 mK in a 20 kHz bandwidth. Four sensors were multiplexed from a single 10 mW laser diode source.

Extrinsic Fabry–Perot sensors have been used as strain gauges in a full-scale aircraft wing in simulated flight conditions [10.64]. The sensors were illuminated by a 1300 nm laser diode via a directional coupler. A minimum detectable phase shift of 1.7 mrad gave a strain resolution of 0.01 microstrains, against 20 microstrains for the electrical strain gauges used for comparison.

Finally, it is worth remarking that fiber sensors are appropriate for the developing technology of *smart structures*. The goal is to produce structures with built-in sensors to monitor variables such as strain and temperature within the structural material throughout its service life, and thus to warn of approaching failure or deviations from design conditions. Lee *et al.* [10.65] describe intrinsic fiber Fabry–Perot sensors for temperature, strain and ultrasonic sensing in such applications.

References

[10.1] A.C.S. van Heel, 'A new method of transporting optical images without aberrations', *Nature* **173**, 39 (1954).

[10.2] H.H Hopkins and N.S. Kapany, 'A flexible fibrescope, using static scanning', *Nature* **173**, 39–41 (1954).

[10.3] K.C. Kao and G.A. Hockham, 'Dielectric-fibre surface waveguides for optical frequencies', *Proc. IEE* **133**, 1151–58 (1966).

[10.4] A.J. Rogers, 'Essential optics', in B. Culshaw and J.P. Dakin (eds.) *Optical Fiber Sensors*, vol. 1 (Artech House, 1988), pp 25–106.

[10.5] D. Marcuse, *Theory of Dielectric Optical Waveguides* (Academic Press, London, 1974).

[10.6] D. Gloge, 'Weakly guiding fibres', *Appl. Opt.* **10**, 2252–58 (1971).

[10.7] A.W. Snyder and J.D. Love, 'Reflection at a curved dielectric interface–electromagnetic tunneling', *IEEE Trans. Microwave Theory Techniques* **MIT23**, 134–141 (1975).

[10.8] A.W. Snyder and J.D. Love, *Optical Waveguide Theory* (Chapman and Hall, London, 1983).

[10.9] D. Gloge and E.A. Marcatili, 'Multimode theory of graded-core fibers', *Bell Syst. Tech. J.* **52**, 1563–78 (1973).

[10.10] D.N. Payne, A.J. Barlow and J.J. Ramskov Hansen, 'Development of low- and high-birefringence optical fibers', *IEEE J. Quantum Electron.* **QE18**, 477–88 (1982).

[10.11] S.R. Nagel, 'Fiber materials and fabrication methods', in S.E. Miller and I.P. Kaminow (eds), *Optical Fiber Telecommunications* (Academic Press, New York, 1988), pp. 121–215.

[10.12] D. Marcuse, *Principles of Optical Fiber Measurements* (Academic Press, New York, 1981).

[10.13] C. Emslie, 'Review of polymer optical fibres', *J. Mat. Sci.* **23**, 2281–93 (1988).

[10.14] Y. Koike, 'High-bandwidth graded-index polymer optical fibre', *Polymer* **32**, 1737–45 (1991).

[10.15] D.N. Payne and W.A. Gambling, 'Zero material dispersion in optical fibres', *Electron. Lett.* **11**, 176–78 (1975).

[10.16] F.P. Kapron, D.B. Keck and R.D. Maurer, 'Radiation losses in glass optical waveguides', *Appl. Phys. Lett.* **17**, 423–25 (1970).

[10.17] S.E. Miller and A.G. Chynoweth, 'Evolution of Optical Communications', in S.E. Miller and A.G. Chynoweth (eds.), *Optical Fiber Telecommunications* (Academic Press, New York, 1979), pp. 1–15.

[10.18] D. Gloge, A. Albanese, C.A. Burrus, E.L. Chinnock, J.A. Copeland, A.G. Dentai, T.P. Lee, T. Li and K. Ogawa, 'High-speed digital lightwave communications using LEDs and PIN photodiodes at 1.3 μm', *Bell Syst. Tech. J.* **59**, 1365–82 (1980).

[10.19] P.S. Henry, R.A. Linke and A.H. Gnauck, 'Introduction to lightwave systems', in S.E. Miller and I.P. Kaminow (eds.), *Optical Fiber Telecommunications* (Academic Press, New York, 1988) pp. 781–831.

[10.20] B.J. Ainslie and C.R. Day, 'A review of single-mode fibres with modified dispersion characteristics', *J. Lightwave Technol.* **LT-4**, 967–79 (1986).

[10.21] P.K. Backmann, D. Leers, H. Wehr, D.U. Wiechert, J.A. Steenwijk, D.L.A. Tjaden and E.R. Wehrhatim, 'Dispersion-flattened single-mode fibers prepared with PCVD: performance, limitations, design optimization', *J. Lightwave Technol.* **LT-4**, 858–63 (1986).

[10.22] F.G. Stremler, *Introduction to Communication Systems* (Addison-Wesley, Reading, MA, 1979).

[10.23] J.M. Senior and S.D. Cusworth, 'Devices for wavelength multiplexing and demultiplexing. *IEE Proc.* **136 Pt J**, 183–202 (1989).

[10.24] J. Lipson, W.J. Minford, E.J. Murphy, T.C. Rice, R.A. Linke and G.T. Harvey, 'A six-channel wavelength multiplexer and demultiplexer for single-mode systems', *J. Lightwave Technol.* **LT-3**, 1159–61 (1985).

[10.25] Y. Fujii, J. Minowa and H. Tanada, 'Practical two-wavelength multiplexer and demultiplexer: design and performance', *Appl. Opt.* **22**, 3090–97 (1983).

[10.26] J.M. Senior, *Optical Fiber Communications*, 2nd edn. (Prentice Hall, Englewood Cliffs, NJ, 1992).

[10.27] S.B. Poole, D.N. Payne and M.E. Fermann, 'Fabrication of low-loss optical fibres containing rare-earth ions', *Electron. Lett.* **21**, 737–38 (1985).

[10.28] S. Ryu, T. Miyazaki, T. Kawazawa, Y. Namihira and H. Wakabayashi, 'Field demonstration of 195 km-long coherent unrepeatered submarine cable system using optical booster amplifier', *Electron. Lett.* **28**, 1965–67 (1992).

[10.29] M. Murakami, T. Kataoka, T. Imai, K. Hagimoto and M. Aiki, '10 Gbit/s, 6000 km transmission experiment using erbium-doped fibre in-line amplifiers', *Electron. Lett.* **28**, 2254–55 (1992).

[10.30] C.J. Koester and E. Snitzer, 'Amplification in a fiber laser', *Appl. Opt.* **3**, 1182–86 (1964).

[10.31] P. Urquhart, 'Review of rare earth doped fibre lasers and amplifiers', *IEE Proc.* **135 Pt J**, 385–407 (1988).

[10.32] D. Marcuse, 'Selected topics in the theory of telecommunications fibers', in: S.E. Miller and I.P. Kaminow (eds.), *Optical Fiber Telecommunications* (Academic Press, New York, 1988), pp. 55–119.

[10.33] L.F. Mollenauer, M.J. Neubelt, M. Haner, E. Lichman, S.G. Evangelides and B.M. Nyman, 'Demonstration of error-free soliton transmission at 2.5 Gbit/s over more than 14000 km', *Electron. Lett.* **27**, 2055–56 (1991).

[10.34] S.V. Chernikov, D.J. Richardson, R.I. Laming, E.M. Diamov and D.N. Payne, '70 Gbit/s fibre based source of fundamental solitons at 1550 nm', *Electron. Lett.* **28**, 1210–12 (1992).

[10.35] P. Cochrane, 'Future directions in long haul fibre optic systems', *Br. Telecom. Technol. J.* **8**, 5–17 (1990).

[10.36] M.C. Brain, 'Coherent optical networks', *Br. Telecom. Technol. J.* **7**, 50–57 (1989).

[10.37] P.K. Runge and N.S. Bergano, 'Undersea cable transmission systems', in: S.E. Miller and I.P. Kaminow (eds.), *Optical Fiber Telecommunications* (Academic Press, New York, 1988), pp. 879–909.

[10.38] T.G. Giallorenzi, J.A. Bucaro, A. Dandridge, G.H. Sigel Jr, J.H. Cole, S.C. Rashleigh and R.G. Priest, 'Optical fiber sensor technology', *IEEE J. Quantum Electron.* **QE-18**, 626–65 (1982).

[10.39] J.W. Berthold III, 'Industrial applications of fiber optic sensors', in E. Udd (ed.) *Fiber Optic Sensors* (Wiley, New York, 1991), pp. 409–436.

[10.40] B. Culshaw, F. Muhammad, G. Stewart, S. Murray, D. Pinchbeck, J. Norris, S. Cassidy, M. Wilkinson, D. Williams, I. Crisp, R. Van Ewyk and A. McGhee, 'Evanescent wave methane detection using optical fibres', *Electron. Lett.* **28**, 2232–34 (1992).

[10.41] F. Barone, U. Bernini, M. Conti, A. Del Guerra, L. Di Fore, M. Gambaccini, R. Liuzzi, L. Milano, G. Russo, P. Russo and M. Salvato, 'Detection of x-rays with a fiber-optic interferometric sensor', *Appl. Opt.* **32**, 1229–1233 (1993).

[10.42] A. Dandridge and G. B. Cogdell, 'Fiber optic sensors for Navy applications', *IEEE LCS* **2**, 81–89 (1991).

[10.43] E. Udd, 'Fiber optic sensors based on the Sagnac interferometer and passive ring resonator', in E. Udd (ed.), *Fiber Optic Sensors* (Wiley, New York, 1991), pp. 233–269.

[10.44] H. C. Lefevre, 'Fiber-optic gyroscope', in B. Culshaw and J.P. Dakin (eds.), *Optical Fiber Sensors*, vol. 2 (Artech House, 1989), pp. 381–429.

[10.45] J.P. Dakin, 'Distributed optical fiber sensor systems', in B. Culshaw and J.P. Dakin (eds.) *Optical Fiber Sensors*, vol. 2 (Artech House, 1989), pp. 575–598.

[10.46] X. Bao, D.J. Webb and D.A. Jackson, '22-km distributed temperature sensor using Brillouin gain in optical fiber', *Opt. Lett.* **18**, 552–54 (1993).

[10.47] A.D. Kersey, 'Distributed and multiplexed fiber optic sensors', in E. Udd (ed.), *Fiber Optic Sensors* (Wiley, 1991), pp. 325–368.

[10.48] P.L. Fuhr, P.J. Kajenski and D.R. Huston, 'Simultaneous single fiber optical communications and sensing for intelligent structures', *Smart Mater. Struct.* **1**, 128–33 (1992).

[10.49] N. Lagakos, J.H. Cole and J.A. Bucaro, 'Microbend fiber-optic sensor', *Appl. Opt.* **26**, 2171–80 (1987).

[10.50] W.B. Spillman, 'Multimode grating sensors', in E. Udd (ed.), *Fiber Optic Sensors* (Wiley, New York, 1991), pp. 157–79.

[10.51] R.W. Huggins, G.L. Abbas, C.S. Hong, G.E. Miller, C.R. Porter and B. Van Deventer, 'Fiber-coupled position sensors for aerospace applications', *Opt. Lasers Eng.* **16**, 79–103 (1992).

[10.52] S. Webster, R. McBride, J.S. Barton and J.D.C. Jones, 'Air flow measurement by vortex shedding from multimode and monomode optical fibres', *Meas. Sci. Technol.* **3**, 210–16 (1992).

[10.53] M.C. Hutley, 'Wavelength encoded optical fibre sensors'. *Proc. SPIE* **514**, 111–116 (1984).

[10.54] A.P. Appleyard, P.L. Scrivener and P.D. Maton, 'Intrinsic optical fiber temperature sensor based on the differential absorption technique', *Rev. Sci. Instrum.* **61**, 2650–54 (1990).

[10.55] Z. Zhang, K.T.V. Grattan and A.W. Palmer, 'Sensitive fibre optic thermometer using Cr:LiSAF fluorescence for biomedical sensing applications', *Proc. 8th Int. Conf. Optical Fiber Sensors*, OFS '92 Monterey USA, 1992, pp. 93–96.

[10.56] M.M. Murphy and G.R. Jones, 'Optical fibre pressure measurement', *Meas. Sci. Technol.* **4**, 258–62 (1993).

[10.57] D.A. Jackson and J.D.C. Jones, 'Interferometers', in B. Culshaw and J.P. Dakin (eds.) *Optical Fiber Sensors*, vol. 2 (Artech House, 1989), pp. 329–380.

[10.58] D. Wong and S. Poole, 'Temperature-independent birefringent fibres', *Int. J. Optoelectron.* **8**, 179–186 (1993).

[10.59] D.A. Jackson, R.A. Priest, A. Dandridge and A.B. Tveten, 'Elimination of drift in a single-mode optical fiber interferometer using a piezoelectrically stretched coiled fiber', *Appl. Opt.* **19**, 2926–29 (1980).

[10.60] A.D. Kersey, D.A. Jackson and M. Corke, 'Demodulation scheme for interferometric sensors employing laser frequency switching', *Electron. Lett.* **19**, 102 (1983).

[10.61] D.P. Hand, T.A. Carolan, J.S. Barton and J.D.C. Jones, 'Extrinsic Michelson interferometric fibre optic sensor with bend insensitive downlead', *Opt. Commun.* **97**, 295–300 (1993).

[10.62] C.E. Lee, H.F. Taylor, A.M. Markus and E. Udd, 'Optical-fiber Fabry–Perot embedded sensor', *Opt. Lett.* **14**, 1225–27 (1989).

[10.63] S.R. Kidd, P.G. Sinha, J.S. Barton and J.D. Jones, 'Wind tunnel evaluation of novel interferometric optical fibre heat transfer gauges', *Meas. Sci. Technol.* **4**, 362–68 (1993).

[10.64] K.A. Murphy, M.F. Gunther, A.M. Vengsarkar and R.O. Claus, 'Fabry-Perot fiber optic sensors in full-scale fatigue testing on an F-15 aircraft'. *Appl. Opt.* **31**, 431–33 (1992).

[10.65] C.E. Lee, J.J. Alcoz, Y. Yeh, W.N. Gibler, R.A. Atkins and H.F. Taylor, 'Optical fiber Fabry–Perot sensors for smart structures', *Smart Mater. Struct.* **1**, 123–27 (1992).

Further reading

G.P. Agrawal, *Fiber-Optic Communication Systems* (Wiley, New York, 1992).

A.H. Cherin, *An Introduction to Optical Fibers* (McGraw-Hill, New York, 1983).

J.P. Dakin and B. Culshaw (eds.), *Optical Fiber Sensors,* vol. I *Principles and Components;* vol. II *Systems and Applications* (Artech House, 1988, 1989).

J.E. Midwinter and Y.L. Guo, *Optoelectronics and Lightwave Technology* (Wiley, New York, 1992).

S.E. Miller and A.G. Chynoweth (eds.), *Optical Fiber Telecommunications* (Academic Press, New York, 1979).

S.E. Miller and A.G. Chynoweth (eds.), *Optical Fiber Telecommunications*, vol. II (Academic Press, New York, 1988).

J.M. Senior, *Optical Fiber Communications: Principles and Practice*, 2nd edn. (Prentice Hall, Hemel Hempstead, UK, 1992).

E. Udd (ed.), *Fiber Optic Sensors* (Wiley, New York, 1991).

Appendix A

LEDs: methods of manufacture

A.1 Introduction
A.2 Substrate growth
A.3 Active region growth: epitaxial techniques
A.4 Wafer processing
     Reference

A.1 Introduction

This appendix gives a brief overview of the methods of LED manufacture. It is not intended to be in any way comprehensive; readers requiring more detailed information are referred to a specialist text, such as Ref. [A.1].

The first stage in LED manufacture involves the growth of a series of micron dimension *active* layers on a *substrate*. The substrate forms a base for the active region to be grown on and gives mechanical strength to the structure. Substrates are grown as single crystals out of, for example, gallium arsenide (GaAs) or gallium phosphide (GaP). Two methods are used for substrate growth, the horizontal Bridgman (HB) or boat growth method and the liquid encapsulation Czochralski method (LEC).

The quality of the bulk crystals produced by HB or LEC growth is not appropriate for the active material of LEDs, which have very thin active layers. It is also difficult to produce ternary materials using these techniques, so there would be a limited number of options on possible output wavelengths. Gallium arsenide substrates can be grown using either of these two methods, but the horizontal Bridgman method is preferred. Gallium phosphide can only be grown using the LEC method.

Once a substrate has been obtained the active layers can be grown using epitaxial techniques, growing single crystals layer-by-layer with each atom aligned exactly with those in the preceding layers. This also allows a certain amount of control over the composition of each layer. The two most commonly used epitaxial methods for LED manufacture are liquid phase epitaxy (LPE) and vapour phase epitaxy (VPE).

Liquid phase epitaxy is used for growing binary compounds (GaAs and GaP) and for ternary compounds like $Ga_{1-x}Al_xAs$ where there are no problems with lattice matching (i.e. matching the lattice constants of the epilayer and the substrate). In this case the *pn* junctions are usually grown straight into the structure. Vapour phase

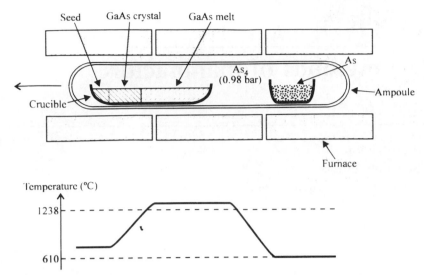

Figure A.1 The horizontal Bridgman method of producing GaAs.

epitaxy is used in cases where the elements used are not the same size, introducing strain which can degrade the quality of the crystal. This is the case for $GaAs_{1-x}P_x$, which is deposited on a GaAs substrate for $x < 0.5$, or for $x > 0.5$ on GaP to minimize lattice mismatch and strain effects. Usually the layers are grown n-type using this method, the pn junctions are formed afterwards produced by the diffusion of zinc into one side.

A.2 Substrate growth

A.2.1 Horizontal Bridgman method

The horizontal Bridgman method is illustrated in Figure A.1. The technique produces D-shaped slices of GaAs with a very low density of dislocation ($\sim 10^3$–10^4 cm^{-1}); this means that for a typical LED of area 10^{-3} cm^2 there will only be between 1 and 10 dislocations. For mass production the round slices obtained using the Czochralski method are more practical but the density of dislocations is higher, $\sim 2 \times 10^3$ to 5×10^4 cm^{-1}. The material is contained within a quartz crucible and heated to form a melt. By either slowly extracting the crucible from the furnace or by using a temperature gradient within the furnace, heated by resistance wire or with an RF field, the melt is allowed to cool gradually from one end of the crucible and crystallization occurs along the length of the sample (Figure A.1). This method of manufacture has the advantage of being cheap, but it is difficult to seed (it is easy to melt the seed crystal if the temperature is not adequately controlled), and it is also

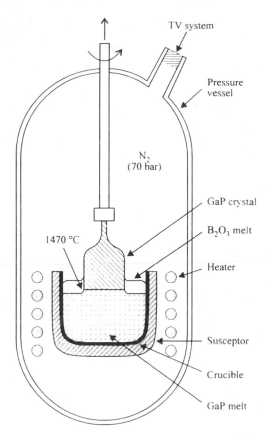

Figure A.2 Liquid encapsulation Czochralski growth of GaP.

difficult to get the material to grow in the correct crystallographic orientation. In addition, because of the high temperatures and pressures required for growth, GaP substrates cannot be produced by this method but are instead produced using the following method.

A.2.2 Liquid encapsulation Czochralski method

This is illustrated in Figure A.2, and we see that in this case the molten material is contained within a melt. A seed with the desired crystal orientation is dipped into the melt and then slowly pulled out, and as this happens new layers of correctly oriented crystal are formed. An inert liquid on top of the melt which is pure, unreactive and nonvolatile stops evaporation. This is required because at high temperatures dissociation can occur and, since elements with different boiling points are present, this leads to an imbalance of concentrations. Boric oxide is used for the encapsulent and a pressure of a few atmospheres is usually maintained to further reduce dissociation.

For use as substrates, the materials have to be doped, usually n-type. Doping is achieved by adding silicon, sulphur, tellurium or compounds of these elements to the melt. Segregation effects can occur so that the density of these elements is not constant throughout the substrate, but the electrical behaviour of LEDs is not critically dependent on the quality of the substrate doping.

Substrate crystals are cut with a diamond saw into slices about 350 μm thick with $\langle 111 \rangle$ or $\langle 100 \rangle$ orientation depending on the type of LED to be produced. The surfaces are polished and etched to remove damage caused by the cutting.

A.3 Active region growth: epitaxial techniques

The techniques of bulk crystal growth mentioned above are unsuitable for the growth of the LED active region. This is primarily because they result in crystals of comparatively poor quality and secondly because large variations in crystal composition over short distances are often required, with which the bulk methods cannot cope. Epitaxial processes allow mixed crystals of any composition to be grown at temperatures low enough to inhibit the formation of intrinsic defects, and the resulting crystals are of high quality. Two techniques are used, one in which the reactants are in the liquid phase, and another in which they are in the vapour phase.

A.3.1 Liquid phase epitaxy (LPE)

The LPE method relies on precipitation of the crystal from a saturated solution, so the type of structures that can be grown are to a certain extent limited by the nature of the phase diagram of the particular binary or ternary material system of interest. Within these restrictions the technique is well suited to the growth of GaAs, GaP and $Ga_{1-x}Al_xAs$ LEDs and generally produces the brightest devices. Here we concentrate on the techniques used for growth in the (Ga,Al)As system. Two different variations are found, each producing 20–50 substrates at any one time, equivalent to several hundred thousand LEDs. These are as follows:

The vertical dipping method. This technique is used in cases where only one melt is necessary, for example in the growth of Si-doped GaAs for infrared emitters. In this process the Ga melt is saturated with As, doped with silicon and placed in a vertical furnace in a crucible. Several GaAs substrates are introduced vertically into the melt at 850°C, the temperature is gradually reduced to 730°C, and the substrates are withdrawn once again. During growth, Si is incorporated as a donor at temperatures above 820°C and as an acceptor below this temperature, so the *pn* junction is grown incorporated from a single solution.

The horizontal slider method. The setup for growth of $Ga_{1-x}Al_xAs$ red LEDs using this technique (Ga,Al)As solutions, one n-type and one p-type, are contained in separate wells of the graphite slider. These are brought into contact with the GaAs substrate at the appropriate times and temperatures to form the pn junction. This method can obviously be extended to the production of multilayer structures.

A.3.2 Vapour phase epitaxy (VPE)

The VPE method has a larger degree of flexibility than LPE since the material is deposited directly from the gas phase and the phase diagram restrictions are no longer present. Thus the technique is particularly useful for growth in the more complicated material systems, the most common of these being $GaAs_{1-x}P_x$, illustrated in Figure A.3. The group V elements are introduced in the form of gaseous AsH_3 (arsine) and PH_3 (phosphine). In the source zone the compound GaCl is formed by reaction of gallium metal with HCl, and this then reacts at the lower temperatures of the deposition zone to form $GaAs_{1-x}P_x$. The crystal composition can be varied by adjusting the partial pressures of the arsine and phosphine, so that active layers suitable for red, orange, yellow and green emission can be grown, together with suitable 'buffer layers'. The latter are layers in which the composition is varied *gradually* between that of the substrate and that of the active layer in order to reduce strain effects.

For P-rich layers the recombination is indirect and nitrogen has to be introduced to increase the radiative efficiency. This is done by introduction of gaseous NH_3 during growth of the active region. In addition, H_2S is added to dope the material n-type by incorporation of the sulphur p-type doping is not possible with this technique; it has to be performed later by diffusion of zinc.

Difficulties encountered with VPE growth include regulation of the gas composition and temperature along the gas flow direction in the reactor. In addition, the reactants involved tend to be highly toxic and explosive. However, VPE is used extensively in industry for the production of low-price red and green LEDs, and it is the only practicable technique for the production of orange and yellow diodes.

A.4 Wafer processing

After growth, the semiconductor wafer, comprising substrate and epilayers, needs to pass through a series of further steps in order to produce an LED, and we briefly discuss these processes here.

The first stage after growth is *passivation.* Surface passivation is used as a diffusion barrier when defining diffusion areas and prevents surface decomposition during diffusion and other higher-temperature processes. It also isolates bonding pads from the semiconductor and may be used as an antireflective coating or as part of high reflectivity contacts. The surface then also acts as a protective coating against chemical or mechanical attack during otherwise damaging fabrication steps.

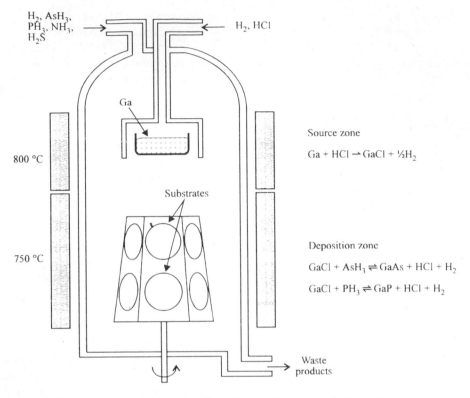

H$_2$, AsH$_3$,
PH$_3$, NH$_3$,
H$_2$S → ← H$_2$, HCl

Ga

800 °C

Source zone

Ga + HCl → GaCl + ½H$_2$

Substrates

Deposition zone

GaCl + AsH$_3$ ⇌ GaAs + HCl + H$_2$

GaCl + PH$_3$ ⇌ GaP + HCl + H$_2$

750 °C

Waste
products

Figure A.3 Vapour phase epitaxial growth of GaAs$_{1-x}$P$_x$ in a vertical barrel reactor.

Materials used include SiO$_2$, phosphorus-doped SiO$_2$, Si$_3$N$_4$ and Al$_2$O$_3$. These materials share a number of properties: they are electrically isolating, stable at high temperatures, chemically relatively inert, relatively hard and optically transparent in the visible and near-infrared. Surface passivation techniques include chemical vapour deposition (CVD) (most common to passivate III–V compounds and of very high quality), plasma coating (similar to CVD but lower temperatures are involved), sputter coating and evaporation.

The next stage may be production of a diffusion or metallization mask whose pattern is defined by *photolithography*. The principle is simple: the sample is coated with a 'resist' which is sensitive to UV radiation in such a way that exposed regions become soluble or non-soluble to an etchant (*positive* or *negative* resists). The desired areas of the sample are masked off, exposed to UV and usually baked for a short time, after which etching reveals areas of the sample which can be *doped* by *diffusion* or *metallized* without affecting other areas. After doping and metallization of the contacts, the wafer is finally *diced* into the final LEDs.

Reference

[A.1] A.K. Gillessen and W. Schairer, *Light-Emitting Diodes: An Introduction* (Prentice Hall International, London, 1987).

Appendix B

Laser diode growth techniques

B.1 Introduction

Fabrication methods for LEDs were discussed in Appendix A; here we turn our attention to growth of laser diodes, mainly in the $Ga_{1-x}Al_xAs/GaAs$ material system. The earliest growth methods were very similar to those used in LED growth, i.e. epitaxial deposition from a melt or a solution (liquid phase epitaxy, LPE) or from a reactive gas mixture (vapour phase epitaxy, VPE). These techniques have been refined considerably over the years in order to reduce the dimensions of the devices produced, but nonetheless only a limited capability exists for the production of the more sophisticated semiconductor heterostructures with very abrupt interfaces using these methods.

However, since the early 1980s, two growth techniques have been developed to meet the special needs of abrupt interfaces. One of these is a modification of the VPE process in which the reactive mixture contains metal–organic compounds and it is therefore known as metal–organic vapour phase epitaxy, MOVPE, or alternatively metal–organic chemical vapour deposition (MOCVD) and the other is molecular beam epitaxy (MBE). The most important aspect of these techniques is that the chemical composition near the growth surface can be altered on a timescale smaller than that required to deposit a single atomic monolayer so that devices requiring very abrupt interfaces can be fabricated.

Because LPE and VPE were discussed in Appendix A in connection with LED growth we leave aside further discussion of them here and turn our attention to the more sophisticated techniques.

B.2 Metal–organic vapour phase epitaxy (MOVPE)

MOVPE has been shown to be a reliable and versatile technique for growing a wide variety of device-quality III–V semiconductor materials and multilayer structures

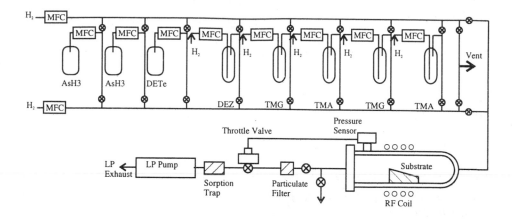

Figure B.1 Schematic diagram of MOVPE equipment used to grow $Ga_{1-x}Al_xAs$ multilayer structures.

with excellent optical and electrical quality. Precise compositional control is possible with both ternary and quaternary materials and doping profiles can be controlled to a resolution of better than 10 nm. A typical reactor for the growth of $Ga_{1-x}Al_xAs$ is shown in Figure B.1. A GaAs substrate is placed in a tube through which hydrogen carrier gas is passed, and carefully controlled amounts of gaseous sources of gallium (trimethyl gallium, $Ga(CH_3)_3$), aluminium (trimethyl aluminium, $Al_2(CH_3)_6$) and arsenic (arsine (AsH_3)) are added. Pyrolysis of the reactive compounds at the surface leads to epitaxial deposition of a single-crystal $Ga_{1-x}Al_xAs$ layer in the overall reaction

$$\tfrac{1}{2}xAl_2(C_3)_6(g) + (1-x)Ga(CH_3)_3(g) + AsH_3(g)$$

$$\xrightarrow{\;H_2(g)\; 600-750°C\;} Al_xGa_{1-x}As(s) + 3CH_4(g)$$

The composition of the deposited layer depends on the partial pressures of the gases involved. Gaseous dopants such as silane (SiH_4) and diethyl zinc ($Zn(CH_3)_6$) are also used to produce n- or p-type material. In recent years, more stringent safety regulations have forced the replacement of highly poisonous materials such as arsine with safer gaseous compounds.

The MOVPE technique is essentially a 'one-way deposition' process and it is this that gives it its advantages over LPE and conventional VPE. The growth rate can be made arbitrarily small by accurate control of the reactants and changes in the gas composition can bring about very abrupt changes in the material composition. In a typical system [B.1], the gas transport is arranged so that the gas composition over the wafer can be changed in a controlled way within 0.1 s. Working at atmospheric pressure at a temperature of 650°C the $Ga_{1-x}Al_xAs$ growth rate can be as low as 0.5 nm s^{-1}, which means that the time required for a change in the gas composition is much less than the time required to deposit one monolayer. This permits the growth of well-defined multilayer structures with reproducible control over composition, thickness and doping profile in the direction of growth on an atomic scale.

The advantages and disadvantages of the technique can be summarized as follows:

Advantages
- The technique can be scaled up to handle many wafers at a time with fast turnaround.
- The equipment is relatively simple and cheap ($\sim$£100 000).
- Ternary and quaternary materials can be handled easily.

Disadvantages
- Because the gas mix flows *along* the substrate, thickness control can sometimes be difficult.
- The source materials (metal-organics) are expensive.

B.3 Molecular beam epitaxy (MBE)

Molecular beam epitaxy is essentially a refined form of ultrahigh-vacuum (UHV) evaporation, with base (residual gas) pressures typically in the 10^{-10} Torr region or better. In this technique, Knudsen cells are used as sources; they provide thermally generated collision-free molecular or atomic *beams* of the constituents. These are deposited onto a heated substrate where they react to form an epitaxial film (Figure B.2).

Knudsen cells are heated crucibles of boron nitride containing, typically, elemental sources such as Ga, In, Al or As, as illustrated in Figure B.3. The molecular beam fluxes are controlled by means of shutters that can be opened and closed quickly enough to control the composition on the scale of atomic monolayers. Growth temperatures are lower than those in other techniques (500°C) and growth rates are typically in the region of 0.01–0.1 nm s^{-1}.

The advantages and disadvantages of the technique are summarized as follows:

Advantages
- The UHV environment is compatible with a whole variety of characterization techniques, such as mass spectrometry (for residual gas analysis); Auger electron spectroscopy (AES, to determine surface composition and purity); X-ray photo-electron spectroscopy (XPS); secondary ion mass spectrometry (SIMS); and reflection high-energy electron diffraction (RHEED), which is an important technique used for the analysis of surface structure during growth.

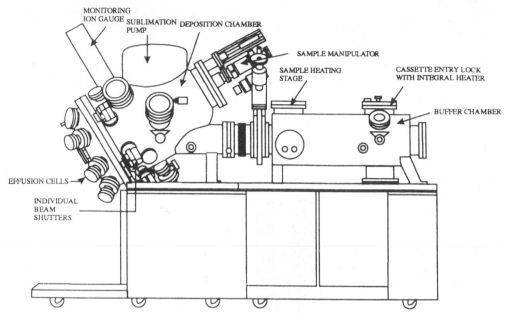

MONITORING
ION GAUGE SUBLIMATION DEPOSITION CHAMBER
PUMP

SAMPLE MANIPULATOR

SAMPLE HEATING
STAGE

CASSETTE ENTRY LOCK
WITH INTEGRAL HEATER

BUFFER CHAMBER

EFFUSION CELLS

INDIVIDUAL
BEAM
SHUTTERS

Figure B.2 A schematic diagram of a solid source MBE system for the growth of III V compounds.

- The growth of $Ga_{1-x}Al_xAs/GaAs$ is easy using this technique and very well characterized.
- Source materials are relatively cheap.

Disadvantages
- The equipment is very expensive ($\sim$ £500 000).
- Turnaround is slower than in MOVPE.

Other techniques exist which are essentially variants of the above, namely metal–organic MBE (MOMBE), chemical beam epitaxy (CBE) and gas source MBE (GSMBE). All of these techniques use MBE high-vacuum equipment but at least some, if not all, of the sources are replaced by gaseous compounds. This overcomes the problem, inherent in conventional MBE, of using exhaustible solid sources by replacing them with inexhaustible gases.

B.4 Scaling up

Scaling up to production quantities has been a problem both for MBE and MOCVD, involving a progression from small (quarter wafer) samples to whole wafers and then multiwafer systems. The problems are essentially those of:

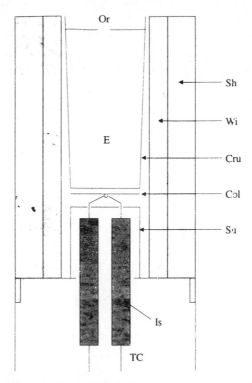

Figure B.3 Schematic diagram of a typical Knudsen cell, acting as a Ga or Al source for the MBE growth of Ga$_{1-x}$Al$_x$As. The evaporation source, E, is contained within a boron nitride crucible (cru) and the heater winding (Wi) is surrounded by tantalum heat shields (Sh). (Or) is the orifice through which the beam leaves the cell. TC is a thermocouple passing through insulators (Is) and connected to a tantalum radiation collector (Col).

1. *Uniformity*: MBE and MOCVD give poor uniformity over large areas in either thickness (MOCVD) or composition (MBE). This is solved by rotating the samples and by more sophisticated source design.
2. *Sample throughput*: This is more of a bottleneck for MBE than MOCVD with transfer from atmosphere to growth chamber originally very slow. This was improved, first, by use of an airlock; second, by cassette (multiwafer) transfer; and most recently by fully automatic transfer with separate entry and exit airlocks.

An excellent example of a contemporary MBE growth system is that of Rohm Corporation [B.2], who have a fully automated custom MBE system which takes seven 2″ wafers in each growth run. The machine is in continuous (24 h) operation and is used to provide 70% of the lasers sold on the open market for use in CD players.

References

[B.1] P.M. Frijlink, J.P. André and M. Ermann, *Philips Tech. Rev.* **43**, 118 (1987).

[B.2] H. Tanaka and M. Mushiage, *J. Cryst. Growth* **111**, 1043 (1991).

Index